AKUNA — A NEW GUINEA VILLAGE COMMUNITY

AKUNA

A NEW GUINEA
VILLAGE COMMUNITY

BRIAN M. DU TOIT
University of Florida

A.A.BALKEMA / ROTTERDAM / 1975

To Sona and Helene
and other fellow Akunans

ISBN 90 6191 004 8 cloth edition
ISBN 90 6191 005 6 paper edition

Printed in the Netherlands

PREFACE

The publication of this material postdates the field research by more than ten years. This is due to a variety of factors, among others a continental change of research focus and three intercontinental changes of residence by the author. While the intervening decade has seen some major political changes in this region the prehistorical and social anthropological pictures have not changed markedly. As is common in ethnographic studies this account will use the present tense and will sketch the sociological picture as it was during the early sixties. For this reason too, I have retained the reference to New Guinea and Papua rather than adopting the present politically correct use of Papua-New Guinea.

A study of this nature inevitably involves an ever-widening circle of friends, acquaintances, associates and persons who assisted in some way or another. For me to do justice to every person who has contributed to this study would be impossible. The persons who will be mentioned here are those who were more directly involved or who have had a more lasting influence. It does not deny the very real and very important contribution of numerous other persons.

The material on which this study is based was collected while I carried out a year's field research for the completion of the doctoral degree. Two persons who were influencial in the early stages must be mentioned, Homer Barnett first whetted my appetite for Oceanic anthropology during a series of stimulating courses at the University of Oregon and Jan van Baal, currently director of the Royal Institute for the Tropics in Amsterdam, opened the field as a real possibility for research. He had just retired as Governor of what was then Dutch New Guinea and his classes as well as private discussions finally lead to the decision to carry out fieldwork in Melanesia. Financial assistance was provided by a grant from the National Science Foundation to the University of Washington under whose auspices this research was conducted. I also acknowledge a grant from the American Anthropological Association — Smith, Kline, and French Committee which permitted me to collect ethnobotanical samples. The greater part of this manuscript was completed under a Graduate Faculty Research Grant from the University of Florida. For these various forms of financial assistance I express my sincere appreciation.

While in transit to New Guinea a very fruitful week was spent at the Australian National University in Canberra where John Barnes and Paula Brown made some valuable suggestions regarding residence and research in the highlands of New Guinea. The whole period of research, but especially my initial settlement and introduction, was immensely facilitated by the assistance and friendship of members of the New Guinea branch of the Summer Institute of Linguistics. Their hospitality and willingness to help in so many ways contributed to the fond memories we have of the field period. I have to single out Chet and Majorie Frantz who also assisted me regarding Gadsup linguistic problems. Appreciation is expressed to Mr John Barrett, entomologist at Aiyura Agricultural Station, for friendship and the identification of plants and insects which were collected. Prior to and during the field work, assistance was received from the Australian Administration of New Guinea, and I would like to refer especially to Dr John Gunther, at that time the Assistant Administrator, as well as Mr J. Donnoly, A.D.O. at Kainantu — our 'kiap'.

In speaking about the year in New Guinea it is impossible to conceive of it without Howard P. McKaughan, a linguist who worked in the eastern highlands during this period. We received all the possible help and kindness from Howard, Bobie and the girls, and their friendship meant a great deal to my family and me. Thanks are due to Bob and Shirley Glass whom I visited while they worked among the Fore, and to Lew Langness who visited us for Christmas. Their questions and discussions stand as luminous points in our period of field work.

This study concerns the Gadsup, and especially the residents of Akuna, to whom immeasurable gratitude is due for their assistance and hospitality to my wife Sona, daughter Helene and myself. It is often easy to criticize them when one forgets that they have no notion of the aims and methods of our research nor of the reasons for our asking so many basic and often simple questions. It is also hard for them to receive an outsider without a bit of restraint until they have formed their ideas about his reasons for being among them. This time of unproductivity is commonly referred to by field workers as that period of 'no returns'. I would like to mention five persons, though I am quite conscious that this is hopelessly inadequate in describing the emotions I feel for the people who were my fellow villagers. Napiwa, horticulturist par excellence, received us with the hospitality and curiosity which could hardly be matched in this country. He offered the land for us to construct a home and additional land for gardens, was always assuring himself that we had the necessary food supplies, and patronaged my research in many ways. To my interpreter and assistant Wa?iyo, and his mother Ara?o I am indebted for their support as informants and for the sense of humor with which Wa?iyo approached life. When he found that I was interested in stone adzes, he formally presented me with one he had made. The total length including the handle was only four inches. His abilities at cricket were stunning to me and his age mates. The hospitality and assistance of Torawa and his mother Wakano cannot be overstated, for they were always ready and available to assist. This is of great importance when we recall that they could hardly fathom the interest which prompted me to ask the same questions twice or often three times. Torawa was invaluable at recalling the exact circumstances during which a person or a topic was referred to, thus avoiding duplication and possible confusion. The research benefited greatly from the sponsorship and guidance by these and many other friends.

After our return from the field I worked at the University of Washington, and am grateful to James B. Watson and K.E. Read, who was chairman of the anthropology department, for their interest and suggestions. During an early reworking of material Theodore Stern at the University of Oregon was of great value. He offered many valuable leads, and his probing questions into every statement were contributory to this synthesis. In the same light I am grateful to A.C. van der Leeden of Leiden, Jan van Baal of the Koninklijk Instituut voor de Tropen in Amsterdam, G.P. Murdock and Barbara Lane, at the time both attached to the University of Pittsburgh, Peter White of the University of Sydney, as well as R. Obeyesekere, Madeline Leininger, Ed Cook and Lew Langness all of whom at various stages gave valuable help in personal discussions and by correspondence. Appreciation is expressed to R.D. Lycan, who at the time was a cartographer at the University of Washington, for the first three maps in this study.

In this account of village life I have used as few labels as possible. Rather than attaching social labels or structural prototypes, and using anthropological jargon, I have attempted simply to present a picture of life in Akuna. The reader can then, after meeting the actors and noting their actions, choices, and decisions, draw his own conclusions regarding labels for kinship, descent, residence and so forth as I have done in the last chapter. Seldom does a student read an ethnography from cover to cover. For this reason a number of subjects have been referred to at more than one place in this book. Thus a subject or a case may be mentioned or treated briefly where the topic arises but will be elaborated in full under the most logical heading.

It remains for me to remark briefly on the orthography which has been employed to record Gadsup words. In earlier publications an orthography was employed which was based more clearly on the phonetic principles of the language. Most of these symbols have been deleted here. Thus differentiation between the middle back vowels -o and -ɔ or the middle front vowels -e and -ɛ have been disposed of. These are now recorded as simply -o and -e respectively. I have retained a differentiation between the normal short -a and -a. indicating length. In this study I have also used the symbol -w to represent the implosive bilabial, though phonetically this was recorded as being closer to a fricative. The last symbol to be noted is -?, indicating a glottal stop. Where pitch or tone was obvious and resulted in possible confusion between items of vocabulary, I employed the marker -' to indicate high pitch.

B. M. d T
Durban, South Africa
June 1, 1974

CONTENTS

Chapter 1

THE FIELD

The island of New Guinea, with its strange bird-like shape lies immediately south of the equator. With its tail and feet straddling Australia and the bird's head looking west into Indonesia it forms a prehistoric link between Asia to the west and Polynesia to the east. It also forms one of the last remaining links with neolithic cultures. Across the length of the island, from east-south-east to west-northwest there extends a central highland zone, primitive in its ruggedness, appealing in its beauty, enchanting in its simplicity, and challenging in its unfamiliarity. At the eastern extreme of these central highlands, forming the watershed for the Ramu River flowing north and the Markham River flowing east, is the Arona Valley.

This part of the highlands is marked by high mountain ranges; and peaks, such as Mt. Wilhelm, reach an elevation of 11,600 feet at its highest point. On the east the highlands drop to the Ramu Fall and the Markham Valley. Here the altitude changes in ten miles from nearly 10,000 feet to a little over 1,000 feet above sea level. Both the Ramu and the Markham Rivers originate in these mountains, growing from mere marshy patches to cascading streams and within a few miles they grow to swift-flowing rivers. During the rainy season these rivers form natural barriers which temporarily disrupt communication between different villages.

PHYSICAL SETTING

Within the central uplift, known as the Bismarck Range, "vulcanism continued into the Quaternary ... faulting, initiated during the orogeny is still taking place, and is responsible for part of the uplift of the Bismarck Range" (McMillan and Malone 1960:13). This volcanic action has led to the formation of three intermontane basins, namely the Goroka Valley (drained by the Bena Bena River), the Kainantu Valley (drained by the Ramu River), and the Arona Valley (drained by the Ramu and Markham Rivers). Prehistorically these basins were

1

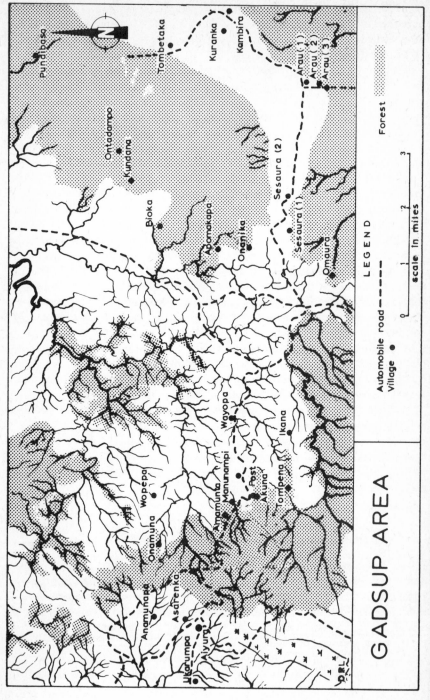

GADSUP AREA

LEGEND

Automobile road - - - - -
Village ●

Forest

scale in miles
0   1   2   3

2

lakes, drained by the rivers which flowed into the Miocene seas pressing up "into the Ramu-Markham areas" (ibid.:52). The beds of these Pleistocene valleys are covered with quaternary sediments forming a fertile valley floor of sand, clay, silt and gravel. This distinctive topography of fertile alluvial lake deposits is covered by kunai and kangaroo grass while the streams are bordered by dense clumps of pit-pit reeds. This valley floor with its deep-rooted grasses and shrubbery contrasts sharply with the surrounding mountains of igneous rock covered in most cases with trees and dense undergrowth.

The Arona Valley itself varies in altitude between 5, 200 feet to 4, 400 feet at the valley floor. It is situated at about 146. 00^0 longitude and 6. 15^0 latitude, comprising an area approximately one hundred square miles. In spite of the fact that it is only six degrees south of the equator, the air is fresh and the temperature mild. While the direct sun scorches the earth during the dry season, the nights are generally cool and the valley lacks the extreme temperature range reported from areas further west. During the period in the field the temperature was recorded an average of twice daily in the shade[1] and no temperature below 58^0 F. or over 78^0 F. was recorded. The average rainfall for the area reported by the Administration is 85. 35 inches,[2] with the pattern being that of crisp sunny mornings being followed by thunder-showers in the afternoon and evening. The year can be divided roughly into two seasons, a dry phase from May to November and a wet phase from December to April. During our year in the field — and this seems to have been the case the preceding number of years — there were two dry seasons, namely November to January and May to July, alternating with two wet seasons.

MacMillan and Malone (ibid.:11) describe the Arona Valley as including mainly two types of vegetation, namely grass which has spread and replaced forests (due to the native practice of burning the bush to clear the land for new gardens), and "lowland and mid-mountain forests". It is felt, however, that the present Arona Valley can be more accurately described as three transitional zones. First

1. Readings were taken at any of the following times: 7 a.m., 12 noon, 6 p.m., and 10 p.m.
2. A rain gauge was not available to us and the rainfall was not measured but simply noted on a calendar. There were 137 days on which precipitation was marked, out of a total of 321 days for which the markings were continued from July 15, 1961, to June 1, 1962. Kainantu, the administrative sub-district headquarters, which is about fifteen miles from the Arona valley, reports the following monthly averages in inches: January 8. 85, February 9. 43, March 11. 97, April 8. 17, May 2. 87, June 2. 59, July 2. 17, August 3. 70, September 3. 85, October 6. 98, November 7. 57, December 11. 61. This gives an annual rain fall of 79. 76 inches (Brass 1964:156).

of all there is the low-lying grassland which is covered with kunai and kangaroo grass, while the alluvial flats beside streams and rivers are covered with dense patches of pit-pit (Saccharum robustum species), a tall reed-like grass with waving white plumes suggestive of ripe sugar cane. This zone extends throughout the Arona Valley, and while an occasional garden is now found there, it was not traditionally used for this purpose due to the danger of hostile attacks, but pigs often root there. The second zone is a transition between grassland and forest with brush and trees covering the ground. It is here, at around 5,000 feet in altitude, that the Gadsup villages are situated and that the brush and trees have been cleared away for garden plots. The latter are generally not on the level plateaus, but against ridges, very often at a sixty-degree angle which hastens erosion and weathering, and makes it necessary to clear new gardens every four or five years. Also in this zone are found many of the privately owned trees such as betel nut, pandanus nut and fruit-bearing pandanus palms, breadfruit trees, together with a variety of banana palms and the clumps of bamboo which are used in house building, for cooking and other needs. Because of the rapid replacement of forest flora, abandoned gardens are soon overgrown and fallow areas can be recognized only because of the immature vegetation that covers them.

Beyond this area of brush and trees is the zone of dense and lush mountain forests, damp because direct sunlight hardly ever reaches the ground long enough to evaporate the moisture. This is the area where colored mosses grow on rotting tree stumps, and where the Gadsup go to collect many of the edible tree fungi and mushrooms, or to spend a night hunting marsupials. It is an area of dangers, too, for on a damp log or slushy pool a person can fall down; and children are forbidden to go into the 'true bush'. This is a man's world and a boy feels very proud the first time he is permitted to accompany his father into this twilight world of unknown sounds and fluttering forms.

PREHISTORY

A recent report on archaeology in the highlands starts with the following general understatement: "The prehistory of Papua — New Guinea is almost unknown" (White 1965:40). This also applies to the Arona Valley. The fact that earlier inhabitants roamed these valleys is generally accepted, but little or no systematic excavations have been carried out. During the past decade the Bulmers from the University of New Guinea, among others, have done important survey work and site excavations in the highlands. Radiocarbon dates from these sites

The village of Akuna

1. Stone adzes collected in the Akuna area. On the left Type I and in the center Type II (see page 5)
2. A hafted stone adze from Tompena
3. Bamboo cicar holders from the Gadsup region showing the variations on an original bird-like figure

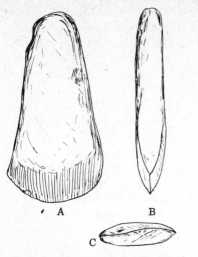

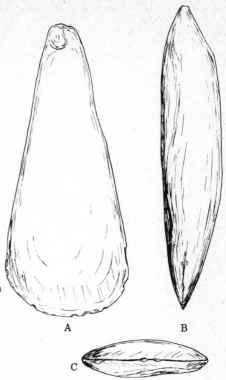

Above (left): Diagram of the 'Vier-kantbeil' reproduced in the center of the plate on the opposite page. Length 11.2 cm; Width at cutting-edge 6.3 cm; Width at butt-end 3.1 cm; Maximum thickness 2.15 cm. A. Side view of flat surface showing gradual bevel. B. Horizontal cross-section showing flattened sides. C. Vertical cross-section at cutting edge.

Above (right): Diagram of the Shovel-shaped stone adze reproduced on the left of the plate on the opposite page. Length 18 cm; Width at cutting-edge 7.5 cm; Width at butt-end 2.05 cm; Maximum thickness 4.15 cm. A. Side view. B. Horizontal cross-section showing shovel shape. C. Vertical cross-section at cutting edge.

suggest human occupation at least ten thousand years ago. Since the dates associated with the first edge-ground axe-adzes are somewhat more recent, Susan Bulmer suggests "between 3000 and 4000 B.C. ... a respectable possible time-depth for agriculture in the Highlands of New Guinea" (1964a: 328).[3]

At this stage of our knowledge regarding the prehistory of the Arona Valley we must turn to a number of surface finds. On the whole these are stone adzes which agree in general with those reported from

3. In a recent survey article Peter White points out that the sixties were significant for the amount of diachronic data which became available and the use of C.14 dating. While well documented sites in the Central Highlands date back to 11,000 B.P. (White 1971:47) this area of the world must still be seen as "a new and largely undocumented field".

adjacent areas (Bulmer 1964b, and Adam 1953). In the Arona Valley a number of these were collected from people who discovered them in their gardens, or had used them in the past and simply stowed them away when metal axes became available. An example of the large, typical Papuan adze-form was acquired in this way. Plate 1 (opposite page 5) shows one of this type described by Adam (ibid.: 414) after a specimen from the Kamano area west of the Arona Valley. I was also able to acquire in the village of Tompena a smaller sample which was hafted (Plate 2) and had originally been used by the owner.

A number of Adam's Type II were also collected. This is the 'Vierkantbeil' or quadrangular adze head with its straight vertical lateral surfaces. It lacks the shovel shape of the previous type.

While in the Arona Valley it was learned that members of the Archbold expedition of the American Museum of Natural History had visited the region in 1959. A number of stone mortars had been collected. According to information these had been found just below the surface in new garden areas and could not be recognized or identified by the present residents of the valley.[4]

The conclusion we must reach at the present time is that this general area has been inhabited for thousands of years, most likely by a people who preceded the ancestors of the present population groups. It might even be possible that this general area of lower montane forest was one of the earliest to be occupied. Based on their archaeological evidence, Bulmer and Bulmer have postulated an early occupation of the highlands by pre-neolithic people. They could have reached this area through gradual population growth and displacement, or "a second possibility is that the lower montane forest was one of the earliest zones of effective occupation by man in New Guinea, and was in fact generally occupied before the less salubrious lowland region" (1964:73). This same differentiation of three zones has been used in an attempt to explain social structural differences in New Guinea (du Toit 1963: 277-283).

Employing a methodological combination of glottochronology and phonostatistics, McKaughan has studied the relationship and divergence of the languages in this part of the eastern Highlands. His conclusions postulated a close relationship between neighboring groups in which linguistic divergence matches geographical distance, the evolution of sub-groups occuring after the present population settled in

---

4. Similar decorated stone mortars have been reported from the Watut River region in the area nearer the coast (Sherwin 1938). In a situation very similar to this, Chinnery (1919:284) reports that when the local population were asked about the objects they were not able to identify them, and he concludes that the objects must have been used by "earlier inhabitants".

the areas they now occupy. "If the latter is the case — and it seems
to be the more likely possibility, then these people have been in the
Highlands area indeed, in this part of the highlands — at the very
least for 1000 years, based upon glottochronological figures which
seem quite conservative" (McKaughan 1964:119). This conclusion
also corroborates Wurm's hypothesis of a single family of languages
in the Eastern Highlands (1960, 1961, 1964). On linguistic grounds
then the people of the Arona Valley (Gadsup) must be seen as part of
a larger family extending northwest (Agarabe) and southwest (Oyana).
It is unfortunate that sufficient ethnographic data on these groups is
not available to confirm this relationship in general. For the Agarabe,
at least, this relationship is recognized by the people in the Arona
Valley.

SETTLEMENT PATTERN

The discussion which is to follow will be based directly on the people
of the Arona valley and even more specifically on the Gadsup of Akuna
village. In general however, a great deal of similarity will emerge
with their linguistic cousins, their geographical neighbors, and even
to a lesser extent with a general eastern highland culture. They will
have closer affinities with neighbors in the highlands, than with trading neighbors in the Markham Valley. Here affinity is more important
than proximity.

Arriving in the valley for the first time, one is struck by the beauty
and simplicity of soft green waves of vegetation covering the valley
floor encircled by dark green, high-rising margins. These dark green
mountains, covered with Casuarine, members of the oak species,
Podocarpus amara and palms such as Pandanus and Areca seem like
a boundary to the valley itself. The softness and uniformity is interrupted as if it were a transition to a similar whole beyond the mountain. From these high areas numerous spurs appear, stretching half-
way into the valley and then tapering off into the soft green uniformity.
It is on these spurs that villages are situated and along their sides the
major garden plots are cleared. The site is always chosen because of
its situation on such a rise rather than because of its proximity to
water sources or level garden areas. For this reason the water is
seldom closer than five hundred yards to the village while gardens
may be as far removed as a mile and a half over rugged and often
hilly country. In most cases a number of gardens will border each
other, separated by fences. A similar fence of split poles forms the
outside barrier to protect the produce from foraging pigs.

The village proper is surrounded by gardens while footpaths connect it with the spring or river from which its water is drawn. Drinking water is never taken from the center or indiscriminately from any stream because of the practice so common in New Guinea of using certain streams to defacate in, lest some hostile person collects the feces for sorcery against one. In most cases the villagers get water from a spring where the trickle of clear liquid comes from the ground. They then dig a slight depression below this place and insert a V-shaped leaf which forms a funnel and from which the clear water runs into the leaf-cup, if persons drink there, or into the bamboo container if the water is removed to the village.

The gardens which surround the village, or which are situated at some distance from it, do not follow any set pattern or form and may be square, round, or irregular, but in all cases a fence up to four feet in height, made of split logs, is maintained around it. This fence, it seems, is primarily there to protect the garden products from pigs, but also serves to denote the fact that the owner is still interested in the garden and is retaining his rights. Within the fenced area one usually sees an old tree, charred by burning, which has been allowed to remain and which supplies welcome shade during the heat of the day. Fruit trees, too, are common in the gardens, as are clumps of bamboo or a few betel nut palms, but this should not distort any statements about the neatness and orderliness of the gardens. Bordering these gardens are always the small trees and immature forest growth, taking over earlier gardens which have been left fallow, and containing many pandanus, banana, and betel (areca) palms over which ownership is retained. Beyond this, and into higher altitudes, is found the real forest of which we spoke above.

It is not uncommon for one to notice a house, usually no longer occupied, which is built in the vicinity of the gardens and which allowed the owner to live by his garden and care for it. These houses it is felt, are the remnants of what we call stage two in a residential cycle which is now nearing completion.[5] Under the traditional system with intermittent fighting going on and with persons always in danger of losing their lives, they all lived in barricaded villages containing a men's house or two — and a number of women's houses (very often houses for unmarried boys and for unmarried girls as well), with all of this surrounded by a strong barricade of fence posts and a gate which could be barred during enemy attacks. This was the period during which the woman would go to the garden to collect food accompanied by her

5. These postulated stages on the village level are discussed briefly on page 11 and more fully in Chapter 9.

8

armed husband, or when the husband would work in the vicinity of the garden with his weapons close at hand in case of need. Approximately fifteen years ago the Gadsup entered the second stage of this postulated residential cycle, when peace was enforced by the White Man bringing superior weapons and cultural objects which were desired. Instead of walking great distances to their gardens every day either to collect garden produce or to feed the pigs, a number of the men constructed houses at their gardens. They could do this because of the enforced peace, but it did not assure safety, since a number of attacks took place up to ten years ago. It was also not very welcome with the Administration, who found it hard to enforce law and order, collect taxes, and — with the help of the Medical Orderly — to improve health and sanitary habits. The result was that a number of these outlying houses continued to operate as residences, but whenever the officials visited a village the residents would be present and the outlying houses would be described as "pig houses". Today we are approaching the completion of the cycle with villages again forming the residential center. Fences are maintained around the village even though these are far-removed from the historically popular barricade. The houses which are still to be seen at the gardens are usually delapidated and have completely fallen into disuse. Should a person be at the gardens when a thunderstorm occurs he will take shelter in one of the houses and occasionally even sleep there, but they no longer function in any way as residences. As was the case traditionally, the village is the social, ceremonial, economic, and residential center, but at present this is partially the result of Administrative pressure — the result thus of external and enforced values rather than of internalized and traditional values. This also opposes that basic Gadsup personality trait of independence and individuality, the right to choose one's place of residence and his associates.

Returning then to the village as it exists today, we can still notice some of the features of this residential cycle. While Wopepa, Onamuna, Akuna, or Kundana, are nearly circular with a fence which surrounds each village, Ikana, Tompena, Omaura, and Pundibasa consists of a number of residential centers with their own fences and often removed by as much as 500 yards from the other part of the village. While the first group is seen, in agreement with the traditional situation, as 'nuclear villages', the second group is seen as dispersed villages. This term is preferred to that of dispersed hamlets, as we are here definitely dealing with a village; the members identifying with each other and interacting as a community. These dispersed villages are seen as a transitional phase between dispersed hamlets (when people lived at their gardens) and the nuclear village phase when people are united in one center.

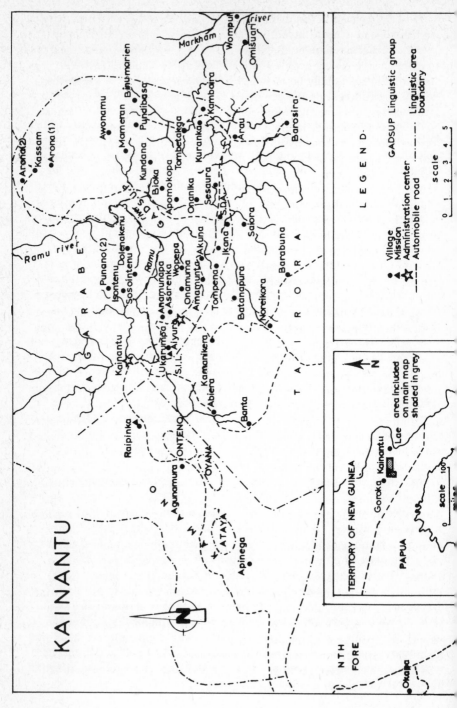

KAINANTU

LEGEND

- Village
- Mission
- Administration center
- Automobile road
- GADSUP Linguistic group
- Linguistic area boundary

scale
0 1 2 3 4 5

TERRITORY OF NEW GUINEA

Goroka • Kainantu
area included on main map shaded in grey
• Lae

PAPUA

scale
0   100
miles

Whatever phase in the evolution of the village we are dealing with, the houses and general layout are the same. In all cases the houses, be they part of nuclear or dispersed village, are organized around an open square where people meet and where announcements are made. The houses may be of the traditional round type in which the floor and walls rest on the ground, but these are no longer very prevalent due to pressure from the Administration that inadequate houses be destroyed. Some officials condemned the traditional houses, preferring the square house with raised floor. This latter type, in which either the whole floor or the part of it which serves as sleeping platform is raised, forms the greatest percentage of house types in the Arona Valley. While traditional houses form the one extreme, at the other extreme and forming just as small a percentage of the total are square houses which stand on piles. Variations on all these types are present, and the village does not give a very uniform impression. All these houses are made of woven bamboo mats which are attached to pole structures while an opening for a door is left, and the roof is thatched. These houses never had windows, but one or two recently erected houses have windows covered with Windowlite or are simply in the form of a board which can be swung open. Among the houses or behind them there is usually a small tobacco garden for each complex of houses. The grounds in front of and around the houses are kept relatively clean and neat by regular sweeping, but this is partly to satisfy the Administration and medical officials. Many Gadsup villages now have a small church, and when these belong to the Lutheran Mission, they are served by a native evangelist from the coast and usually by a native teacher, too, but the latter nowadays is frequently one of the Gadsup from a nearby village. Those villages which have a church ascribe a fairly important position to it, and it occupies an important place, as the old church in Akuna which stood in the center of the village. The houses of the luluai, the government-appointed village headman, and tultul, the government-appointed interpreter, do not differ from those of the other villagers, nor do they occupy any important position. On the outskirts of the village are now always a number of latrines which have been erected due to the insistence of the Administration, and also a women's house where women go into seclusion during the menstrual period, during postparturition, and, often after some domestic squabble. Latrines are built more to satisfy the Medical Orderly than for use, and while long speeches are made in the village square about the presence and danger of "gems" the people still prefer the bush and kunai, and men especially seldom use these outhouses. Referring to what he calls "pre-village" conditions, Rowley states: "Often the official 'village' is still mainly a place of assembly for census and taxation purposes. The inhabitants may live for most of the year in family or clan groups at the garden sites, and from their point of view the 'village' may be no more than a device which keeps the Administration at a distance; a place where the patrol officer may discharge the formality of inspection, and where the latrine, spic and span and unused, is the symbol of conformity with the reforming urge of the white man" (1966: 33-4).

THE PEOPLE

Since the bulk of this study deals with the people who live in the Arona Valley, this brief discussion is used to introduce the actors. It introduces them not as individuals or persons, but as milieu — as the context which always has to be kept in the back of the mind as the reader visits villages and houses, or squats with the women in their gardens. For this is my ideal, that a reader who has not shared in the pleasant experiences I now recall, may never the less be able to place himself in the situation. He may lack the smells, the sounds, the pulse of the Arona Valley, but in closing out his present setting, his mind should take flight. It is very difficult, and some scientists feel unnecessary, to do this in an academic study and even more difficult to maintain this communication throughout. But it is the best way to communicate more than words and sentences.

The visitor who is new to the highlands and especially to the Arona Valley, is immediately impressed by the mildness of the climate. Due to this temperateness the people who live there need very little in the way of dress. That which they do wear satisfies primarily the cultural requirements of modesty and aesthetics, and only secondarily any need for physical protection and comfort.

Most of the women in the valley wear the traditional grass skirt — manufactured by attaching a great number of strands of dried grass fiber to a girdle of bark or string. This basic form of dress differs between young girls, post-pubertal girls, and married women, — each variation marking a change of status in the life cycle. A female, whatever her age, will never appear naked and even little baby girls emerge at two weeks of age from the women's house with a small grass skirt firmly attached around the waist. During the next two or three years the skirt may frequently shift,while children play, sometimes even covering the side of the thigh, but girls never appear without them.

Boys on the other hand may go naked until they are seven or eight years old, at which time they receive a pubic covering, and it is thicker, as if built up in tiers of various layers of grass fibre. This grass fibre, 'yienni' as it is referred to, will be discussed below.

12

In addition to this basic covering, a variety of decorative features appear although they are usually preserved for festive or ceremonial occasions. Many of the older people still braid their hair into dozens of little strands which are rubbed with rancid pig grease. The young people, and especially boys who have returned from the coast where they worked, comb their hair into fluffy tops covering the head. A woman may appear with a short pointed stick about the thickness of a match, protruding from the front of the nose; she may wear the dried foot of a fruit bat, a piece of fluffy fur of a tree kangaroo; or under modern conditions of trade store contact, may be resplendent in a colorful necklace of beads. When she is on her way from or to her gardens she always has a string-net carrying bag hanging from her head. The men in nearly all cases have a woven bark armband around the biceps of the upper arm. Generally they carry tobacco leaves or pepper leaf attached to this but during ceremonial gatherings it is used to attach decorative leaves of many colors, or bunches of grass which aid them in camouflage during an ambush. On these latter occasions too the men wear their decorated sandstone nose plugs which are passed through the septum of the nose, or the tusks of boars specially sown together "to look fierce". An adult man is rarely seen without his string bag hanging round his neck and stocked with betel nut, gourd and spatula for lime, and a bamboo cigar holder. Even today males do not leave the village without their bows and arrows. Also preserved for festive occasions are necklaces of dogs' teeth or head-dresses of cassowary, cockatoo and other feathers.

At the time of this study an increasing number of people, especially males, were changing their dress due to increased contact with the Administration and trading stores. A very small number of women have acquired either cloth, which they wear in the form of a lap-lap (wrap around the waist), or in some cases loose hanging maternity dresses of cotton. To a major extent these foreign dresses are reserved for festive occasions associated with the mission churches — both Lutheran and Seventh Day Adventist. Partially too these materials are used to remove the constant irritation of a bark and grass skirt which irritates the waist, often causing open sores on the women's hips. On the whole then, women's dress is still the traditional grass skirt with variation appearing either to protect the body or to make them more acceptable to Christian missionary gatherings. A variation on this is a lap-lap which is loosely draped across the shoulders in the form of a shawl to cover the breasts.

A decreasing percentage, perhaps ten per cent, of the males still wear the pubic coverings discussed above. Most adults, however, have one which they don for ceremonial occasions. About the same number of men, mostly the young men who have been out of the valley on indentured labor to the coast, now wear short pants, occasionally accompanied by a neat shirt or tee-shirt. The remaining eighty per cent can perhaps be equally divided between those who wear the lap-lap either with nothing under it or over a grass pubic covering, and those who wear the lap-lap over a pair of shorts which is often worn and torn.

One frequently observes women, and less often men, whose hands have been mutilated by severing the joint of a finger. For women this marks the loss caused by the death of a beloved child or husband. Men would show sorrow in this form for the death of a child but never for a wife. Occasionally though this was only observed in the eastern part of the Arona Valley, women would be decorated by scarfication.

Beyond the Arona Valley a number of villages are found which agree basically in language and culture with the Arona Valley people. There are four of these village clusters each named after the largest village within the cluster. On the west, across the mountains from the Arona Valley, are two large villages and one small one which will collectively be referred to as the Aiyura villages, taking their name from the Aiyura Agricultural Experimental Station on the eastern border of the Kainantu Valley. In a southwesterly direction from Aiyura, but separated from it by a number of Tairora villages, are the five Onteno villages with a population of 380 people, and beyond the mountains further southwest are the Oyana (counting 318 people), and finally the Ataya villages (238 people).[6] While the Aiyura Gadsup maintain strong ties with the Arona Valley Gadsup, they are no longer a homogeneous group, and intermarriage and acculturation have followed the contact with laborers from all over the highlands who live at the agricultural station. Due to their geographical locality and history, there is a relatively high degree of intermarriage with the Agarabe, their linguistic neighbors. The Aiyura Valley Gadsup also maintain strong ties with the Onteno villages, and marriage is frequent. The Onteno Gadsup, in turn, maintain these same ties with the Oyana Gadsup. The latter, in turn, visit and intermarry with the Ataya Gadsup. While all of these Gadsup groups have intermarried with Agarabe, Kamano, and Fore speakers, and have undergone acculturative influences from them, their culture, kinship terminology, and language are closer to Gadsup as spoken in the Arona Valley than to any of the neighboring groups.

Although the Gadsup of the Arona Valley stand as a unit when contrasted with those of other clusters or other language groups, among themselves they recognize linguistic and cultural differences between

6. These figures were kindly supplied by Mr. John Fowke of the Sub-District Administrative Office in Kainantu.

what we have called Eastern and Western divisions of the Gadsup, the border being somewhere down the middle of the valley. Though a river flows the length of this valley, it is rarely of such force that it would cut off all communication. Social ties and interaction between the two groups are very weak, while intermarriage is nearly nonexistent.

The Gadsup border to the north on the low-lying areas of the Ramu and Markham, on the east and south live the Tairora, and on the west and northwest, the Agarabe. Within the linguistic boundaries of the Gadsup are two non-Gadsup-speaking villages, namely Binumarien and Kambaira. Both of these linguistic groups, each being no more than 150 persons, have their affinities with the Tairora. The total land area occupied by the Arona Valley Gadsup is approximately fifteen miles from east to west and nine miles from north to south, or 135 square miles, with a population density of fifty-eight persons per square mile.

The field research on which this study is based was conducted primarily in the village community of Akuna. While this is the field of concentration attention will be given to extra-village and inter-village relations, where Akunans interact with members of other communities.

SYNOPSIS OF THE COMMUNITY

The village of Akuna is situated in the western part of the Arona Valley. On one side of the village extends virgin forest, which is gradually being pushed back to make way for new gardens under the system of swidden agriculture. Since coffee has been introduced to the economy, the need for new gardens has arisen and increasingly larger areas are being cleared. The eastern part of the village leads off into the valley which extends for many miles north-east and offers valuable land for pig grazing. The major garden areas are southeast of the village and every day the women can be seen travelling between village and garden or bent over a digging stick as they clean their gardens or dig into fertile brown earth to uproot the tubers which form the staple of their diet.

The village of Akuna is compact, organized around a central village square which is clean and neat. The houses are made of plaited bamboo with thatched roofs, while the whole village area is enclosed by a fence of split logs. Since each house opens into the village square, there are a number of earth ovens which seem to unite those houses closest to them. In the background, between the semi-circular outlay of houses, and the village fence, are a number of latrines which have been built primarily to satisfy an administration which believes in

15

the germ theory of disease. In this transitional zone, too, are a number of seclusion houses where women go during menstruation or after the birth of their children. At the western extreme of the village, removed from the houses and yet within the village fence is the Lutheran mission church. The evangelist and his assistants live outside the village. Akuna has fifty-five houses and, while we are speaking here of a village, we are in effect dealing with two parts of the village of Akuna for while this fissive force is noticeable in the physical layout of the village, it is even more noticeable in the association of the people.

There are 245 people who live in Akuna. This is fairly average for this part of the New Guinea Highlands and does not include the missionary appendage to the community. These people spend relatively little time in the village proper, for while women have their chores which take them to the gardens, males visit the forest area for hunting and other semi-economic activities. When the members of the community are in the village, women usually are involved in food preparation, or the basic processing of natural materials which they will use for making grass skirts, mats, string bags, or similar objects. The males congregate in smaller groups or as a body, if need be, to discuss communal activities like a pig hunt, or to discuss court cases. Children are always present, running, singing, playing, or assisting a parent. While life is rarely a hustle, some activity is always taking place. Group interaction centers around the house, the earth oven, the court area, the hearth, the women's house or the village square. Each of these localities is the centrum for a particular variety of association, for a special sub-field of interaction, for a limited or more extensive number of persons to meet and to discuss. This is the primary group of people we will be discussing in these pages — people who belong to Akuna because they live there and pledge loyalty to the community members and the leaders. These are the places where we will visit them; a family in their house; a group of neighbors around an earth oven; the adult members of the community in court; a number of people talking quietly as they sit around the hearth while smoking or chewing their betel nuts; those who retreat to the women's house to cool down, or recover, or to become neutral; or the children playing in the village square.

But this community of Akuna does not exist in isolation. Around it are other villages which are formed into a district and among whom certain attitudes and loyalties exist. There are furthermore ties of friendship, of kinship, or of affinity which link the members of these various communities. They all speak Gadsup and so their relationship will be seen as they all contrast with persons and communities which differ from them along these lines of division.

16

Chapter 2

SOCIO-HISTORICAL BACKGROUND

The eastern extreme of the central highlands lies in the administrative sub-district of Kainantu. The villages in this area, and particularly those which we will discuss had their first known contact with Europeans in 1920,[1] and possibly shortly before that date.

FOREIGN INFLUENCES

There is some possibility that Hermann Detzner, a captain in the German Defence Brigade might have visited this region during his wanderings, but this could not be confirmed. It will be recalled that this northern part of New Guinea, prior to World War I was German Territory, known as Kaiser Wilhelmsland. With the outbreak of the war against Germany, the area was occupied by Australian forces who administered it until they were replaced by the League of Nations mandate. Between 1914-1918 Detzner fled from Australian forces and penetrated into the central part of the island as far west as Mt. Hagen.[2] During these four years he journeyed this unknown and often hostile island, finally giving himself up after armistice in 1918 and being deported. By way of Rabaul and jail in Australia he finally reached Germany, somehow still retaining notes, observations, and rough sketch maps made during the previous four years. His map (reproduced in Rhys 1942: 40-41) suggests that he passed over the watershed of the Markham and Ramu rivers. But whether he in fact came near the Arona Valley, thus influencing the culture of eastern Highland villagers, is not clear. Missionaries who entered this region a few years later do not report any recollections of an earlier visitor.

1. This pre-dates by nearly a decade the commonly accepted supposition that Leahy and Dwyer, and later Taylor (representing the Australian administration) were the first to explore this region. See Chinnery (1934: 113-121), Leahy (1936: 230, 238), Leahy and Crain (1937: 42-74), R. Berndt (1962: 2-3).
2. Ruhen (1963: 130) claims that he "set up his standard on the peak of Mount Hagen ..."

There also seems to be some question as to the factuality and authenticity of both his map and his claims of travel and discovery. Some persons have gone so far as to call his claims no more than fiction, while Souter states: "The awkward fact remains, however, that valleys resembling those he described in 1920 were actually discovered in the early 1930's. Certainly his description of them was vaguer than one would have expected from a surveyor" (1964: 122). One has to keep in mind though that missionaries were entering the eastern part of these high valleys shortly before his writing, and had been exploring the Finisterre, Saruwaged, and Rawlinson Ranges for two decades. During this time of his wanderings too, Kaiapit in the Markham Valley was already established and missionaries must have had reports or actual glimpses of the grassy plains and towering peaks. One could expect that the lonely German would have numerous friendly exchanges with the German-speaking Neuendettelsau missionaries during his wanderings, and while they would not have assisted him due to their declaration of neutrality, he nevertheless learned from them.

It seems that the same reserve was felt about Detzner's claims by the Gesellschaft für Erdkunde in Berlin. They forced him to retract all fictitious claims regarding his discoveries and to file corrected copies of both his map (1919) and his book (1920) in the offices of the society. At this time too, either on his own initiative or due to pressure from the officers of the Gesellschaft, he resigned his membership in the Geographical Society. For the sake of clarity the explanation given by Detzner is quoted here in full. The unsigned official pronouncement appeared in the Zeitschrift der Gesellschaft für Erdkunde zu Berlin (1932: 307-308) under Brief Communications, and reads: "Detzner's Explanation. After discussion resulting from initial examination of the book Four Years among Cannibals by Hermann Detzner, leader of defensive troops, the Board of Directors and the Advisory Board of the Geographical Society in Berlin held it their duty (without prejudice to the completely acknowledged achievement of Herr Detzner in carrying through of his border expedition and the annexation of the Saruwaged in New Guinea) to publish the following explanation of Detzner in order to create scientific clarity, in conjunction with which the account published in the Zeitschrift der Gesellschaft für Erdkunde (Newspaper of the Geographical Society) in 1919 (page 371) in the corresponding parts becomes weak. The explanation reads:

"I must explain at this time that I have given occasions for false interpretation with my book Four Years among Cannibals (Berlin, Scherl, 1919) relative to my trip into New Guinea. This book is only in part a scientific report of fact. In the main, it is a romanticized presentation of my stay there, originating from the press of special

circumstances as they acted upon me after my return home after the
World War. Several of my sidetrips are not discussed in the book; on
the other hand, it contains sections which do not correspond to the
fact. I cannot thus maintain with certainty that I myself on my trip
through Kaiser-Wilhelm Land reached Josefsberg in the Fall of 1914.
The breakthrough attempts which I described did not actually take
place. The journey to Saruwaged took place in 1916 (not in 1915, as
given), and not alone or independently (there were several cogent
reasons on the writing of this book for this presentation). A copy of
the book containing the necessary deletions and alterations in text and
maps has been deposited in the library of the Geographical Society.
The parts containing no deletions I will stand by as being correct. I
request that in the future these and my scientific publications in the
Mitteilungen aus den deutschen Schutzgebieten be relied upon. I intend
to complete them through further scientific work."

MISSIONARIES

The influence of various missionary contacts and undertakings in New
Guinea is of two kinds. We are dealing with both historical influences
due to the date of contact as well as the policies of missionary groups.
Since it is of primary importance for the people of the Arona Valley,
we have to single out the Lutheran Mission for discussion here.

The first Lutheran missionary to start work in New Guinea, except
for Otto and Geissler who landed at Manokwari on the western side of
the island in 1855 was Dr. Johannes Flierl. When Queensland annexed
the part of New Guinea immediately opposite her shores in 1883, and
Britain ratified the occupation of the southern half of the island — the
present Papua — it left only the northwestern section of the island.
This was occupied by Germany later in the same year (1884) and
called Kaiser Wilhelmsland, while permission was granted to the
Neuendettelsauer Missionsgesellschaft to start work. And so on July
12, 1886 Flierl landed at Finschhafen. He established his station at
Simbang and soon the influence of this new settlement and its teachings were spreading beyond the Huon Peninsula. A year after this
first mission denomination, named for the sponsoring town of Neuendettelsau in Bavaria, a second Lutheran denomination entered the
territory. Eich and Thomas representing the Rheinische Mission,
which is also alternately named for the sponsoring town of Barmen in
the Rhineland, established their headquarters a hundred miles west
at Bogadjim. In 1921, these two denominational missions united as
the Lutheran Mission and the new undertaking in its post-war phase

was heavily dependent on American funds and personnel for the continuation of its work.

Especially during the earlier phase of this work, though it has now become characteristic of the Lutheran Mission, there was a very strong emphasis on the involvement of the whole community. Keysser, one of the important members of the Neuendettelsau Mission and author of such books as Anutu and Eine Papuagemeinde "advised young missionaries not to preach at all at first, but to get into touch in a human and neighborly way with his people and establish vital relations with them" (Knak 1938: 20). The Lutherans also required that the person who became a Christian should not become isolated from his community. "When a solitary man from the Hube, a neighboring tribe came to become a Christian, expecting to settle among the Katé (where the first Christian communities had begun), since life would be impossible for him among his own people, it was the congregation — who insisted that such a thing could not be" (ibid.: 6). Thus the new convert had to remain in his community, causing changes among those with whom he had daily intercourse. Flierl also followed the policy of using vernacular languages for both education and mission work. The first groups to be contacted by the Simbang located missionaries were the Jabim and the Kai people and their languages, Jabim and Katé, became the first languages reduced to written form. By the first decade of this century these two languages were used as tools in the mission schools while textbooks and parts of the New Testament were printed in these media. Soon other languages appeared which were used for mission work, but Katé became identified with the Lutheran Church and today is the Lutheran lingua franca in the highlands.

Associated with this very important aspect of policy and modus operandi, the Lutherans were of great importance due to their exploration of the unknown regions and their establishing mission stations in these territories. The earlier phase is recounted by Johannes Flierl (1931:67), but it was his nephew, Leonhard — who had since joined the mission undertaking — who aimed at extending the mission field (du Toit 1966). By 1918, a station had been established at Kaiapit on the upper Markham, but on the northern side of the river. "Two years after Kaiapit had been established an exploratory party of missionaries went into the Eastern Highlands as far as Pundebassa" where they met with considerable hostility. "A year later in 1921, Missionary L. Flierl went into the same area. Contact was established and maintained so that the first evangelist station could be established in the Highlands ..." (Frerichs 1957: 59).

Their experiences on these expeditions and a detailed account of nearly a decade of explorations in the Eastern Highlands is contained

1, 2, 3. Examples of stone mortars
collected in the Arona valley by the
Archbold expedition in 1959 (see page
5). (Courtesy of the American
Museum of Natural History)
4. Pigs teeth nose plug from Akuna
5. Valued decoration made of dogs'
teeth. The smaller object was worn
on a man's temples, one on each side
while the larger decoration was tied
to the nape of his neck

1. The traditional
Gadsup house with
plank or woven walls
and horizontal boards
to serve as door
2. A beautiful example
of 'koma' weaving
3. One form of woven
wall, using flattened
bamboo slats

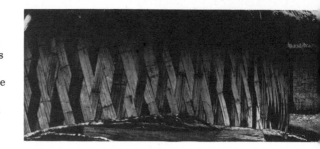

in Leonhard Flierl's little book, Unter Wilden.[3] The account starts
with an expedition from Kaiapit in January 1920, across the Markham
River to the Kratke Mountain ranges which were nearly completely
unknown. Once again we find reference to the Neuendettelsau policy
"that its young communities in New Guinea become the supporters of
independent missionary undertakings. Each congregation has a
special missionary territory, for which it is responsible. The community sends proselytes out, pays them itself, and exercises super-
vision over them as much as possible" (1932: 7). Since the Sattel-
berg congregation was wanting to extend its field of work it "was
advisable to establish friendly relations with the Gazub,[4] the mountain
dwellers, in order that it would be possible in a not too distant time
to begin missionary work" (ibid.). Shortly after leaving Kaiapit the
expedition crossed the Markham and ascended the steep mountain
wall on the south side from which they looked down on a wide valley
crossed by a stream which they later discovered to be the Ramu River.
They proceeded to the village of Binumarien. This village where they
established camp and which in fact became a gateway to the highlands
is important for a number of reasons. "The inhabitants of Binumarien
already stood in a trade union with the Uzera people", in the Markham Valley, Flierl tells us, but this was also the village where a
missionary had spent a night a year before, in 1919. This was Missionary Lehner (ibid.: 8 and personal communication 1962), the same
man Frerichs referred to in the quotation above. It is also most
important to keep in mind that the Binumarien are in the Highlands
and surrounded by Gadsup speakers, yet belong to the Tairora linguistic group (McKaughan op.cit.: 118 and Wurm 1960: 127). It is therefore not unlikely that some of the changes which were now being introduced might have filtered through to the Tairora Valley, perhaps by
way of Kambaira — another isolated little Tairora-speaking village
(Wurm and Laycock 1961: 138). This is all the more likely since
evangelists who accompanied the missionaries shortly afterwards
spread into the Arona Valley while a young man named Singeroa established a Lutheran station in the Tairora Valley.

The missionary expedition had a 'warm', though not friendly

3. I was most fortunate to learn that Missionary Flierl, though retired, was
still living in Germany. He was kind enough to send me a copy of his book which
was not obtainable elsewhere, and on two occasions (1962 and 1964) correspon-
ded at length about his experiences and reminiscenses — also sending two early
photographs. One of these letters starts with the emotionally laden statement:
"Anfangs 1929 musste ich Neuguinea aus gesundheitlichen Grunden für immer
verlassen."
4. These are the Gadsup speakers with whom this study deals.

reception in the highlands. From their headquarters among the Binumarien, the party decided to press further west in an attempt to establish contact with the community at Pundibasa.[5] It seems that these Gadsup were not on very friendly footing with the Binumarien people who warned the missionaries that the forementioned villagers "were completely bad people, with whom they continuously warred". The missionaries would not heed this advice, nor react to warning signals in which the Pundibasa scouts deposited fern bundles in the footpath. The result was a somewhat unfriendly reception which resulted in two of the missionaries receiving arrow wounds and in turn fatally shooting Makakayu, the strong man — as we will refer to the New Guinea equivalent of the village chief in other parts — of Pundibasa.

Among the Binumarien greater success was enjoyed by the missionaries who were able to establish a station which was manned by Miwong, an evangelist from among the coastal Hube. At this time too, a station was successfully established in Wampul which is on the border of the highlands and marginal to the Gadsup. They, however, speak a dialect of Atsera which is spoken in the Markham Valley, and would then have less significance for changes in eastern highland cultures. While the latter station was permanent, manned at first by evangelist Gapecnuo, and thus at least serving as a door to the highlands, the Binumarien undertaking was less happy. On one or two occasions Miwong withdrew to Wampul, and after about a year he felt so harassed that he withdrew his household and assistants.

Missionary Flierl visited this eastern extremity of the highlands every year between 1920 and 1928. He reports on frequent contacts he and the evangelists, who remained permanently, had with the non-Gadsup villages of Binumarien, Wampul, and smaller settlements along the Wanton river, as well as with Gadsup speakers in Omisuan and Arau.

During the last year he was in New Guinea, Flierl undertook an important expedition into the eastern highlands in an attempt to open permanent communication with village communities in this area. Some important observations appear in the diary he kept during this July-August expedition. The first part of the journey, as they were coming from the northeast, took them over the tip of the mountain which borders the Arona valley on the east. They saw Kundana, which he describes as a "large group of villages", and were descending to Sesaura when they looked across the valley westward. "On the grassy hills diagonally opposite us we see many fields; that is where the

5. Flierl refers to this village as Bundewateno or Buntibaza as the Uzera people in the Markham Valley refer to it.

Afunakeno[6] live" (1932 op. cit.: 42). Having spent the night in Sesaura the expedition departed but "not one of the Sesaura was willing to guide us, not even for a stretch of the way. They merely opened the exit through the high wood fence, let us go through, and barricaded it again right away, for they were expecting an attack from their enemies". They then left alone and travelled to Omaura where much the same condition greeted them but once they had identified themselves the palisades were opened and they remained for the night. Rather than going into the Arona Valley from here the expedition turned south into a small depression between the mountains, and arrived in the Tairora Valley. This is a neighboring linguistic group to the Gadsup but one with whom the latter do not have cordial relationships. Here Flierl refers to the country of the Beza who were enemies of the Omaura, but it is not quite clear who the Beza are other than being Tairora speaking. That same afternoon they arrived at Barabuna.

It seems that two days were spent here visiting a number of village communities. In addition to Barabuna, he mentions the people of Noraikora and Apaira before arriving in Kainantu. Now the party made visits to a number of the Agarabe speaking villages in the vicinity, and the author remarks the preparation of a feast in which the food was cooked in bamboo containers and "grossen Töpfen".[7] After three days in this region the party decided not to go any further inland but to return to Omaura and become better acquainted with the area and people around it. That night they spent in Isontenu[8] where the people were reserved and shy though not hostile. "This morning (it was August 6, 1928) we went further through a broad valley in which the contours of old fields were clearly visible. The owners of the land probably were driven away by their enemies. Before us lay the territory of Afunakeno; we wanted to arrive today" (ibid.: 49). We learn a number of things about the Akunans of the day, the dismembering of

6.   Flierl consistently used Afunakeno instead of Akunakeno. Both these names consist of the name of the village or place and the suffix -keno. It is also the same suffix he mentioned earlier when he spoke of Bundewateno. This suffix denotes the relationship between a person or people and time or place. Thus, the first mentioned are the people of Akuna (or in his statement of Afuna) while the latter are the people of Bundewa". It seems as if Flierl heard and used the f where we now use k. From his description of locality — as we shall see later too — there is no doubt that this village is in fact Akuna, the community with which this study deals.
7.   The making of clay pottery and trade for these by the Arona Valley peoples will be discussed under the village economy.
8.   In most cases the names have been changed to the form in which they are used today. This reads Isoteno in Flierl.

a finger for a deceased and covering the body with clay; the great popularity of sugarcane and a native bean; the status of women; the fact that sweet potato is the main crop while taro, yams and banana are second in importance, and similar aspects of life which could be readily observed by the outsider. Flierl was greatly impressed with the expanding fields of these people and he states that "they make a good impression because they are so surely divided and kept so clean". From here they returned to Omaura, it seems by way of Tompena and Ikana, for the author states that "we climbed down to Ramu (River)". This then would be through the stream Yondonanomi which feeds into the Ramu further down the valley. The villages which they then visited as they passed by to Omaura, were four in number, he says, and later we shall see that Ikana is in fact divided into two or three subgroups; he could then be speaking about these villages.

In the early thirties the missionary Wilhelm Bergmann, who had been associated with the work in the eastern highlands since the beginning, received instructions from the mission-directorate to establish a European station[9] among the Gadsup. This it seems was something more permanent and better equipped than the ordinary station manned by evangelists from the north coast. Bergmann was told that his new post in the highlands was to serve initially as provision-post and to be followed later by a permanent settlement which would be established further into the highlands. The evangelists, many of whom are still working in this area found this a great moral support. Kevengu, the evangelist in Akuna, recalls with great clarity how Bergmann's presence supported their work, and it was shortly after this that they spread out. At this time Singeroa entered the Tairora Valley, Kevengu settled in the Arona Valley, and stations were permanently established in Kurangka, Binumarien and other villages to the east of the Arona Valley. In response to his instructions, Bergmann now established the permanent European station in Kambaira — it will be recalled that this is a Tairora group surrounded by Gadsup speakers. Shortly afterwards he pushed further west across the Ramu River and established a station at Onelunka. In 1935[10] the Lutheran Mission headquarters for the eastern highlands was moved from Kaiapit to Raipinka, four miles south-west of Kainantu. There a Katé community gradually developed with teachers and evangelists, relatives and assistants — and increasingly also hangers-on from the neighboring regions —

9. "... unter den Gazupleuten eine Europaerstation anzulegen ..." (Flierl personal communication, 1964).

10. This is the commonly accepted date, also mentioned by Frerichs (1957: 64) who at that time was head of the Kaiapit station. Flierl (personal communication, 1964) has suggested that the change took place in 1938.

which forms a unique linguistic and cultural enclave. Its presence is also of great importance to the whole of the eastern Highlands since each of the evangelists in the village where he teaches is now united as if by an umbilical cord with the headquarters. This "cord" also allows two-way influence — there is a continuous visiting by evangelists and villagers to Raipinka, and the latter frequently sends white missionaries to the villages on special religious occasions.

By the time Bergmann was asked to establish a permanent base in the Highlands, explorers and administrators were also entering this region and pressing westward. Thus it was a relatively short while before Kainantu started developing as the administrative base for the eastern highlands. In the late thirties, shortly before World War II, the Seventh Day Adventist Mission established a station and school in' Kainantu. Some time later they extended this to Omaura in the Arona Valley where a school and small hospital have been established. As in the case of the Lutheran Mission, such headquarter villages also serve as base from which native evangelists are sent out to surrounding villages. Few are the settlements that do not have their own resident native evangelist belonging to one of these religious denominations.

We should also mention the fact that at the time of this study the Salvation Army had started work in this district. They had at the time two nurses stationed in Kainantu who would visit the Arona Valley once every six or eight weeks. Visiting basically at the administration's medical aid posts, they were aiming their interests at the general health and childcare for expectant mothers and their children.

We have dwelled at some length on the work of the Lutheran Mission for two reasons, the most important being the early date at which they initiated their work in the eastern highlands. It would be folly to expect that the presence of a mission representative, be he white or an evangelist from the coast, had influence only on the supernatural beliefs of these people. These in fact seem to have been little affected for at least three decades and it may be plausible to suggest that it is only the generation that is born and matures under the influence of school and church that really changes. The early decade at least was marked by requests, exchanges, threats, attacks, intrigues, defrauding, and ambush by members of the local population to acquire western cultural objects, especially metal axes, knives and the like.[11] Flierl relates numerous cases of this nature but one especially ingen-

11. McCarthy, at present Director of the Department of Native Affairs in New Guinea was one of the first patrol officers to visit J. L. Taylor (see below) at Kainantu in 1931. He refers to the Yongi (obviously an Agarabe group) who "knocked down a carrier and ran off with two knives he carried" (1963:87).

ious device by which the Gadsup acquired metal tools. On one occasion they had killed numerous pigs and presented these to the missionaries as a gift. Gift giving, however, requires reciprocity and thus the missionaries were placed in the precarious position of not being able to refuse the gift for fear of insulting the local leaders, and by accepting the gifts they were placed in a position of obligation — this obligation quite clearly being satisfied by one type of gift only. We have also given more attention to the Lutheran Mission than to other denominations working in the Arona valley, because the village of Akuna has for many years had close ties with Lutheran evangelists from the north coast.

ADMINISTRATION

In addition to the search for new congregations by the missionary undertakings, the opening up of the eastern highlands resulted from the search for gold.

The first gold discoveries were reported in 1877 but these were of relatively little importance. In 1913 an Australian prospector, Arthur Darling, found gold on the Watut River and he was followed by "Shark-Eye" Park who commenced mining operations in what became famous as the Morobe gold-fields. Life and activities along the Bulolo River, at Edie Creek, and at Wau have been vividly described by Boothe (1929) and Rhys (1942), while Taylour and Morley (1931) discuss the mineral significance of these undertakings.

Every prospector, however, was hoping that he would be able to make a significant discovery of a rich source. The result was that people were always exploring further inland, looking for signs of alluvial or reef gold. Flierl mentions that after their return from the visit to Tairora country, Kainantu, Akuna and Omaura in August 1928 they met at Wampul "our fellow-country-man, Herr Baum" (1932 op cit: 52). He wanted to know from the missionaries whether on their inland tours they had come across any gold in the Ramu or its tributaries. As early as 1929, E.W. (Ned) Rowlands had crossed the Bismarck Range and set up camp near the watershed between the Ramu and Markham rivers. This grassy plateau country where Rowlands started his settlement was fairly densely populated and the prospectors which joined him in time, must have had an influence on the surrounding Agarabe speaking peoples. When Michael Leahy made his first visit to the central highlands in April 1930, he visited Gaitluren[12]

12. He identifies this village as Kaimantu, but this latter term had already been used by Flierl on the tour discussed above.

before returning to the Papuan coast. In October 1930, he returned to
the headwaters of the Ramu river, where at a place called Onapinka,
he visited Ned Rowlands (Leahy 1936: 238). Ted Eubank, a Kainantu
resident who is still living, came in at the same time as a prospector
for gold (Watson 1964b: 2), as well as the Peadon brothers and Kar-
Kar Schmidt (McCarthy op cit: 86). Shortly afterwards Leahy and
Michael Dwyer bypassed these diggings by travelling north of the
Arona-Kainantu valley area, and there started their famous explorations of the Central Highlands. The Leahy-Dwyer expeditions (see
Leahy and Dwyer 1937), joined later by a younger brother, Daniel
Leahy, opened up the Central Highlands, especially since they were
followed by a government party. The latter was led by an Assistant
District Officer of the Australian administration, J. L. Taylor, and
included a surveyor K. L. Spinks. Joint expeditions by these persons
established a government station at Mt. Hagen and explored the Purari
and Sepik rivers. This party had been "dispatched from the Ramu
Administration post" in March 1933 (Official Handbook of the Territory of New Guinea 1937: 281).

The Australian administration seems to have established a district
headquarters at Kainantu shortly after 1930 (C. Berndt 1953: 112; R.
Berndt 1962: 3). It was served by a small aircraft which called
"fairly regularly" to provide supplies for the patrols — J. L. Taylor
was then stationed here — and prospectors. A year or two later Mc-
Carthy was sent in to prepare a better airstrip (op. cit.: 86). While
the influence of an administrative post was not overly important, it
did influence the traditional culture in neighboring villages. Fighting
was halted, trading started, the missionaries received indirect support from the presence of officials of the government. "By 1935,
there were altogether ten mission posts occupied by Europeans between
Kainantu and Mt. Hagen, and seventy-four occupied by native mission
teachers" (Souter, op. cit.: 185). With the outbreak of the Second World
War, civil control was relaxed and most of the Europeans and missionaries had to return to the coast. While intervillage fighting flared up
again and there was a periodic return to the hostilities which preceded
European administration and patrols, changes which had occurred
could not be undone in a few years.

During the war Japanese parties infiltrated to the north of Kainantu
and also briefly entered the Arona Valley at Kassam Pass. This was
in 1943 and it reverberated throughout the Arona Valley, as far as
the Gadsup-Tairora border on the south.

Dexter describes how the bombing took place around Aiyura in June
1943, and a month later there were reports of Japanese at Arona. The
Australian troops were also crossing the Arona Valley and on August
27, of that year a platoon under Captain McKenzie killed six Japanese

after leading them into an ambush (1961: 251). The major forces were only now being assembled as the Australians met at Arau on September 6, while on the 24th of that month about one hundred Japanese were reported at Arona (ibid.: 416- 431). Whether these two forces clashed is not known but if they did it most likely occurred in the Markham and not in the Arona Valley. A number of the villages assisted or were forced to assist this new type of contact but it was both marginal and very temporary. The major influence still originated from civil and military personnel in Kainantu, and the associated settlements which were appearing as traders entered the region to set up bartering undertakings and later trade stores as money became more commonly used.

Since 1938, an Agricultural Experimental Station has been functioning at Aiyura. This settlement is on the eastern border of the Kainantu Valley and in fact forms the western extreme of the Gadsup cultural area. The station was briefly closed down for eight months during 1942-43, when the Japanese were north of Kainantu and bombing took place in the general region. The experiments with agricultural products and especially the growing of coffee has had a very important influence on the Arona Valley. Not only do ideas and changes spread from this settlement, but opportunities for labor have increased. This was also true at Kainantu where local labor was utilized after the re-establishment of administrative authority, rather than drawing on peoples in the western regions, such as Chimbu, as had been done earlier.

Many of the older people still clearly recall the first time they saw a White man and the first aeroplane which flew over the Arona valley. The 'luluai' of Akuna, with all the acting skill and oratorical gifts he possessed, dramatized his re-enactment of this great occurrence. "We were working in the gardens down where Wayopa now stands", he explained. "I was a young man like Yifoka.[13] We were clearing a garden and all were working when suddenly we heard this thunder approaching out of the clouds in the sky, but the thunder did not quiet down — it was unlike anything we had ever heard. We were scared and yet wondered what produced such strange noise. Then we saw this large 'pisin' (pigeon) going past through the clouds and disappear. When we saw it we ran for cover under the brush in the garden. That night when we returned to Akuna — the village was still at the old site against the mountain — we killed a pig and smeared the blood on our hands and arms and over our faces so that it covered our eyes that had seen this large pisin". From Torawa, who is much

13. This young man was about twenty two at the time, and while married had not yet produced a child. Later that year his wife had a baby.

28

younger than the luluai, the following description was obtained as his father, Inunti, had recounted it innumerable times. "We were at Yankapuki, near where Wayopa now stands, and all working in the gardens picking peas — you were not yet born. Then while we were working one man remarked on a strange sound and we put our ears to the ground and listened for it sounded as if the rumbling came from the ground. Then suddenly we heard it overhead. The women all ran for cover under the brush and trees around the garden and the men gripped their bows and arrows for defence. They did not want to stare up at this new sight. When we got back to the village that night we killed a pig and took the grease and smeared our eyes — everybody who had seen this new thing had to protect his eyes. That night the men all gathered in the men's house and prepared strong bows and arrows, for if it should return, we were going to bring it down and eat it". At this point in the recounting of the history there is much laughter for the people say how could they have eaten this big bird that is made of metal. They called it 'anda nu?mi' composed of 'anda' a corpse and 'nu?mi' the generic term for bird (differentiated from 'numi' meaning lice) — because it was like a large bird.

"About three years later" (the luluai in fact counted three Orande ceremonies) the luluai continued, "we saw the first white man. It was shortly before the end of the day when we saw the white man coming — he had crossed over that hill (pointing northwest of Akuna)[14] from the Agarabe and came across where Wopepa now is. We thought the ancestors were returning. As he came closer we grabbed taro, yams, sugarcane and tobacco, and holding them in front of us we ran backwards for cover. He motioned to us with matches that he wanted to buy food in exchange for his gifts, but we said "You desire food? Why don't you take everything" and with this we threw tobacco and sugar cane at his feet and ran for the bush. The white man was a Lutheran Missionary. Later a Lutheran native evangelist came and he and his wife lived at Onampa. " This in fact was Kevengu who relates their history from this point on.

For some years there had been an interest in official circles regarding the possibility of raising livestock in the obviously healthy highlands. The Australian millionaire Mr. E.J.L. Hallstrom did in fact start raising sheep in the Wahgi Valley, western Highlands. Thus in 1948, the Arona Livestock station was established in the northern part of the valley and shorthorn cattle and saddle horses bred. Simultaneously experiements were made to improve the natural pasturage. These experiments with breeding seem not to have been very satisfac-

14. It should be remarked that this was indeed the mountain Flierl would have to have crossed coming from Isontenu among the Agarabe.

tory for the station ceased operation in 1959,[15] and at the time of this study two years later few signs of the station were left. There were no cattle or horses in the southern part of the valley and the local population lacked recollections of any important influences which originated in this setting. It is fairly obvious that since we are dealing with a horticulture-based economy, changes in this and related spheres of the culture will be of greater significance and more acceptable than such innovations related to livestock.

In 1956, the first members of the Summer Institute of Linguistics arrived in New Guinea on their way to Ukarumpa. This little settlement is only two miles south of Aiyura and has become the Operational Base of the New Guinea Branch of the S.I.L. Additional to the influences which naturally flow from such a settlement, and the fact that they buy timber from the natives of the Arona Valley, a family from the group has been living in the Arona Valley while doing language study. This, then, is another influence in the changing culture of the Gadsup speakers (du Toit 1964 (b)).

The operations of this latter group, as well as influences due to traders and a number of white settlers in the Arona Valley are all facilitated by a road which was completed in 1953. It leads from Goroka by way of Kainantu to Lae on the north coast, after passing through the Arona Valley and thence through the Kassam Pass and down the Markham Valley. While the road is closed during the rainy season it allows a free flow of heavy vehicles during the greater part of the year. It has also become the road which is travelled on foot by people from the Western Highlands who are going to the coast. The road follows this general course of traditional trade routes, but has taken a much greater significance because it has widened the horizons and increased the contacts of every person in the Arona Valley.

While these new contacts have most certainly introduced new relationships and associations, they must also have influenced the worldview of the Gadsup. The pacification by white administrators has made for a widening of the social field of interaction and at least suppressed traditional hostilities. While new alliances might not have been formed, old ones have lost their political and military significance. The administration in Kainantu, where a small hospital is maintained, has Medical Aid Posts at Arona, Pundibasa, Akuna, and Aiyura. Each one of these is served by a medical orderly — a position which has become

15. The Department of Agriculture, Stock and Fisheries of the Territory of Papua and New Guinea, Konedobu, Papua informs that the Arona Livestock Station was sold on 25th April 1961, to a Mr. Oxlade (Personal communication, 1967).

known as a 'dokta boy' who serves a 'haus sick'. Throughout the valley, government rest-houses have been constructed within walking distance from each other. Those are the places where a patrol officer and his assistants stay while they patrol different areas on inspection and to collect taxes.

These, then, are the foreign influences; those factors which affected and changed the local population members. It is our subject, however, to concentrate on the people themselves as they live and work in the Arona Valley.

THE PEOPLE OF THE ROAD

This discussion will introduce the people who surround the Arona Valley, and, in detail, those who follow "the road" and who live in the valley. They will be treated on a village and district basis, for neither of these categories is sufficient in itself. The village, as was shown in the previous chapter, refers to a number of households which are geographically separated from other such clusters, and which are regarded by the people to constitute a distinct unit. Its size allows for personal contact between all of its members and is small enough so that they can all be personally acquainted. This community is the largest economic, political, and social unit of any significance and includes a linguistically homogeneous population whose members are not necessarily related by kinship ties. The village, then, is not a localized kingroup. In speaking of the district we refer to that geographical area occupied by a number of villages with gardens, pig-grazing area, and bush which surround them, with which the people identify and from which they extract their livelihood. The concept of district is a fluid one as the people utilize various criteria and associations, depending on the tranquility of social relations. While the Gadsup definitely point at the division between Eastern and Western Gadsup, they also (though less stringently) distinguish between three districts within the geographical limits of the Western Gadsup. In this study we are utilizing both of these classificatory concepts in the way the Gadsup conceive of them, and allowing the fluidity which they recognize for particular times and under particular conditions.

Before going into this subject any further, we must distinguish· "Gadsup" from non-Gadsup, and define the geographical limits to which this study is confined.

Conception of their own group

As is typical in the greater part of Melanesia, and especially in the Eastern Highlands, the Gadsup are only a numerically small group amidst many others that differ in language and culture from them. These differences, in many cases, are not very significant and suggest relatedness between the groups concerned, and, either due to this historical relationship or because the people themselves notice these similarities, we find a gradation in the degree to which a group identifies with neighboring groups.

The Gadsup concept of "self" and "others" is a linguistic one, graded to allow for closer or more distant affinity. The universe is defined by what they know of the Arona Valley and its immediate surroundings, and the linguistic groups that live there. The traditional conception of how all of this came about is related in one of the Gadsup tales about the culture hero Mani?i.[16] This legendary character is said to have lived so long ago that kinship ties can no longer be traced with any modern kingroup. The tales told of him are often summoned to justify the practices or taboos which they relate or to give antiquity to modern practices. In this particular tale it is related how the Arona Valley and surrounding regions were peopled.

Very long ago there was an old man named Mani?i and his wife. They were very old and did not walk about, but just sat with their arms folded around their knees, in their house at Maropa?i. These two old people had one daughter who had a son. All the time the little boy would play outside and show his grandfather what he did or what he was playing with.

A little distance from the old man's house was a large 'warim' tree with a great number of its fruit, and every day a 'kapul' would come and eat the fruit. When ever the old man visited the tree he found that the fruit had been eaten. One day Mani?i called his grandson and said: "Tentinai, I want you to go to the large 'warim' tree and hide with your little bow and arrows, for there is a 'kapul' that steals the fruit and you must kill it." So the boy left with his bow and arrows and hid near the large tree. At dusk he heard the rustle of leaves and saw a young 'kapul' which he shot and took to his grandfather saying, "The 'kapul' which you spoke about, I have shot it." But his grandfather looked at it and said that it was not the one he had wanted.

The following evening as the moon was rising, the boy again shot a 'kapul' and took it to his grandfather, but he was still a long way off and carrying the young 'kapul' when his grandfather called to him that it was not the one he had in mind.

16. A more complete treatment of Gadsup Culture Hero Tales will be found in du Toit, 1964(a). Appreciation is expressed to the editors of the Journal of American Folklore for permission to reproduce the various tales in the context of this study.

Some days later the boy again went into the pit-pit reeds close to the 'warim' tree and hid. He saw an 'ironi' approaching the tree and shot it. Again his grandfather told him that it was the wrong one.

This went on for some time. Then one evening during the full moon the boy was watching the tree listlessly, scratching the ground and thinking of all the tree-kangaroos he had shot when he saw yet another one approach. Quickly he placed an 'awemi' arrow in his bow, aimed and shot the 'kapul' in the chest between the two front legs. He jumped up, and as he was used to doing, he took it by the hind leg and started dragging it home but he could not move it. He tried any way he wanted to but could not even move it, so he took his bow and arrows and returned home. "Were you not lucky this night?" his grandfather asked, but when he heard what the boy had to tell, he and his old wife accompanied the boy, and, together, they dragged it home. After they had placed it in the house, the old man said: "This is the kapul I told you about. Do you not see how large it is? One man cannot carry it. And here, do you see the lines down its back just like the leaf of the 'Imi' tree?"

Now the old man, Mani?i, and his wife started to cut the meat. They spent all day cutting the pieces and piling them up in the house until the whole house was filled. They did not skin it, but cut off the legs and then cut up the body also, cutting the skin and hair on the back (with the meat).

When they had cut up the whole 'kapul', the old man called to his grandson, and said: "Tentinai, go up to the top of Yana?anuka?i and look around in all directions." The boy climbed the mountain and looked all around him, but in no direction did he see any people nor were there any fires or smoke rising into the sky. He went back to his grandfather and reported, saying "Tentinapu, I looked in every direction and saw grass and trees and the ground and rivers, but there are no people and no smoke from their fires." Then the old man and his wife took all the meat, went to the top of Yana?anuka?i and with great power threw one front leg away from him; it did not fall very far away but fell close by at Abinapa?i on Yapana?u mountain and from that place came up the Agarabe. He took the other front leg and cast it toward the place where the sun disappears; at that place came up the Kamano. A hind leg was cast away and became the Tairora. The other hind leg was taken; it fell where the river Upanomi breaks the mountains and at that place appeared the Arona people who also live down to the big river. Then the body was left and Mani?i took the neck and it fell in the direction from which the sun appears, and there arose Pundibasa, and he took the heart and tossed it in the same direction, but it fell against the mountain side and here Kundana grew up. Then he took the shoulder and it fell close by

33

against a mountain and it became Akuna. Then they returned to the house. That night the boy climbed the mountain and he told his grandfather that he looked in all directions and saw people and there were fires.

Thus came all the people to this place and they lived here.

Yana?anuka?i which is referred to, is a butte which rises near the present village of Onamuna. The important distinction is made here between the torso and the limbs in which the latter form non-Gadsup speakers, but yet neighbors who are included in the world-view and with whom there are relations. The body itself is subdivided to form the different Gadsup villages in the Arona Valley.

The degree to which this tale corresponds to the linguistic picture has already been suggested. For the Gadsup, the center of strength, truth, and courage is the heart, while the shoulder forms an important body part, partly because of its proximity to the heart and partly because it is the place where all men carry their burdens. The villages of Pundibasa, Kundana, and Akuna[17] are pointed out as the truly old Gadsup village, as is Maropa?i (from where the meat had been cast). Maropa?i has been vacated and two villages have since developed, the older Onamuna and Wopepa which has reached its maximum size only recently. All other villages are of recent origin and have come about by fission of an older village. Closest to the heart and shoulders are, of course, the front legs. These became the Agarabe and Kamano with whom amiable ties and occasional intermarriage exist. The one hind leg became the Markham people with whom a minimum of social intercourse occurs, except for the occasional trade of earthen pots from the river lowlands to the Arona Valley. The other hind leg became their Tairora neighbors, who live much closer, geographically, than the Kamano, but with whom social intercourse is extremely sporadic and limited and with whom sexual intercourse is taboo. While no tale was recorded which explains this social distance between such close neighbors, informants maintained that sexual intercourse between members of the two groups proved physically detrimental to the parties involved, citing as corroborative evidence two examples of present Gadsup cripples (a man and a woman).

The simple fact, however, of belonging to the same linguistic group has not in the past assured peace, social interaction, or co-operation, nor does different linguistic-group membership prompt hostilities. In spite of this general statement, the logical condition of course is that

17. This does not refer to the present village of Akuna. The old Akuna was situated about one mile east, but it was vacated in 1941 and the new village constructed.

34

a person feels more at home and will more readily assist a member of the same linguistic group, than a member of another group.

The point, then, is that we lack any form of group consciousness on the political level, let alone allegiance to a central authority, and we lack anything which we can call a tribe.[18] The villages of Onamuna and Wopepa have much closer ties and less strife with their Agarabe neighbors across the Ramu River than with their own linguistic brothers in Kundana. When outsiders first entered the Arona Valley they found that among the Gadsup a person was safe among friends and relatives, and very rarely did he venture a great distance from his village of residence or a village in which ties of friendship and kinship offered safety. It was also primarily within the village that a person found those with whom he interacted most frequently. These were people of his own choice and they were persons whom he called 'tikono'. This term does not denote institutionalized friendship, but is used for any person that a Gadsup considers his very close friend. There may also be more than one of these and they are all members of that person's interest group. This is a group of persons, of both sexes, who may be related to ego and who associate with and assist him. We will discuss it in greater detail in the context of descent groupings below. In this respect we can differentiate various degrees of closeness a person feels to others. The widest, most inclusive and generalized group is the 'ayukam waya', which theoretically includes all the Gadsup and, in fact, all persons who speak the Gadsup language. Then there are all the members of one's 'ankumi', namely his bilaterally related kinsmen. Both of these groups, however, are general and simply denote the fact that the persons under discussion have much in common. Every person also belongs to a patrilineal clan but it is of less importance than the 'ankumi' and certainly of much less importance in Akuna than the interest group.[19] More restricted and of greater importance to any Gadsup are the persons who comprise his residential group, namely fellow villagers and men and women who have many of the same interests and values and who for that reason

18. If this term is used as "a politically or socially coherent and autonomous group occupying or claiming a particular territory" (Notes and Queries on Anthropology, 6th ed. 1954: 66). See also Hogbin and Wedgwood (1953: 252), Read (1954: 11), and Reay (1959: 43). Our village community as discussed in these pages in fact coincides with the criteria of 'tribe' mentioned by Williams (1930: 156) namely: (1) a common territory, (2) certain idiosyncrasies of custom, (3) a distinct dialect, and (4) its common enmities. In both cases we lack any central authority (ibid: 325) usually associated with 'tribes'.

19. For a discussion of the apparent contradiction in this statement, see our discussion of the community structure in the tenth chapter and especially the last chapter where the social structure of the community is discussed.

associate together. Usually within this group of villagers, but not restricted to it, a person builds up his interest group and retains their interest and backing by constant effort. None of these groups (except of course the clan and 'ankumi') is limited to kinsmen, and in fact is usually equally distributed between cognates, affines, and friends. The smallest and most loyal of these associations is a small group who usually form the core of "one-talk-villager-interest-group-member", and these call each other 'tentikono' — my friend.[20] This author was so considered and addressed by a number of males about his same age group, mostly from the same village but not limited to it. While these groupings still exist, they seem to have been of greater importance in the past and were the bases on which traditional Gadsup society rested. Only with the coming of the missionaries and later the Australian administration was this universe widened, as the Gadsup became aware of more distant places and persons differing physically from them. Although the first intensive and prolonged contact with outsiders was not with Europeans but with evangelists from coastal groups, in this study we are using 'White Man' to refer to the generic influence. Changes that have affected the Gadsup have been a direct result of Whites seeking converts, wealth, or subjects to pacify. This does not in any way deny the internal dynamics of the Gadsup culture and the fact that changes in one set of ideas may affect other ideas and actions to a greater or lesser degree. Whites, like Agarabe, Tairora, or Kamano, form part of the Gadsup social universe, and are set off from those who follow 'the road'.

Based then on the degree of relatedness and the linguistic intelligibility as they see it, the Gadsup set themselves off from others. While this is not based on any feeling of kinship ties (descent), alliances against outsiders, or 'tribal' feeling,[21] it is an equally strong tie which distinguished between 'real-road people', those who are fairly close or related to them, and those who belong to different speech communities.

All speakers of what we now call Gadsup designate themselves as 'ayukam waya', i.e. 'road real/true talk' (Frantz 1962: 49-50). As they say: "Others have their talk, but ours is THE talk." This linguistic identification then is the basis on which people distinguish them-

20. This term is not related, nor is it derived from the same root as tentiko — my taboo, or avoidance relative. This latter term will be discussed in the context of kinship terminology in the tenth chapter of this study.

21. Oosterwal (1961:17) prefers to speak of the 'tribes' of the Tor Valley in the former Netherlands New Guinea. This term has not been used because we lack the group feeling and the allegiance to a central authority which are usually associated with the term. See also footnote 18 above.

Weaving bamboo
slats in the 'koma'
pattern for the
wall of a house

Preparing a grass skirt

selves from others, and while all the 'ayukam waya' are by no means
friendly and periodic hostilities arise between villages, one feels
closer to and can expect help and food more readily from a speaker
of the same language than from an outsider. The criterion of language
is interpreted so that any person who uses the Gadsup language fluently,
who identifies with the group as a social and residential unit and later
marries into the people of 'THE road', is counted among the 'real
road men' ('ayukam wainta'). These include individuals captured[22]
during infancy in raids against the Tairora or Agarabe and who grew
up as Gadsup. "They ate our food and are fluent in THE talk, they
are one of us", an informant said.

The members of this group with which one identifies are contrasted
with others on a linguistic basis. The Agarabe are referred to as 'para
waya yumuno', "they who turn THE talk", and the Tairora are 'opam
waya', "another talk".[23] The term applied to the Markham peoples,
who are called 'Datama', is obscure. Informants gave two explanations
for this form of reference: the first being that the Markham Valley
people are much taller than they and while a Gadsup can lie and sleep
next to a fire at night and be warm, the tall 'Datama' must curl around
the fire to heat their bodies, causing them "to be crooked". The other
explanation was that during traditional fighting, people would warn each
other "Do not go near that place, for you will be killed". The reference
used was always an exclamation of 'datayo'. The suffix -yo, however,
occurs in many Gadsup exclamations, and it is likely that the latter
term is merely a variant (possibly the vocative) on the basic form
'Datama' referring to the crooked position during sleeping. It might
be noted that the bamboo bent into a tweezer, by which women handle
the hot rocks of the earth oven, is called 'data'. This too is bent
double like the 'Datama' sleeping by the fire.

The term 'Gadsup', like most of the group names in the area, has
been applied to this people by the Administration. They do not use it
among themselves nor did they traditionally have such a common den-
ominator for themselves. It is said to have been introduced by the
early Lutheran missionaries after a small village, called 'Gasupu',
on the eastern entrance of the area where they had spent some time

22. This seems to have been a common practice, especially in inter-village
fighting among the Gadsup speakers. Numerous cases were recorded in the field,
and Flierl (1932:22) refers to adult women being captured and carried off in
1926.
23. This linguistic relationship is also recognized by researchers who classify Gadsup and Agarabe as one language while they are differentiated from
Tairora, vide McKaughan (1964:119) and Wurm (1964:84) for a discussion of
linguistic distance and glottochronological data on these languages in the Eastern
Highlands of New Guinea.

after their arrival from the Markham. This report of Kevengu, the evangelist in Akuna, does not seem to be tenable. In his discussion of the first contact with these highland peoples, Flierl simply introduces them as "den Bergbewohnern, den Gazub" (1932: 7). He does not refer at all to a village by this name and it seems that one would have to seek the meaning of this name in the Azera language of the Markham Valley. Only now are the people themselves starting to use the term in contradistinction to such terms as Agarabe, Tairora or Kamano, but among themselves they still use the traditional term 'ayukam waya'.

Demography[24]

Population figures for the eastern highlands of New Guinea are not plentiful. Those that are available are not very accurate since they depend on general taxation censuses or village averages rather than detailed household analyses. A comparison of our own figures with government census figures suggest the latter to be invariably higher by 5-7%. They will be given in this discussion simply as a control for our own figures based on household censuses in each of six villages and comparable data extracted from genealogies which were collected in the field.

In the introductory discussion above it was suggested that the Gadsup may be divided into at least two major groups, an eastern and a western sector, with some outlying villages spread southwest from the Arona Valley. The Neuendettelsau missionaries who first moved into this part of the highlands have given us some figures for different villages. These are all in the eastern sector and include most of the villages in that part of our ethnographic area. These figures, I would suggest, are rounded out rather than being individual counts.

I would suggest Kulaka to be the village known to the administration as Kuranka as one finds in various contexts that the Gadsup speaker will replace the 'r' in European words with an 'l'. Mamerain is listed as including two settlements, referred to as 'kleine Dörfer', while Kundana and Sasaura include three small village settlements each. In the table they have been listed separately as if we were dealing with

24. See also 'household families' in Chapter 9 when Akuna will be treated in greater detail. My appreciation is expressed to Mr. R.D. Lycan who is primarily responsible for the computation and tabulation of Tables II-VI from demographic data we extracted from the genealogies collected by the author among the Gadsup. James B. Watson also gave valuable assistance in the final form of these tabulations.

Table 1. Village populations in 1933*

Village	Population
Kulaka	150
Tombetaka	150
Mamerain	150
Mamerain	150
Urauna	100
Kundana	100
Kundana	100
Kundana	100
Sasaura	100
Sasaura	100
Sasaura	100
Omaura	250
Average per village	129. 2

* These counts were made by missionary Bergmann and submitted to the
Neuendettelsau offices in Germany. They have been kindly sent to us by
L. Flierl (personal comm. 1964).

individual settlements. One of Kundana's daughter villages might very
well be Apomakapa which will be discussed again in this study and
even during our visit was seen as a relatively recent off-shoot from
the larger and older Kundana. In his Unter Wilden, Flierl remarks
on Pundibasa which consisted of three 'Dörfergruppen' or village
groups (op. cit.: 8) but unfortunately he does not give any estimate of
the population. While we are counting each of these settlements as an
individual village community, and have also done so below when dealing for instance with settlements like Wayopa and Manunampi, the
administration has tended to group them together with a mother village in some cases as those just mentioned, while in others they have
recognized the sovereignty of each separate residential community
for instance in Ikana.

During the time spent in the Arona Valley a census was made of
each of the five villages with which Akuna had closest ties. This concentration of social ties agreed with geographical proximity. In doing
this we recognized distinctions which the people themselves emphasized, namely the distinction between Amamunta and a daughter village,
Manunampi. Also the distinction between Akuna and the daughter village Wayopa which will be discussed below. These distinctions are
not made in government census listings.

These population figures agree with the suggestion of Hogbin and
Wedgwood (op. cit.: 242, 244) that village communities vary from
seventy to three hundred persons. As soon as a village community in
the Arona Valley approaches the upper limit, we find schism taking

Table 2. Village populations in 1961-62

Village	Government census* 1961-62	Genealogies** 1961-62	Household census 1961-62
Amamunta	286	171	196
Manunampi	-	-	73
Wopepa	290	57	301
Tompena	243	323	229
Akuna	355	233	245
Wayopa	-	79	82
Average	293. 5	156. 0	187. 7

* From: Village Population Register 1961-62, Department of Native Affairs.
** During the period of fieldwork extensive genealogies were collected. These contain the names of approximately 4,000 persons, living and deceased, but are on the whole "shallow genealogies" and include references to villages throughout the Arona Valley. Genealogies obviously refer to past and present generations and for that reason older villages will be much better represented than recently settled ones. In this regard it is of interest to compare the figures for Wopepa and Tompena. The latter is an old and well settled village, the former only recently settled but having grown to its maximum in population.

place and a group of persons moving away. This new village will be associated for some time with the mother village. Kinship ties, visiting patterns, and ceremonial exchanges will assure bonds of friendship, and only very gradually does the new village acquire complete independence. This will require novel associations and neolocal residence by people who join the community. For Akuna, at least, it was not always possible to ascertain a trend in terms of the person who initiates residential change. In terms of the somewhat more dominant and certainly more mobile role of the male, it could be suggested that males initiate change in residence and thus also new loyalty associations. A number of examples will be discussed below.

The western Gadsup show a higher percentage of females than males, though this is primarily due to fewer adult males than females — a condition quite possibly brought about by the warfare and hostilities which were characteristic of this region up to fifteen or twenty years ago. These figures are best represented in Table 3 as they have been deduced from the genealogies.

When dealing with demographic figures which have to be extracted from genealogies, and when these apply to a community such as we are now dealing with, we can only deal very generally with age. Anthropologists have devised a variety of means for dating the ages of persons in illiterate societies or in communities where no records are kept. For the western Gadsup it was possible among other things

Table 3. Sex and marital status of living population*

Village	Male			Female			Male population as ratio of whole	Ratio of married female to married male
	Married	Single	Total	Married	Single	Total		
Amamunta**	39	46	85	51	35	86	49.7	1.3
Wopepa	21	4	25	25	7	32	43.9	1.2
Tompena	69	86	155	81	87	168	48.0	1.2
Akuna	56	55	111	76	46	122	47.6	1.3
Wayopa	24	17	41	23	15	38	51.9	1.0

* While the status of single and married does not necessarily denote age, since bachelor and spinsterhood may be important factors, the Gadsup are expected to ' marry and by the early twenties are or have been married. This married/single breakdown would then allow us some general statement about the age categories in each village.
** Amamunta here and in the tables which follow includes Manunampi.

to refer to the Second World War when bombing took place at Kainantu and Aiyura. This was around 1943. Persons could be dated in terms of the stage of the life cycle they had reached at this time. We are then dealing with relative age, and even more so when persons on the genealogies are dated relative to persons who are known. It might be more accurate to speak of generational level, but the breakdown of ages also occurs within generations. Errors no doubt have occurred and been included in the calculations, but it is trusted that the genealogical averages will roughly coincide with the actual cases.

Table 4 points at a number of very important facts. It will be recalled that Wopepa is a very young village settlement, as is Wayopa, which split off from Akuna after 1948. We notice that neither of these villages are represented by persons in the upper age category, though in the case of Wopepa a number of old persons were met. We might be dealing here with a case where generational age and chronological age do not coincide. To balance this, however, there is a very high percentage of the population who moved to these villages as young (possibly post marital) individuals. They are now in the third generation level, thus within the group that have not yet completed their reproductive cycle. Informants stated that Tompena, too, was one of the older and earlier settlements. The major competition for people to settle, due to similarity in size and possibly in age — though Akunans claim to be the community which gave rise to Amamunta — was between Amamunta and Akuna. The figures for generational levels one (I), four (IV) and five (V) are nearly exactly the same, while differing only slightly in the other categories.

We find this same differentiation between the three older villages

Table 4. Relative age of population (percentage)

Village	Generational level				
	I 1–17	II 18–23	III 24–39	IV 40–59	V 60
Amamunta	31.8	16.1	33.9	17.0	1.2
Wopepa	10.3	8.6	62.1	19.0	0**
Tompena	41.3	12.8	36.4	8.6	0.9
Akuna	31.8	12.4	37.8	16.7	1.3
Wayopa	30.4	8.9	50.6	10.1	0
Average	29.1	11.8	44.1	14.3	.7

* The five relative age levels, or generation levels, agree in general with the following phases in the life cycle:
I.   infants and children
II. unmarried youths, or persons recently married but as yet childless
III. married persons and parents of young children
IV. people who have completed the reproductive cycle and represent the generational level whose children have children
V.   great-grand parents. These are 'old people'. Elderly childless couples are assigned the same age as siblings and colateral generation mates.
** Contrary to the case of Wayopa where no person lives who can be seen as 'very old', the genealogical data in this case of Wopepa do not coincide with observations.

as contrasted with the two young ones when the genealogies are analyzed for fecundity. We find again a lower number of children per woman of childbearing age, quite possibly because the women are on the whole younger in the younger villages and the village has not been in existence long enough for large families to have been completed. These same tendencies may be reflected in the 25% infecundity recorded for Wayopa.

A count of bride and groom migrations points at a high rate of local brides, thus females who resided natally in the same village community as their husbands. In Amamunta only 42% of the brides and 56% of the grooms are native to their village of residence, while all

Table 5. Fecundity in Western Gadsup village communities

Villages	Number of children									Total no. of children	Percentages of women childless	Average no. of children/women of childbearing age
	0	1	2	3	4	5	6	7	8			
Amamunta	11	9	11	7	7	6	0	2	0	124	20.8	2.95
Wopepa	4	8	4	3	3	1	1	0	1	56	16.0	2.67
Tompena	9	16	14	12	13	8	5	1	3	233	11.1	3.24
Akuna	11	11	15	14	7	6	8	2	1	211	14.7	3.30
Wayopa	6	6	7	2	1	2	0	0	0	40	25.0	2.22

the other village communities show appreciably higher rates. This in fact confirms accusations which were repeatedly heard in Akuna, that Amamunta people were attempting to popularize their village, to attract settlers, to accept persons who were shunned from their natal villages. Informants gave the impression of a Statue of Liberty at Amamunta's entrance which offered a haven for all the tired, weary, poor ...

Nearly the opposite reputation is ascribed to Tompena. Here is a community which cares little about its neighbors, a community with a great deal of pride, a great deal of drawing on its own resources both ceremonially and socially. Wives are not generally sought beyond the community boundaries nor are prospective grooms encouraged to visit Tompena and thus we find a high rate 94.2% of the brides who are native, even higher than the husbands in the same community. The fact that Tompena borders on a neighboring linguistic group with whom traditional hostilities existed may in part be reflected in these figures.

Wayopa once again proves to be an interesting case. It will be recalled that Wayopa split off from Akuna after 1948 (this process to be discussed below) and for that reason we would expect a small percentage of persons native to the village settlement. When a new village is formed it is the 'strong man', the leader in the community, who must make his village attractive to others. Since neolocal residence is generally accepted as a possibility, one would expect young and recently married persons to look for a place where they can best satisfy their personal needs. Granted the fact that Wayopa is very young, it is still important to note that less than 13% of brides and grooms who settle in the village are native to it.

Once again it cannot be emphasized too strongly that most of these figures have been extracted from genealogies. They are therefore dependent on genealogists who might not have known each of the persons he reported on or who might not have remembered the detail very well. All the genealogies were checked with other informants,

Table 6. Inter-community immigration

Villages	% of resident males who are native	% of resident females who are native	% of husbands who are native	% of brides who are native
Amamunta	85.9	66.3	56.0	42.0
Wopepa	92.3	81.2	88.5	76.0
Tompena	98.1	95.8	88.3	94.2
Akuna	91.9	86.9	84.7	68.3
Wayopa	43.9	36.8	12.5	12.5

but the recorder might have overlooked many important details. Together we most likely gave too little attention to childless couples for only by having a family does a person grow to full adulthood. The same is true for infant mortality since an individual who has not been named, a child who dies at childbirth might as well never have been born — he was never a person. While we have already remarked on the lack of genealogical depth, it certainly must have affected these figures for the third and even the second ascending generations.

Since most of the subsequent discussion in this study will be dealing with the village community of Akuna, an attempt was made to compile statistics as far back as possible. Mr. John Fowke who patrolled the Arona Valley during our stay was willing to assist by supplying figures kept in the offices of the sub-district administrative center in Kainantu. It will be recalled that we suggested earlier that government population figures are higher than our actual counts in the village. This is true as well in the last line of this table which applies to 1962. The number of births are nearly twice those that we recorded. These figures, as all government statistics, apply to Akuna and Wayopa.

The figures in Table 7 show a gradual population increase in these villages. In most cases the decrease in the totals as for males in 1949 and again in 1953 and 1960 can be explained by a larger number of persons away on indentured labor. From information I could

Table 7. Akuna and Wayopa demography 1947-1962

Year	Deaths M	F	Births M	F	Migrations in	out	Labor potential M	F	Absent at time of census M	F	Average family size	Totals M	F	Grand total
1947	3	1	5	6	1	0	-	-	-	-	-	111	136	247
1948	-	-	-	-	-	-	-	-	-	-	-	-	-	-
1949	4	3	6	6	11	16	64	95	15	0	-	97	134	246
1950	1	0	9	7	14	4	69	81	0	0	3.0	128	142	270
1951	2	0	6	2	1	3	69	85	0	0	3.0	134	140	274
1952	2	1	3	9	-	-	74	89	0	0	3.0	134	149	283
1953	4	1	9	5	-	1	75	88	23	0	3.0	115	153	291
1954	-	-	-	-	-	-	-	-	-	-	-	-	-	-
1955	3	5	3	11	3	0	78	94	10	1	3.0	133	155	299
1956	2	2	5	6	-	-	82	99	11	1	3.2	138	161	311
1957	2	1	11	4	-	2	82	96	6	5	3.4	152	158	321
1958	6	4	5	13	3	4	86	101	5	0	3.5	152	171	328
1959	4	3	8	8	3	1	89	92	1	1	3.3	163	169	334
1960	5	3	5	6	1	10	86	95	21	0	3.2	143	166	330
1961	-	-	-	-	-	-	-	-	-	-	-	-	-	-
1962	2	5	19	18	1	3	95	90	7	0	3.0	161	187	355

gather it seems that people from this part of the highlands have been going to the coast for quite some time, possibly as early as the years immediately following the Second World War. The exact meaning of the fifth column was not ascertained, but it would seem to refer to adults, thus excluding the young and the infirm due to age. The picture one gets from these figures is that Akuna has been growing gradually in numbers while the family size has not changed a great deal over the past decade and a half and the female preponderance has been maintained. We will return later to the demographic data related to Akuna and associated villages when we deal with the household and also with the ties which exist between the members of various village communities.

A typical day in Akuna

Traditionally one day was much the same as the previous, nor would the next one differ essentially from them, unless there was some feast or an enemy attack which had to be dealt with. The men had their meetings and their hunting activities to occupy them while the women had children and gardens, food preparation, and the making of clothing — meagre as it may be — that required their time and interest. Weeks were not divided into a certain number of days, and the day of rest occurred at any point when it was needed. The coming of missionaries and the White Man has changed this, and today we find Akunans thinking in terms of a seven-day week, ending with a day of rest when visiting can take place and people often wear the items of European clothing they have acquired primarily for this reason.

Even the lunar changes which have been noticed are not tied in with any regularity, nor are names given to the successive bright or dark nights. The waxing and waning of the moon is noted, for on the nights when it is light, the men can set out for the deep bush and spend the night trapping and hunting the tree kangaroo, and their return will be a happy occasion in a land where meat is scarce, and that which is available is subject to strict rules. The moon is also important (as is the sun) because at a certain time the old men inform the younger people that the moon is setting over a particular place and is retiring over a well-known tree, and that it, therefore, is time to prepare the gardens for an abundance of food, and to start setting traps in the bush, for the time of the 'Orande' is close by. The calendar has now been introduced into most villages by missionaries, teachers, or progressive youths returning from a few years of labor on the coast, and it is well known that the 'moons' have names, and that by counting the days, and finally the months, it is possible to know when to expect the patrol officer collecting taxes, the medical officer who destroys inadequate housing, or even to know when a 'brother' who has been away for a long time at 'nambies' (Neo-Melanesian word for the coast) will be returning. In this way Gadsup life is slowly forming itself into a Western mold, with certain days set aside for Government work, certain days for activities connected with missionaries, and the knowledge that money will have to be saved, for though they have very little need for it, the kiap is sure to arrive and taxes of one pound per adult man will be required.

Though the people recognized a difference between certain times of the year in terms of the amount of rainfall, they did not classify it into dry and rainy seasons. The time when the 'Orande' was close at hand was a time when plenty of rain would fall and an abundance of food would be harvested. There was more of an emphasis placed on the amount of sun which was visible, and for that reason they preferred the months of rain when nearly every day had a thundershower, but the rest of the day the sun would shine clear and warm. This matter of natural heat is important to a people who live in nature as much as they do and lack any form of bodily covering except that which is worn for reasons of modesty. The amount of rain that fell and the duration of this precipitation had a great effect on the activities of the day, and any group activities which could be undertaken.

In speaking then of a typical day for the Akunans, we are using a term which seems more appropriate than the one which is more widely used, namely 'the daily routine'. It seems that routine is and was lacking in Akunan life due to their dependence on natural conditions, and also that routine as such is avoided as far as possible. Two days have to be excluded from the following discussion. Friday has become known as 'road day', for each Friday the people have to go out onto the road, clean it, and keep it in good shape for vehicular travel. It is also the day when the police constables come through the villages checking on those who do not turn out to do their share and the only people seen in the villages on these days are pregnant women, those with small babies to care for, and the very old, who, because of their physical condition, are exempt from such trying activities. This is also the day when teen-age boys take off for the bush as they can hide there all day without being noticed. The road work is hard on the people especially during the rainy season when roads and bridges are swept away, or the people are called out in the middle of the week to clean the road of some landslide. Road work receives priority over regular economic activities, and people often return home late at night from their gardens having gone there to get food after doing the regular day's work for the Government. The other day which we have to exclude here is the weekly day of rest which is expected by the

46

missionaries. About a third of the Gadsup converts belong to the Seventh Day Adventist Mission, and the other two-thirds to the Lutheran Mission, and these respectively spend Saturday and Sunday in the way they have been taught to, at rest.

On a sunny day one is awakened shortly before sunrise (at a latitude of six degrees north of the equator, this usually occurs around 6 a.m.) by the squealing of a young pig being carried from a house by the woman who owns it, and who places it outside the house, but often outside the village enclosure. She returns briefly to her house to blow on the fire and to clean out excreta of children and pigs, before emerging with a string bag hanging from her head, filled with small or left-over sweet potatoes, sweet potato and taro peels, and while calling her pigs a short distance from the village fence with a deep grunting glottal 'ma, ma' (here, here) she squats down and feeds them, striking at other pigs that come too close, often picking up a young pig onto her lap and delousing it with all the loving care she gives her own child. Meantime the men emerge from the smoke-filled houses to stand around in twos and threes, discussing the events of the day or a pig hunt which is being planned. Children are already sitting in small groups around the rocks used in the earth oven and beating with monotonous regularity at the hard shell of 'puyami', or 'imi' nuts, the contents of which they enjoy with smacking lips. While the boys go off to start a game of marbles or cards, the girls will join their mothers in sweeping around the house and preparing the morning meal. This is usually sweet potatoes roasted in the fire, but taro is often roasted in addition. Occasionally when the sun breaks through after a period of rain, the women will prepare for a long day in the gardens and will have their main meal of food cooked in the earth oven in the morning. If the day is to be spent in the gardens, a mother and her older daughters will leave the village at approximately 8:30 a.m. or thereabouts, often leaving the baby with a sibling or a relative, and with string bags over the head and shovel in hand they set out for the garden anywhere from a few hundred yards to two miles distant. Should a baby accompany the mother, it is laid down on a folded mat of pandanus leaves in the bottom of a string bag and carried, hanging from the mother's head, either fore or aft depending on the direction of the sun. As in all societies, domestic duties require a certain amount of the housewife's time, and should she be needed in the village to make string bags or grass skirts, or to repair rain gear, she will sit around after the meal sharing in the gossip with friends and relatives, and then join them in some activity. The first or second day after a rainy spell will usually find most women in the village repairing their flimsy rain gear made of pandunas leaves which show signs of tearing after a few days of constant wear.

Any day, including Friday or a day of rest, may be interrupted by an informal court session, over which the luluai, a government-appointed village official, presides, often assisted by the tultul (the government-appointed interpreter). The occasion may be that a pig has been killed, some person's garden destroyed by foraging pigs, or a couple has suggested or committed adultery. Cases often result from women leaving their husbands, being then accused of either adultery or prostitution. Such periods are spent by the men and often some of the older women, gathering round the luluai and the accused in the village square if weather permits, or in the luluai's house, while chewing betel and smoking, offering suggestions and comments based on previous cases like this; they make lengthy orations which are more often than not a repetition of something which everyone knows or which has already been stated. Such gatherings usually take place in the morning while the men squat with the pleasant morning sun warming their backs, and while some chew on a cold roasted sweet potato left over from the previous evening's meal; others wait for their womenfolk to roast some sweet potato or taro for the family. The daily activities, furthermore, depend on the weather. During our stay in Akuna one could not really speak of a dry and a wet season, both of them being somewhat irregular, and the dry season being described by the local people as freak weather.

Should a court case not require their presence, the men will soon emerge with their bows and arrows, which they always carry, and set out for the bush to hunt pigs, repair a garden fence, or visit their various clumps of betel nut palms and other trees. The younger adult men at present spend about two or three days a week cutting trees which are pulled to the road and then taken to Ukarumpa by a representative of the Summer Institute of Linguistics, who buys them for timber. This gives these men a regular income, and also offers a new kind of incentive for the younger men with enough initiative, who organize the 'lines' which work for and with them. During the heat of day, the women work in the cool places of their gardens, often taking shelter under one of the old trees which were allowed to remain standing when the garden was cleared, while the men go out into the bush, and the village is relatively deserted. Men will often enjoy an afternoon nap in the shade of some forest tree, shoot at birds that perch too low or too long, visit snares set up for the rats and tree kangaroo, or split fence posts from half-dry tree stumps. As a general rule, however, very little is done which is constructive by men during the afternoon. While men may return with large stumps of wood to be cut up as firewood, they are not expected to supply this as the case seems to be among the Kuma (Reay op. cit.: 16).

The women start to drift back into the village at about four-thirty in the afternoon, their carrying bags loaded to capacity with sweet potatoes, hanging from their heads, with another balanced on the head with garden produce and greens, and finally a load of firewood on the top. Should the day be clear with little prospect of rain, the woman and her daughters may cook their meal in the gardens where provisions are made for earth ovens, while still working and then return to the village at dusk with the husband's meal. Usually, however, the late afternoon is spent preparing an 'Onakareno' (Gadsup term for earth oven, or 'mumu' in Neo-Melanesian) in which a number of neighboring and/or related women will co-operate. This is usually cooked by six o'clock, and should the men be engaged in some topic of conversation, the wife will take the food to her husband. As a rule, however, the members of a household eat together, a share being offered to ány person who might be in attendance. Children are not expected to join the family, and take pieces of food at any time, returning to their age mates and continuing the game in which they were engaged. This is a leisurely time when the activities of the day are rehashed and the members fill each other in on the news and gossip they received. After the meal they move inside, or remain there if the meal was taken indoors, and chew their betel or smoke while discussion follows its course and people relax.

After dark has set in, the members of each household sit around their hearth in silence or discuss activities which are to follow. A man might wander over to visit with a friend or relative, but usually each is back in his own house by half past nine. Older men may beat a few stanzas on a drum, or sing to themselves the story of some past event, while children curl into their blankets around the fire. Hardly an evening passes without someone making an announcement from the village square or informing the villagers to keep their village clean and the latrines in order. While everyone remains in their houses, they may reply or argue fiercely, very often drawing forth bursts of laughter that spread around the village like ripples on a pond. While there is nothing formally taking place, one hears intermittent talking in low voices up to about ten-thirty. At any hour of the night a baby may start crying, waking the members of the household, which gives a chance to put some logs on the fire, and as the soft voice of the mother is heard 'ma' (giving the baby the breast) calm settles in again and the village is left to sleep.

Chapter 3

THE VILLAGE AS A COMMUNITY

The previous discussion was primarily aimed at outlining the eastern highlands and locating the Gadsup speakers and the community of Akunans within this larger area. In the present chapter the larger regional unit, or district is seen as consisting of a number of local village communities. Each community includes the village proper together with the lands, bush, and gardens owned and worked by various households. As will become clear, these divisions of the society are not sufficient in themselves, as the social divisions based on kinship overlap both of these categories. In the last analysis, the local group, the place of residence, the persons with whom an individual interacts, and the locality of his garden rights are the important aspects which define his loyalty and his identification. This point has been made at this juncture because it is all-important to keep in mind that locality and decisions in this respect play as important a role as does kinship. It is a nexus for daily association and interaction, the group of persons with whom happiness and sorrow are shared.

THE VILLAGE COMMUNITY

This study pertains in general to the Gadsup of the Arona Valley, but throughout the focus will be on Akuna. As a village in the Arona Valley and even in the eastern highlands, many things will be typical in Akuna, but the question should be posed whether we are allowed to turn our fieldglasses on our village — whether there is justification for viewing this small group of 245 persons as a community and whether this then could be viewed as typical.

Recently Arensberg and Kimball (1965: 13-14) have suggested that such questions about a community be answered on the basis of four key issues. These will be discussed, not according to the order in which they appear, in deciding the representativeness of Akuna. It is important to know whether we are dealing with a whole community or simply a part of a larger one. It is obvious that communities form parts of larger fields of interaction, that they are in the last analysis only sub-systems of larger systems, and in turn contain systems and sub-systems of interaction. But the question remains whether the community we focus on forms a relatively independently viable unit which articulates with neighboring communities. In this respect Akuna stands as a community. Some contacts certainly overlap with other communities, but the village here is an independent economic and social group which forms an autonomous political and warring unit. Even on the basis of exogamous relations we find that villages may supply brides to their own males.

A further issue which needs to be answered is the degree to which the institutions and cultural traits common to the whole group or area, are included in the culture of the community. Once again it can be stated that every village community forms a part of a larger whole, the linguistic unit of district. Yet, every village community also is unique to some degree as regards the particular way a ceremony is performed or with reference to elements of vocabulary or the phonetic system. These differences are minute, though, when compared with the culture of a neighboring linguistic group; we may decide, I think, that every village community applies its own filter to the beliefs and actions of its members, producing slight variations from one community to the next. Here again Akuna shows some variation from the culture of Pundibasa or Aiyura, but these differ in turn and yet all three have a greater part in common with each other than with the Tairora village, Noreikora, across the mountain.

The third point that Arensberg and Kimball mention can best be stated in their own words: "When is a community integrated enough, common-minded enough, co-operative or sharing enough first to be a community at all? Second, to mirror, but not over- or under-represent, the fissions normal to its society?" (ibid: 14). In this regard Akuna stands as a typical example of not only the Gadsup but also the eastern highlands in general. It was fortunate that we were able to study the process which led to fission when Wayopa separated as a new village, and also to observe the process at work in Akuna during our stay. These forces of co-operation and factionalism, of integration and fission, are typical of the area we are discussing.

The last question which we need to answer really hinges on the other three, namely the degree of representativeness. Throughout the discussion which is to follow repeated reference will be made to neighboring villages or to Gadsup in general. It will become obvious, in terms of the answers already supplied and from reading the material, that Akuna is largely representative of the Gadsup linguistic group.

There is however more to our decision to select this village as a community sample. Nadel has suggested that the concept of community applies "to the sort of unity entailed in common religion, language, or culture in general" (1958: fn. 153). Now, while this may certainly be a necessary requirement it is not sufficient for the segregating out of a particular group and area which we call a community. There must be more than simply common culture (which logically includes religion and language), for persons must also feel and recognize a bond. Park and Burgess, though writing more than a quarter of a century before Nadel, seem to have been much closer to the crucial point. They explain that "an individual is not, at least from a sociological point of view, a member of a community because he lives in it but rather because and to the extent that he participates in the common life of the community" (1921: 163). The fact of residing in a village called Akuna is not enough for inclusion in the community of Akuna and for that reason it will become clear that the community in fact includes two sub-systems, two interaction fields. It can be expected that in time this will lead to two communities, geographically separated.

We must also include in our conception of a community two further components. The first has been alluded to above in terms of space which sets off one group or one interaction field from another. But this is not crucial for geographical space is only one aspect of space, the more important being space in terms of a field or setting of social relations. Here one is able to define clearly, isolate out and study fields of interaction. Here one is able to study and analyze the organization and structure of a community. This, however, requires the second component, namely, that of time. A community exists in space but through time and also at a particular point in time. It is of primary importance to include the time aspect in the study of a community to locate it on a continuum, but also to separate events from each other and to denote continuation and/or repetition. For the village community being discussed here it would make a vast difference whether it was studied during the thirties, forties, or sixties. Even since we left the field a government school has been built just outside Akuna, national elections have been held in New Guinea (including among these people), and we are now dealing with a different community. This is time as a continuum, but it is also necessary to discuss events in the village as interaction network — each a smaller field for analysis. These are the patterns which we observe, the interactions we record, the associations we analyze. Only very infrequently is it possible to observe the community as a whole while the members gather in the village square for some major event. The rest of our studies are dependent on observing a group working, a family enjoying a meal, a court session in progress. We are in fact observing,

Variations in
female dress.
Small girl has
pubic skirt of
grass. Post-
pubertal girls
wear long grass
skirts in front
and bunch of
leaves in back.
Married women
wear skirts that
go around the
body but are cut
shorter in front

1. A court case. The 'luluais' of Akuna and Amamanta are seated on the pandanus leaf mat. The 'tultul' Apoma?i is in the center behind them, while Ma?e looks suitably remorseful (see page 104)

2. The orator Opura reminding the people of Akuna of his ancestral ties with their village (see page 77)

studying, analyzing a sub-system and reconstructing from a number of these the system. We are observing functioning parts which represent the whole. In this process of abstracting and synthesizing two lines keep moving, an axis of time and an axis of space which permits us to make positive statements about place, time, duration, repetition and similar subjects which really are basic to anthropological studies.

HAMLET CONCENTRATION

When discussing the village pattern in the previous chapter, we were primarily interested in the physical aspects of the village structure and its layout. At this stage we will look at these same villages in terms of the human beings who constitute the population and who are responsible for intervillage ties and supravillage organizations, temporary as these may be.

The village constitutes the social and ceremonial center, as well as the political unit, of traditional Gadsup society. Co-residence, linked as it is with the other aspects of economics and politics, is of the greatest importance, overriding in many cases kinship obligations and other loyalties. It is on this level that we may speak of a discrete unit in which the members know, associate, and identify with one another and where a person may operate as leader of the group. In most cases such a village is part of a cluster in which these same forces operate but to a lesser extent. Such village clusters are geographically close to each other because of historical affinity, but also because of the lay of the land and the availability of level stretches. Both of these sets of reasons are responsible for a relatively cordial relationship between the member villages of such a cluster, partly because they are kinsmen or have strong kin ties with members of the other village, or because of the fact that they interact much more than with persons who are further removed from them. The cluster of villages was traditionally the only possible alliance during war, and is even today the most frequent festive unit when intervillage prestations, visitations, and 'sing-sings' take place.

Gadsup villages come about by fission and alignment, two tendencies which emphasize the individual's freedom of choice and which will be treated in more detail in later chapters. What duration of time is involved before a village is ready for fission is hard to say, for just as some families or households reach maturity sooner and allow members who wish it to move out and start their own households, so some villages may require more time before certain persons wish to move out. In a strongly unilineal system sons may choose to remain with the father and his lineage, and this gives greater stability to the village structure. Among the Gadsup freedom of choice and action is highly developed, and as persons make the decisions to start their own households when they are physically, socially, and economically in a position to do so, in the same way fission on the village level often follows when the residents of a village have become too numerous.

The separation of one group of persons from another may be termed 'calving' when it is associated with amicable relations and when the fission results from overpopulation or shortage of land. Fission which is the result of a quarrel can be referred to as 'schism' (Hogbin and Wedgwood 1953: 70). The latter term is also employed for cases in which a leader moves off because he finds it impossible to satisfy his aspirations in the home village. Following both 'calving' and 'schism' there is a slow process during which individuals and families align themselves with the new village and its leader. An example of each process among the Gadsup will be given.

During the previous generation the village of Akuna was too large for the level stretch of land on which the village was built, and because gardens had to be laid out at increasingly greater distances from the village, a number of people decided to leave and start a new village, Amamunta. Who the leader was that instigated the 'calving' we did not inquire into, but it may well have been Uyatima, who was appointed the first luluai of Amamunta. Uyatima's son, Pano, returned to Akuna. Nokame, whose parents lived in Akuna, moved to Amamunta and is at present the luluai.

A more recent case and clearly one of 'schism' involves the origin of Wayopa. The first luluai of Akuna, who was appointed supreme luluai of all Gadsup (a practice which has since been discontinued), was Pumpua. He was accepted by all and recognized as the 'yikoyumi', or strong man and leader. He had four wives and aligned himself with many people in various villages. During his last years prior to his death in 1948, two potential strong men, namely Nori and Opura, emerged, both of whom started seeking the prestige and the recognition of the status. Upon Pumpua's death the Administration appointed Nori as new luluai, and within a month he had married Wainano, the elder sister of Opura, and the widow of Pumpua.

Opura, frustrated in his political ambitions, decided to leave Akuna and moved to a new site about two and one-half miles east where Wayopa soon started to grow. Here we have a number of definite examples of the freedom of choice which the individual has, for not only did Nori in this lose his first wife's brother, Makada, and his third wife's brother, Opura, but also his sister's son, Kampuma, all of whom took their families with them. The move of Kampuma is of

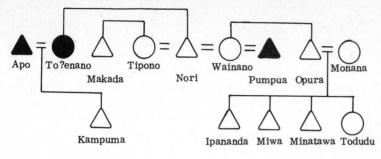

Diagram 1. Schism in Akuna

special significance in the light of the part played in a man's life by
his mother's brothers. At the present time the relations between the
two villages are very cordial, and while Akuna likes to point at Wayopa
saying 'piccanini bilong mipella' (our offspring), the latter insists
that it is an autonomous village. During our stay in the field there
was a heated court case between the two villages, and just to prove
the superiority or at the least, equality of Wayopa, Opura made a
lengthy oration in the best Gadsup tradition to prove this point.

Leadership in a village does not depend on kinship, for in the
present Akuna hierarchy we find Nori at the top, closely aided by his
brother Naniyu, and some older men like Munowi but luluai Nori's
other brother, Anokara, is of no significance in village affairs. A
person's position is not assured because of his relations with the
leader, as is clear from the relative unimportance of Yajima, eldest
son of Nori and his first wife. Also Yayo, eldest son of Pumpua, is
of little importance in village or intervillage affairs. Individual status
rests upon personal actions and achievements rather than kin ties.

The sketchmap of Akuna (and this fact is also true for Tompena,
Pundibasa, and other villages) reveals some striking features. Although the village is elongated in part because the only level ground
is the shoulder on which the village is situated, clear fissive tendencies are also at work here. Akuna has fifty-five households in addition
to other structures, and a population figure that is approaching 250.
In the field notes, the referents 'upper Akuna' and 'lower Akuna' were
used because the villagers themselves see the difference. The lower
part, smaller in area and population, is very neat, clean, and organized and is led by the tultul and the husband of his wife's sister, a
man named Napiwa. They have their own feasts and visit chiefly with
each other, while they clearly set themselves off from the upper part,
led by the luluai.

From the foregoing it will be clear that a person always had rel-

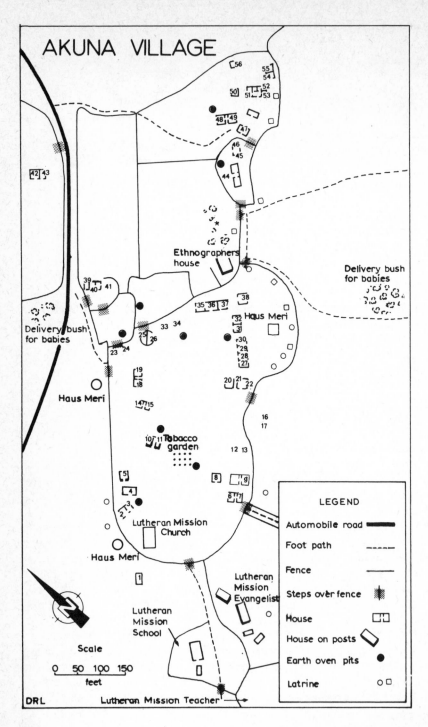

AKUNA VILLAGE

56

55
54

50
51 52 53

48 49
47

46
45

42 43

44

Ethnographers
house

Delivery bush
for babies

39
40 41

38

35 36 37

Haus Meri

Delivery bush
for babies

33 34

32
31

30
29
28
27

25
26

23 24

19
18

20 21
22

Haus Meri

17 15

16
17

10 11 Tobacco
garden

12 13

5

8
9

4

6
7

2 3

Lutheran Mission
Church

Haus Meri

1

Lutheran
Mission
Evangelist

Lutheran
Mission
School

N

Scale

0   50 100 150
feet

DRL

Lutheran Mission Teacher

LEGEND	
Automobile road	▬▬
Foot path	-----
Fence	———
Steps over fence	▨
House	⊡
House on posts	◇
Earth oven pits	●
Latrine	○ ▢

atives in other villages, though he often lives without kin in a village that he joined for personal reasons. We have to emphasize, therefore, that while a village is the largest visible unit within Gadsup society, the residents do not belong to it alone. In each village there are individual families who left the main body of kinsmen to settle elsewhere. In this context, too, our choice of 'interest group' to refer to the persons that surround and aid a person, will become clear. This can best be clarified schematically as it appears in Diagram 2.

Such villages have ties with a number of other villages. Akuna is tied very strongly with Wopepa, Amamunta, and Wayopa, while a few individual families have kinship ties with Ikana or Tompena. In the same way Wopepa is strongly tied with Akuna and Onomuna while members live in Aiyura and Amamunta. Loyalty fluctuates between kingroup and village, but women never turn against the husband and his family in time of hostilities or steal his semen for sorcery as is reported for the highlands to the west of Gadsup. Although a recent bride might return to her group in time of fighting, once children have been born this rarely occurs. This is also true for widowed persons. Older persons who have spent a number of years in a village identify very strongly with their residence group. Loyalty then grows

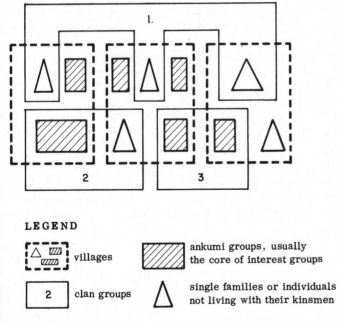

LEGEND

△ ▨ villages

▨▨▨ ankumi groups, usually the core of interest groups

2 clan groups

△ single families or individuals not living with their kinsmen

Diagram 2. Diagrammatical representation of village composition and increases between members of residence and interest group,[1] while it becomes more objective and unattached for the distant kingroup. The kingroup is the first and most durable of the two in terms of its historical composition, but it is always changing and adjusting to the choices of the individuals and their place of residence, their locality of economic expression, and their preferences in terms of interest group membership and activity. For an Akunan, then, the village and its local organization of persons becomes the group of primary significance.

Intervillage contact and relations, too, take place both on the individual and village basis. While persons are continuously travelling back and forth to visit with relatives or discuss matters regarding pig hunts or gardens, the village frequently acts as a unit in preparing a feast for another village, usually within the same district. Traditionally such intervillage festivities always followed some hostilities where two villages, or members of two villages, allied against an enemy. These are in fact one of the prototypes on which the present 'sing-sing' dances and decorations are based. As mentioned, these intervillage activities attended by the total population of a village usually took place within the district; Akuna then would invite Amamunta or Wayopa, while Ikana might invite Tompena, and Wopepa invite Onomuna. Slightly less frequent are those occasions when a village invites members of a village who reside in a neighboring district, while Wopepa would never invite Ikana, i.e. a village separated by another district. In such receptions it was always the 'yikoyumi' who extended the invitation after discussing the matter with his own villagers, and after the women had been given ample time to prepare food and gather enough wood.

KINSHIP AND ASSOCIATION

The Akunan strikes the observer as being extremely individualistic in his everyday life and as free and unrestrained in his relations with others. Each person exercises those choices regarding kinship affiliation as best suits him, for not only are they limited by memory, acquaintance, and place of residence, but they are exploited according to the needs which arise. Traditional rules and structural prerequisites are not only vague but lack the tightness found among many groups with unilineal descent. The lines of affiliation differ from one indiv-

1. The interest group, and the ways in which it differs from the Ankumi and the residential group, is discussed on page 64.

idual to the next depending on the depth of his genealogical knowledge, or on the depth he wants to recognize at the particular time.

While a person often has his strongest ties with members of his own kingroup, this is not primarily due to the fact that they recognize ties of relatedness between them, but quite possibly because of common residence and constant association, for a person may have similar ties with others who are not related but live in the same village. Every person simultaneously belongs to more than one group, and the decision he makes regarding alignment depends on which of them will be most favorable to him.

On the whole, the division between the sexes is a clear-cut one with each occupying its own ascribed status, but these are not irreconcilably separated. While the realms are different, the statuses are not at such opposite poles as is the case of Kutubu (Williams 1940; 48-51) or among the Kuma, of whom Reay speaks in one breath about "women and pigs" (op cit.: 22; see also 23 and 44 for the status of women). Indeed, among the Gadsup there are actually a great many points of overlap and informal contact between the sexes, with certain spheres, e.g. that of shaman, belonging to the females. In this context it is significant that females may initiate courtship or divorce. When a young woman initiates courtship and shows her interest in a boy, this might in fact be interpreted as the result of much love magic. Private meetings are often held in the bush, and in later stages these might result in sexual relations between the sexes, but for such courtship to be carried on in public as among the Kuma (ibid.: 176) would cause great 'shame'. Such informal contact between the sexes continues throughout a person's life, and might in fact be at the root of all the adultery accusations which are made, many of which seem to function as warnings rather than actual accusations following some act or attempted act.

The most important relationships for a person are those which involve village-members, and nearly as important are relatives. Gadsup seem to fluctuate in loyalty between relatives living elsewhere, and good friends in the same village. This is not strange, as a person's whole life and all his social interaction takes place in the village setting, and the place of residence is a choice which each adult person makes for himself. This is frequently motivated by his loyalty to a friend or a trusted leader. These choices are open to all, and decisions they make in this respect affect their families. However, not only males make such decisions. Should a young woman decide that she prefers living elsewhere, she moves; should she decide she likes another man, she leaves the first and offers herself to the second; should she feel that it might be more advantageous to live in her natal village, she suggests it to her husband and he will in many cases consent, especially if there is more land available than in his natal village.

Tradition does not prescribe the individuals with whom one associates nor where to find strong ties of amity. In the village, visiting between members is frequent and does not require any special reason or any return on either side. It is quite common for a relative or someone else to drop by in the evening. The men will sit around the hearth smoking and chewing betel nut while discussing some village happening. Women join in the conversation, and when food is prepared, offer a share to everyone present. Visits to other villages involve more time and some purpose, even though they are usually made to the houses of relatives or affines, and therefore presuppose greater hospitality on the part of the host. Such visits are frequently connected with feasts or some ceremony in the life cycle of a sister's child.

Schematically presented, an individual has the strongest, most intimate and enduring relations with persons in the shaded areas of Diagram 3, namely relatives in his village or residence and affines

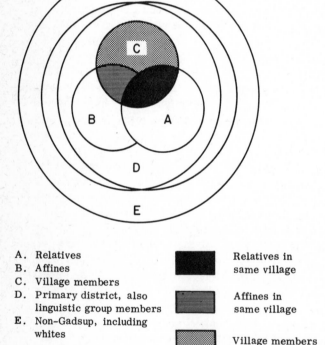

A. Relatives
B. Affines
C. Village members
D. Primary district, also
   linguistic group members
E. Non-Gadsup, including
   whites

Relatives in
same village

Affines in
same village

Village members

Diagram 3. Schematic representation of Akunan sociological universe

in his village of residence. The intensity of their social interaction (measured by duration and frequency and resting on informant statements and observatory recordings) is greater than that with relatives or affines living elsewhere. Fighting never takes place with members of categories A and C, and a person maintains good relations with those of B because of their usefulness to him in everyday living and especially in leadership aspirations.

Persons in category D include all members of the linguistic group, who are not included in the interlocking nuclei. This does not mean that a member of the Eastern Gadsup has stronger ties with all Eastern Gadsup than with any Western Gadsup. Simply because the frequency and duration of contact within the district is greater than outside it, we find that the person's sphere of interaction is greater within his district, but overlaps occasionally to other districts, just as it does in rare cases to E, namely non-Gadsup. This latter category includes the occasional traveller passing through the Arona Valley from Chimbu or the Markham, intermittent ties with Agarabe or Tairora, and weak recent ties with non-Melanesians.

During the field study, Torawa and his mother, Wankano, travelled to Noreikora (a village in the traditionally hostile Tairora country) and offered gifts of betel nut to the luluai while requesting him to give hospitality to Kama and his family. The latter, a younger brother of Torawa, is teaching under the auspices of the Lutheran Church among the Fore and was due to return home through the Tairora country for a period of leave. Similar treatment is given travellers through the Arona Valley. In Akuna there are two men known as 'Awoku Wainta' (good men), both of whom make a point of showing hospitality and offering food to travellers. This conduct, traditionally of such great value when dealing with kinsmen or group members, is now becoming of great value when dealing with outsiders. Friendship and hospitality are two very important characteristics of the culture of this community, and this is especially true in any dealings which involve, though are not necessarily limited to, 'People of the Road'.

Within a person's social universe, however, very strong ties of friendship develop between friends who are not related.[2] These ties are not institutionalized nor formalized in the manner found elsewhere (e.g. Pouwer (1955: 55) mentions that among the Mimika close friends use kinship terms and this leads to exogamy between their children, and Watson (person communication. 1962) mentions institutionalized friendship among the neighboring Agarabe). Nevertheless they are sufficiently strong and enduring to lead to changes in resid-

---

2. See the discussion of tikono below.

ence, adoption of children, farming out of pigs, and modification in established rules of inheritance.

While it is true that the relations between members of the same village or district are strong and generally amicable, this is not always the case. Hostilities may flare up between them and domestic quarrels are nothing rare, but in all these cases Akunans will explain that they are quarrels and not fights. In other words, they see themselves as fighting to punish, but not to kill. In this same context it was true that the district was the unit of interaction, both as regards amity and enmity relations. Thus Akuna intermarries with Wopepa but fights with the latter's neighbor, Onamuna. While Ikana and Tompena are neighboring villages, Akuna intermarries with the first but considers the latter an enemy. The point here is that geographical proximity does not imply any particular social relations with other villages.

The village community is based on a set of what we may call age-groupings rather than being an age-class system or a system of age-grades as these are discussed by Prins for East Africa (1953: 10). By the term 'age-grouping' it is here understood that the society is divided into people roughly five to seven years apart. We can then distinguish young children, those approaching puberty, post-puberty children, young people who are courting, and approaching marriage, recently married adults, people with one child, those with a number of children, and so on up the line. While these groupings are not formalized or institutionalized, they are generally recognized, and form the basis for the Akunan practice of teknonymy.[3] This form of name usage is but one expression of the way in which juniors address and refer to seniors and a way in which seniority is formally recognized. It is not suggested that seniors or adults are honored, but due to the fact that adulthood is one prerequisite for leadership, children are taught the correct form of address. They are not taught to venerate or even respect adults in the way we think of it, and children can often be observed disobeying their parents or grandparents, running away from adults after arguing with them, or engaging in horseplay with an adult. This treatment is not resented by grownups, but rather encouraged, and on this basis the very close relationship develops between grandparents and grandchildren.

From childhood boys and girls are separated, not by any formal injunctions which regulate their actions and behavior, but simply because boys, once they are free of the mother's care, follow the 'monkey' bands and spend the day playing. Girls, on the contrary,

3. For a more specific discussion of this and examples of its use, see Chapter 9.

while they spend much free time in the company of their age-group members, are gradually introduced to the statuses they will occupy and the roles they will perform, as adult women, namely homemaker, horticulturist, and mother. In this way the young folk operate in two separated universes, meeting from time to time in the village, or when the 'monkeys' pass through the garden; but interaction is otherwise limited.

Although we will treat this subject in some detail in a future chapter, a few remarks are pertinent at this point regarding these informal groupings of youths. The most important point in the light of the foregoing discussion is the fact that they are based on the age groupings outlined above. Since the earliest years a child finds himself toddling around the village square in the company of other children of the same age and the same unpredictable physical condition. Together they fall and together they rise. Soon they start to imitate the older children in the games they play, and boys spend hours playing 'marbles' with fruit seeds while girls sit around a hole in the earth tossing fruit seeds in the air and attempting to catch an increasing number every time. By age five or six, the boys spend their days alone in the village while playing games, descending sharp declines on sliding boards, or playing 'King-o-the-Mountain' on clumps of pit-pit. While the girls associate in groups from time to time, they are primarily in the company of a mother or elder sister until puberty. Boys grow increasingly more independent up to this same stage in their life cycle. Groups usually are constituted of about four or five persons, but may grow or decrease depending on the kind of activity that is participated in; this, then, implies that there are more than one per village, and in fact very often more than one per age-grouping. These relations and friendships are carried over into the post-puberty years where they become stronger and recognized by outsiders because of constant association. In time they form the basis for friendships which so greatly influence the choice of a person where more than one possibility of action or behavior is open to him.

Puberty for both sexes is marked by ceremonial introduction to the adult realms of the particular sex, and for girls it opens the way to marriage. While each sex allows considerable freedom for post-pubertal children, and opportunity exists for informal contact between individuals of different sexes, the females are closer to adult status and perform roles more congruous with that expected of adults, than is the case with males. With the male and female role well outlined, and each person conscious of the behavior and choices expected of him, courtship follows, and the next stage in the life cycle, with its roles to be performed, can be entered into.

When we are dealing, however, with filiation and affiliation, we are distinguishing between kinship groupings and groupings in which option plays the primary role. Persons belong to one or another group because of birth and to a certain village either because they chose to identify with it or because certain villages are traditionally identified as the loci of particular clan groups. While a person may change his place of residence or may align himself with a different group by association or by adoption, he can never change his membership in his natal clan. For the duration of his life certain obligations and bilaterally extending exogamous rules apply to him due to this membership, and on the basis of these reciprocal expectations he must perform the roles prescribed by the members of that society.

By birth every person is ascribed the status of member in the patrilaterally extending clan group, but beyond this he recognizes his own personal associates and in time there develops about him an 'interest group'. This group is largely selected and dependent on the personality and prestige of the persons involved. While inheritance always follows recognized lines, e.g. from parent to child, it may be extended between persons of the interest group — or conversely may be extended to increase membership in the interest group. This latter group, ties in very closely with residential and other decisions such as association, land tenure, and other considerations leading to identification and alignment with a certain person on whom one can depend in times of need.

While outlining genealogies, informants repeatedly used the term 'manankumi' when referring to their relatives, or 'wenankumi' when it pertained to another person who was not related. This 'ankumi', it seems, referred to the group of relatives. It was not used for all Gadsup ('ayukam waya'), nor those living in one place only. The 'ankumi' includes all the persons related to ego through both parents as far as relationship can be traced and also extends down to his children and children's children, but has a preference in the patrilineal line. Marriage is prohibited between all persons who can trace relatedness between them, thus all persons of the same 'ankumi'. The term 'ankumi'[4] is the same word that is used to refer to a bundle of sugar cane shoots. While in the bush one day, Munobi was speaking of the bamboo shoots at the base of a bush and he spoke of these as 'ankumi'. We have the same symbolism here, namely that of 'offshoots' where people who sprout in various directions are all related

4. See in this respect the use of the tree metaphor referred to by Firth (1936:371) and Read (1954:11). Due to this strong bilateral leaning I was inclined originally to ignore the patrilineal tendency. It now seems, however, that the clan is at least of nominal importance to the person under consideration.

at the base. Most of these persons who are included in ego's 'ankumi' also form part of his interest group, an important group of persons who associate together and include bilaterally extending relatives, affines and persons who live together in the same village community.

Every Akunan, then, is a member of a wider group extending pat-rilaterally and matrilaterally, with whom he has cordial relations and with whom he exchanges kinship terms. Though he can always expect hospitality and assistance from members of this group, and though his rights find expression through membership in this group, the group as such never acts collectively and does not have any corporate rights. This fact became very clear in the collection of genealogies. A person, depending on his age, place of residence, and often his aspirations, might know most of the persons related to him or only a very few. An old woman, Ampino, who lives in Akuna with a friend of hers, was able to mention thirty persons in her ankumi and did not seem upset when unable to extend this. In contrast to her is Darayi, an old man in Wopepa who without much trouble listed 400 persons and their exact ties with him. The average, in forty-two genealogies collec-ted from males and females, is ninety-five named persons. In this light Freeman's statement that "at any given time the sociological limits of a kindred are set by the recognition extended to cognates by the individual at its centre" (1961: 207) becomes meaningful. But even at this level of analysis, namely by focussing on the individual at the center, we are still faced with the problem of delineating the ankumi in terms of its value to the person concerned. This problem has been formalized in the following way: "The difficulty posed by cog-natic or non-unilineal descent is that collateral cognates (from first cousins onward) belong to more than one cognatic stock. This means that cognatic stocks, at this level, overlap; and consequently, unless some criterion other than, and in addition to, descent be brought into operation, it is impossible to achieve the division of a society into discrete groupings. No account of a bilateral or non-unilateral system can be considered complete until the way in which this difficulty is solved has been demonstrated in detail" (ibid.: 200). In this study we will treat only part of the ankumi, namely that part which has any value for the person at the center. In fact, it seems as though only those persons are of importance and need consideration, for those members that are not considered to belong to the ankumi, for all intents and purposes do not exist. Unless the person at the center recognizes certain rights and privileges in relation to another, that other person might as well not be a relative. The importance of cog-nition and alignment are the two primary factors with which we will deal. While discussing the Sinhalese kindred, Yalman points to this whole kindred as forming a "protective wrapping" around the individual (1962: 553). For Akunans, and this seems to be generally true, only a part of this ankumi can be said to be any form of "wrapping" around the person at the center. For that reason we will define those persons who are of interest to the person involved and which assure him security and personal contact. Lawrence speaks of this core as ego's "security circle" (1955: 12), and Jules Henry discusses what he calls the "personal community" while showing for the Pilaga that in spite of the importance of kinship, this group is based primarily on co-residence (1958: 828). In much the same way but closer to Lawrence's approach, Leach introduced the valuable concept of "personal kindred" (1950: 61-2). Among Akunans, however — and this agrees with the approach of Henry — the members of the ankumi do not function alone but in all cases are joined by friends and affines, in other words, persons who are not related but who associate together because of free choice. Freeman has proposed the referent "a cognatic category" for this kindred group (op cit: 202), while Held recognizes the functional importance of these groups and calls them "feestkernen" (1951: 122), i.e. festive cores or nuclei, namely a group of persons who associate with and assist one another on ceremonial and festive occasions. We have chosen to speak of interest group, as it clearly defines the justification for identifying with a person. The interest group is constituted of ankumi members, a number of selected affines and friends, and is usually based on and expressed by co-residence. Important is the fact that this grouping is not static, that it will keep changing as persons are born and die, and furthermore that the grouping can also be applied while focussing on a number of persons. By changing the focus on an individual the membership of this group will change, too. It should be kept in mind that the interest group differs from the clan, the ankumi and the residential group, even though it always includes members of each. It should also be kept in mind that the interest group is fluid, always changing and adapting to the individual's needs at the particular time, nor does the interest group of one person ever entirely agree with the interest group of another, for while the ankumi group is ego-oriented this is even more true of the interest group which each person builds up for himself.

Above it was pointed out that the ankumi, and even a person's interest group, in all cases extends beyond the village of residence. This leads to the ambivalence and the fluctuation in loyalties which a person feels, often resulting in his preference for the residential group over his kinsmen residing elsewhere. Kroeber has suggested this to be a general characteristic, when he stated that "it seems doubtful whether sex lineage reckoning as such is as important a factor as residence" (1942: 210-11).[5] Among Akunans they are of about equal weight, and while a person will never turn his hand against his fellow villagers during fighting, he will simply remain neutral should they attack a village where he has kinship ties. As a general rule, informants stated, villagers did not attack to kill in these villages where their own members had kinship ties.

In the light of statements above regarding settlement and village compositions, we can look more closely at the kinship and affinal ties within Akuna, as well as those which link Akuna with the other villages. The results of a detailed analysis of genealogical material and a village survey of Akuna produce some interesting facts: Household 1 through household 41 are associated with Upper Akuna, while Household 42 through household 55 form the association which we above designated as Lower Akuna. Our count shows that the upper part of Akuna has more ties per household which link it cognatically and affinally to the other houses in Upper Akuna, than is the case in Lower Akuna. From a total of 722 ties,[6] only ninety-five are with and between members of Lower Akuna. While the average number of ties per household for the whole of Akuna is 13.1, the number for Upper Akuna is 15.2 per household as against only 6.7 for Lower Akuna. While a single significant tie, either on the cognatic or the affinal level, may be of much greater importance than a number of more distant ties, we have no way of establishing the primacy or importance of these ties, and have therefore made the only possible division. Within each category all ties are considered of equal weight and importance, while kinship ties are seen here as of relatively greater importance than affinal ties. This will substantiate the statement made above that in the process of settling a new village, members join it for reasons other than cognatic or affinal ties with its members. It is significant to note that Upper Akuna has no cognatic ties linking it with Lower Akuna except for seven househeads who are related to the young woman who lives in household 52, and four househeads who are linked affinally with her. Furthermore, there are no households which are linked cognatically and only eight that have affinal ties with Lower Akuna househeads. This substantiates the suggestions made in an earlier section that Lower Akuna represents the initial stages of schism.

One household must be excluded from this discussion, namely household 40. The reason for this must be blamed on an oversight on the part of the ethnographer. The household was included in the village

5. This same point was of course made nearly a century ago by Maine and later by Morgan (vide du Toit 1964(d):87, and Lowie 1937:50).
6. The writer is conscious of the fact that many of these are reciprocal, and are therefore duplicated.

Table 8. Ties between Upper and Lower Akuna

Village section	Upper Akuna Cognatically	Affinally	Lower Akuna Cognatically	Affinally
Upper Akuna	143	150	11	23
Lower Akuna	0	8	8	14
Household 52 in Lower Akuna	7	4	0	0

survey, but was never cross-checked with genealogical data, although it does seem significant that none of the other genealogies, covering the total population of Akuna, claims either cognatic or affinal ties with any member of that household, consisting of a nuclear family. This suggests a correspondence with the case of household 12, but in the absence of specific information it is denoted as 'unknown'. In the light of the suggestions raised above, namely that we are dealing with an absence of rigid requirements regarding residence and with a freedom of association, it might be significant to look at two groups of households. Those to be discussed briefly form the upper and the lower limits on our household count referred to above namely those with more than twenty-seven ties (cognatic and affinal) and those with less than two ties with other households in Akuna. Under the first category are households 1, 3, 8, 10 and 14. Household 10 is that of the luluai of Akuna, while households 8 and 14 are occupied by his brothers. The first two households are the sons of the previous luluai, Pumpua. This shows two things, firstly that the government-appointed luluai in these cases (and this is generally true in the Arona Valley) have the cognatic and affinal connections upon which leaders can draw and thus the accompanying social backing which would have aided them to become 'strong man' ('yikoyumi') in the traditional system. It will be recalled that the interest group includes persons who are linked by kinship and affinal ties, as well as those who associate with a person because of common residence, and that grouping would therefore include members from both these groups discussed here. Secondly, the cores of certain kingroups are usually localized and certain clans are associated with certain villages. We are conscious of the fact that common residence may be an expression of ties as well as a cause for these, and both of these reasons are present in Akuna. At the opposite pole in terms of social ties represented in our count are a number of examples; household 12, whose occupants come from Tompena and have settled in Akuna for personal reasons; household 43, whose occupants also came from Tompena while a married daughter has since married into Upper Akuna. The occupants of household 47 came from Amamunta and have very strong ties of friendship with the tultul and Napiwa, the potential leaders of Lower Akuna. The woman who lives in household 49 comes from Wopepa and was married to the househead of household 42, but is divorced and has decided to remain in Akuna. Topide, who lives in household 54, has relatives in Wayopa (patrilaterally) and in Wopepa (matrilaterally) but none in Akuna, and his wife comes from Ikana. Karempo, resident of household 25, comes from Ikana, and E?epuro, his wife, from Wopepa. Both these families have chosen on personal grounds to reside in Akuna.

The previous discussion dealt with those households which would form the two poles on our kinship bond and sociometric table when each tie, cognatic and affinal, is given equal weight. This discussion, it is felt, clarifies two points which have been discussed and which form the core of this study, namely the fact that there is a positive correlation between residence or locale and the place or group of primary interest to the individual concerned. Regarding the lower limits of the table, it is clear that these persons settled in the village where they reside for reasons other than any strong kinship or affinal bond. This, in fact, is clearly stated by informants when they remark that Akuna is a 'gudpella ples' (good place) and that their decision was based on 'ting-ting bilong mipella' (our thoughts or decision).[7]

The upper limits of our table show that the leaders on the village level are the persons with the greatest number of cognatic and affinal ties. It would also be possible to recheck this by looking at the households of other influential men and to see how they compare on this level. During the field study, a set of questions was asked to a number of informants of different ages and thus from different status positions. The questions dealt with the following general topic: "If we now were living under the traditional system, where there was no kiap and no government-appointed village officials, who do you think would be the strong man ('yikoyumi') in the village?" In each case three names were asked for, though not insisted upon. It could also be remarked that characteristics which lead to leadership and which were valued in leaders were recorded. For Akuna there are four names which were mentioned most frequently, namely the luluai, the tultul, Napiwa, and Munowi. We shall briefly look at each of these persons, in terms of the total number of ties counted and referred to above.

1. The luluai (household 10) has already been discussed, as he forms the upper limit with a total of thirty-two ties.

2. Munowi (household 16) moved to Akuna shortly after his marriage. He is the person who introduced coffee into the Akuna economy

7. See Tables 14 and 15 in Chapter 9.

and is by now quite wealthy. Also, and this is important, he is a very skilled orator and speaks with power and conviction. He has only nine ties, of which a number are with his married children in Akuna.

3. Napiwa (household 44) was born in Akuna but has cognatic ties with Ikana. His mother was captured during intervillage fighting with Tompena, and he married his first wife in Kundana, an Eastern Gadsup village. He is a wealthy man and one who can speak in the best of Gadsup tradition, but linked by only thirteen ties within Akuna.

4. The tultul (household 53), too, finds himself in the lower limits with a total of twelve ties, of both cognatic and affinal nature. Once again, two of the factors which are involved in his choice are his wealth and his oratorical abilities.

In all four instances the choice was partly dependent on the fact that the men named could make peace between quarreling villagers and that they were sympathetic in their dealings with others and 'felt sorry' for others.

This once again gives weight to the suggestions that a leader may draw other persons to him by characteristics which he has, that there is a great latitude in choice between localities of residence, and also that personal factors may influence the person who is making the decision. Leaders may attempt to manipulate their followers, but the latter are free to change their loyalty and their place of residence. It also shows that leadership is not based on kinship or the number of cognatic ties alone, but allows for personal abilities to be cultivated, and for the matter of common residence to be utilized.

Regarding intervillage relations, we note in Table 9 that Wopepa (the mother village) and Wayopa (the daughter village) are strongly bound to Akuna by both cognatic and affinal ties, while extradistrict ties are negligible. There are no ties with non-Gadsup, except villagers who are on labor.

It should, of course, be kept in mind that we are dealing with villages of various sizes and that we cannot simply compare the total number of ties of two villages irrespective of their size. Wayopa and

Table 9. Ties between Akuna and other Gadsup villages

Extravillage ties	Cognatically	Affinally
A — Wayopa	21	29
B — Tompena	15	15
C — Wopepa	26	28
D — Amamunta	19	19
E — Ikana	9	12
F — Onamuna	2	6
G — non-Western Gadsup	2	5

Amamunta are both "daughter" villages of Akuna, with Wayopa being of very recent origin. This village has a total population of eighty-two persons who are linked by a total of fifty cognatic and affinal ties to Akuna. In the case of Amamunta (and its daughter village Manunampi which is included with it) there is a total population of 269 which is linked cognatically and affinally by a total of thirty-eight. The present population of Amamunta is thus less homogeneously of Akunan origin. The other persons which make up the population-total have settled there, among other reasons, because it is close by the main road between the highlands and the coast or because a White couple (members of the Summer Institute of Linguistics) has lived there for three years and offered store products for sale to their villagers only. Also, being a young village, there is still a great deal of land available for cultivation and persons are encouraged to join it. All of these villages, too, belong to the same district. The high percentage of ties with Wopepa is clearly explicable in the fact that Akuna originated from repopulation by Wopepa people, and also the important fact that the latter is a much older village — though not by that name[8] — and that cognatic ties can therefore extend further back in time and wider consanguineally, to link more persons together. This is not the case in a young village where the grandparents lived in Akuna and their siblings' children still do. The percentage for Wopepa is especially high when we compare its fifty-four ties with its total population of 301.

The strong tendency for intradistrict marriage, or at least marriage within the group outlined as Western Gadsup, is also clear, pointing at the group identification and association.

SUMMARY

While there is ample justification for speaking here of a village community in contrast to other such communities, it would not be correct to suggest a further subdivision at this stage. At the time of this study the members of upper and lower Akuna had enough interaction to justify grouping them together in contrast to other village communities. We should of course not loose sight of the fact that we are dealing here with interaction networks and sub-systems. In this regard a number of factors have been considered to ascertain the nuclei of such networks. Above we already mentioned and gave evidence for cognatic

8. It should also be remarked here that villages frequently change the locality of settlement. Wopepa had been located for some time in the valley next to the Ramu and moved to its present site in about 1949: Patrol Report, No. K. 4, 1948-1949, by D. Whitforde.

and affinal bonds which link households in one part of the community with households in the other and also with other village communities.

We can go further than this and plot the locality of earth ovens as has been done on the sketch map of Akuna village. The question then would be which different households use the same earth oven, and whether any pattern emerges from this. With the exception of the earth oven which is located directly behind houses no. 10 and 11, each earth oven would be used by those families living closest to it. The oven which is excluded is that behind the house of the luluai and while it usually served family members in adjoining houses, it would frequently be used when visitors arrived or on some village occasion like a feast or a court session. If the oven is used by those households proximate to it, we have to agree that it serves basically as a unifying factor for neighbors, rather than being used selectively by persons from different parts of the village.

Very closely tied to this subject is the question of who resides in adjoining houses. We stated earlier that Akuna does not have any one group resident in the village, but that various kingroups have been joined by a number of unrelated individuals or nuclear families who settled in the community for reasons of their own. While this is true in a number of cases in Upper Akuna it is especially true in the lower part of the village. Our counts furthermore show that while the members of a household are frequently related cognatically or affinally to persons in adjoining households this need not be the case. An example is household No. 12 as well as others. While the luluai as head of the strongest clan lives in household No. 10 and his two brothers reside in No. 14 and No. 8 and his sister's daughter and her husband in No. 11, unrelated families occupy other houses in that part of the village.

During warfare in traditional times the males from one village would leave the barricade for action and return to it. When a pig hunt is organized the men from one village will cooperate. When work teams are formed to cut logs which are sold, the men from one village join the same 'line'. In times of initiation, of the Orande, of ceremonial feasting and dancing it was always the residents of one village which joined, which shared, which interacted. We may speak then of Akuna as an example of a village community among the Gadsup. It could be suggested that this picture holds true for the eastern highlands. In this regard Read has pointed out: "The village of the east is a closely knit group. Its members are in close daily association with one another, and they share the widest range of common interests and common life ... The village, moreover, is the center of ceremonial and ritual life. All important group activities take place either at the men's club house, in the village street, or in the surrounding gardens. While, therefore, a man's closest bonds are with

72

the members of his patrilineage, his daily association from infancy onwards with members of other groups has an obvious importance in the formation of wider ties and group loyalties. Village life gives rise to a continually interweaving pattern in which the independent identity of each segment, though never lost, becomes absorbed in the common identity of the whole. On the organizational level, the village is the core of community. It stands in relationship to the social structure as that center on which all the most important group ties converge and from which they radiate. The common life of the village, the give-and-take of daily intercourse, and the manifold corporate interests and activities give rise, moreover, to strong emotional bonds, to that sense of belonging, that idea of togetherness which underlies and upholds the group organization" (1954: 14-15). While the members of a person's clan, his 'ankumi', or affinal group may reside in neighboring village communities, the interest group which every Akunan developes for himself will primarily reside in his community.

Chapter 4

COERCIVE AND JUDICIARY FORCES

For social order to exist within a community, and it seems that this is even more so in the case of these small village communities, there needs to be a high level of agreement regarding what is permissible. The members of such a community grow through a process of social- ization and enculturation to the point where they know what they may do, and also what forces may be brought to bear on them for miscon- duct. For this reason, it seems we may treat these two concepts not as dealing with the same realm of behavior but as being related. Coercive forces might be applied in the context of Akuna primarily in the realm of inter-clan and inter-community relations; while the jud- iciary forces establish and maintain common agreement on rights regarding people and with relation to property. Both of these have been subsumed in the arrival of a White Administration, and so a social order has been brought which differs in kind but not in degree from the social order which characterized the old Akuna.

THE STRONG MAN

It will be commonly agreed that social order requires some kind of leadership in the community. I am not thinking here of an officially elected or hereditary leader who occupies a position of chief or cap- tain. The fact that New Guinea generally lacks any kind of ascribed leadership or system of rank[1] is well known. While it frequently hap- pens that the strong man in a community comes from the largest clan in that community, this is not necessarily so nor will his son have any prior claim to the position of his father.

Traditionally the 'strong man's' influence hardly extended beyond the clan of which he was a member and the village community which

1. As more information becomes available, e.g. from Wabago (Read 1954: 10) exceptions may appear but at the present time we lack examples which con- tradict this statement.

recognized him. We should then see him as village leader for this was the primary realm of his influence. He might be dependent on clan members in other communities for assistance, or at least neutrality, but he looked basically to the people in his village. This 'strong man' in Akuna was called a 'yikoyumi' and achieved his status on the whole by gaining prestige. Every culture recognizes a number of clearly definable aspects which give prestige to the persons in whom they are found. Every culture also recognizes avenues for acquiring these valued characteristics or objects, and where status is not specifically ascribed to an individual, these avenues must be explored by those who qualify and desire to achieve that status.

The status position of 'yikoyumi' was restricted to males, and in reality open to any man within the village. On the whole, as has been suggested, the leader was a member of the largest clan in a village community. One may even speak of 'primus inter pares' for the adult males belonging to his and other clans were also eligible for this position and in any case voiced their views, often vociferously, in any discussions. Nor did the son of a leader necessarily succeed him.[2] In fact accession to office was one of the basic reasons for schism in village communities. The leader's role was informal. He did not have distinctive regalia or any symbol of his status, and no ceremonies marked his assumption of office. His behavior or treatment by others did not set him apart from his fellows, and his house blended with those surrounding it. In court sessions or on war parties he would not be noticed except for his behavior, his initiative, and above all his oratory.

In achieving his status the leader depended on prestige, and this was based on two primary factors, namely personality characteristics and wealth. It should also be kept in mind that only a strong, healthy, virile, and physically active person would be in a position to manipulate those two factors in achieving status. Akunans explain that their traditional leaders had the same personality traits which they still value today. These differed whether a clansman or fellow villager was involved or whether he was dealing with somebody from another group and thus a potential enemy. The 'yikoyumi' had to attract attention and loyalty for this was the only way in which a community would grow in numbers and strength, and it was the way to assure alliances in times of warfare. He was therefore in addition to, or perhaps because of the personality type he was, also a manipulator

2. This seems to be one difference between eastern and western Highlands. Berndt (1962:175) sees only a "slight tendency toward heredity", while Reay (1959:114) found 64 per cent of her cases where a son succeeded his father as leader.

of social relations. He was kind and just, and filled with compassion for clansmen and fellow villagers. He assisted them where possible and was continually working at maintaining peace, cooperation, and goodwill within the village.[3] As long as the members of a village community lived in harmony, the village would prosper; but when feuds broke out between different clans, when competition was too well developed between competitors, or when people were not satisfied with their leader, schism could develop and new villages be formed, or individuals and families drift away to attach themselves to other villages or to profess loyalty to other leaders. The 'yikoyumi' then, was one who wanted to do right and wanted to be liked by the members of his community.[4] When dealing, however, with strangers and potential enemy, he was still fair but ready to retaliate and outwit.

To a great extent the leader could manipulate social relations because of his wealth. It is good to inherit many pigs from your father, it was said, but once pigs and land rights are inherited, it was up to the individual to use them. The 'yikoyumi' was in all recorded cases, a polygamist. By having a number of wives — up to six, perhaps — there would be many gardens which produce tubers and greens, and since one household cannot eat all the food produced by a garden, much of it would have to be given away or used in feasts. The transfer of a gift, even as small as a couple of sweet potatoes, does not end there for it must be returned at some future date. Before it is returned, the receiver is obligated to the giver; after it is returned, there usually has developed a relationship of reciprocity. The good leader then was one who had reciprocal ties with many people and in many villages for in time of fighting they might be counted on to form an alliance, or at least to remain neutral. Another aspect very closely related to this one was the fact that the 'yikoyumi' tried to marry wives from different villages. If for instance the leader of Akuna had five wives, representing the villages surrounding her, then Akuna in fact had a regional alliance and could expect assistance or neutrality from the whole district. This was an ideal situation, but genealogies do not confirm that it was consistently practiced.

3. Lowie (1953: 162) emphasizes these same abilities in the Crow chief who "would exert himself as a peacemaker" for it strengthened the tribe in inter-tribal relations and competition.

4. Leadership among the Kamano, as described by Berndt (1962: 174), calls for "a dominant, aggressive personality, ability to override others in argument, quickness to take and to give offense, and proficiency and courage in warfare". Reay (op. cit.: 116) states that Kuma leaders show "wealth and reknown" and a "flamboyant personality".

In warfare the 'yikoyumi' was the fearless fighter who could lead his men, who could lie in ambush, and would be quick to retaliate if one of his villagers were killed. He was the orator who could use his hands and his tongue in flamboyant speech, rich with symbolism.[5] I clearly remember the present strong man of Wayopa, a tall muscular man with a voice as clear as any in the community. Men from Akuna had shot a pig, being under the impression that it belonged to one of them. It was in fact the property of Opura. He arrived in Akuna, taking up position in the village square and started an oration that lasted nearly two hours. The men of Akuna looked and listened as he recounted his whole history; how he had been an Akunan and then moved away; how he had started his new village and its history; how some people kill pigs without even looking and that such action could cause trouble. As his voice rose and his arms spread, he turned; and pulling a piece of wooden picket about three feet in length from the village fence, he drove it into the hard ground at his feet next to a pandanus leaf blanket. "I belong here; this is where my fence posts and those of my family have always been. Your parents conceived you while they were having intercourse in the dust of the kunai, but my father placed this mat on the ground for my mother to lie on ..."[6] And then, when the Akunans thought it best to discuss the possibility of paying Opura or replacing his pig, he shrugged his shoulders and dismissed the whole matter. When he had needed help these people had always been kind, and who will take care of him when he is old and not wealthy any longer? He thus gave them the pig and the Akunans felt very guilty and appreciative after first the harangue and then the gift.

There were two realms, that of inter- and intra-community relations, where the 'yikoyumi' was of particular importance. The first will be discussed here primarily in terms of warfare, the second in terms of court cases and the rights on which they are based and from which they flow.

WARFARE

No Gadsup community — and certainly not Akuna — ever started a war or initiated hostilities, at least this is what informants maintain. Hostile actions, an attack or an ambush, are always an act of retaliation.

---

5. See also under death payment — regarding the tultul of Ikana — page 128.
6. `Opura had moved away, but his wife was still working gardens near Akuna which they had claim to. The suggestion in his address here is that he is at least their equal and his parents might just have been superior to theirs.

They always represent an action which balances matters between two communities.

It will be pointed out in greater detail in the next chapter that death in Akuna was never natural with the sole exception of some very old people. This then called for discovering who caused the death and of taking vengeance. Since the other community will maintain its innocence, it is necessary to retaliate to balance matters and so a continuous — if intermittent — state of warfare existed. To complicate matters arguments frequently developed around women and pigs. As fighting gave prestige, young men were often grateful for the opportunity of breaking the monotony of everyday living and acquiring prestige in the act.

Men never left the safety of the village without their bows and arrows. Even today one will never see an adult male walking the valley paths or emerging from the bush without these weapons. To a certain extent these weapons are carried for use if they happen to meet something which can be hunted; partially, too, the long bow and accompanying arrows are a mark of adulthood and therefore seniority; but basically the idea is one of self-defense. Traditionally surprise attacks, ambushes, and retaliatory murder by stealth were the primary forms of hostilities. The latter two forms endangered every man while he was going about his daily affairs. The first also entailed the wife and daughters laboring in squatting position in the gardens, and for that reason a man was usually close by in case of need.

In Akuna I collected thirty-five named varieties of arrows, but they can be classified into three different categories. First there are the bamboo arrows referred to in the discussion of village economy. Here a piece of bamboo has simply been split to give a slightly hollowed out form which is then sharpened and hafted by means of intricate weaving. The small (about nine inches long) pointed arrow is called 'puyani' while one with a point of thirteen inches which is the more important of the two, is called 'kayemmi'. These are 'pig arrows', for they can be used at short distances when the aim is to kill a pig in the bush or near the village. They were hardly ever used in fighting because the irregularity of the surface affected the accuracy of this kind of arrow. The second variety of arrows to be excluded from our discussion of warfare and fighting are the so-called 'bird-arrows'. This category includes all those three — and five pronged arrows described to me as 'ayantakumi' which are used to hunt birds. The area of effectiveness has been increased in these arrows by making a number of prongs, each of which is separately carved, and attaching them by means of bark-weaving to the shaft. Instead of an arrow point there now are three or more points about two inches apart. The last category is that of 'fighting-arrows', and it is within this category that the greatest number and greatest elaboration takes place. Each of the arrow heads is carved out of a single piece of hard wood and varies in length between thirteen and twenty-four inches. This solid point is now hafted by means of bark wrapping or weaving and forms a firm unit with the shaft. The shaft in all three categories of arrows is about thirty to thirty-six inches and is made out of a strong bamboo-like variety of pit-pit. Solid sticks are never used.

Regular arrows in this category such as 'oma?e', 'aipara?e', and 'wara?e' were used for long range fighting when shots were accurate in spite of the fact that no arrows used by these people were supplied with wings or feathers nor was there a notch for the bow string. The, more elaborate arrows, as well as the pig-arrows when these were used, were employed for in-fighting when loss of height and direction was not critical. In an attack each man would carry twenty-five to thirty arrows in a storing bag over his right shoulder. Should a person's supply be exhausted and could find no enemy arrows to return, or when his bow string broke, he would retire to the rear of the group.

This category of fighting arrows is of particular specialization and attraction (to the onlooker) since they are delicately carved and decorated, painted and stained, giving a symmetrical product. The arrow heads are notched and barbed, and some have prongs of up to two inches pointing backwards and away from the body and point of the arrow. One of these, called 'awemi', has a point eighteen inches long and barbs two inches long and are carved from the central part and extend all around the arrow — a total of twelve such barbs. When I asked an old man, Daweka, about this arrow, he gave his usual chuckle and explained that in the time of fighting a man who had been hit by an 'awemi' was as good as dead. The other arrows could be pulled back and cut out, but the 'awemi' with its deadly prongs would have to be forced through the body. In somewhat the same category was the 'ate?na' with its point of twenty-four inches in length carved into numerous shorter barbs, and while the arrow could be pulled out it left a gaping wound which was often fatal to the victim.

As has been mentioned above these arrows were cut from the body of the wounded. The victim would lie on the ground while one male stands over him, holding the arrow and pulling very gently thus causing the victim's flesh to bulge around the barbs and notches. A second person will kneel immediately over the wound with a sharp bamboo knife slicing with small delicate strokes to free the arrow as it is pulled upward and finally removed. Many victims of course suffered damage to internal organs and for them, as well as those who lost too much blood while awaiting treatment, death came soon.

The bow ('itanda') is made of a very rigid shaft while the bowstring ('inami') consists of a thin and narrow strip of bamboo which

79

is periodically dipped in water to retain its pliability. A man always has an extra bowstring in his carrying bag in case of need and may in fact use it to start a fire.[7]

Arrows, as our discussion of village economy will show, are frequently included in all kinds of payments. They were an important part of the value system and even today are still highly valued. While they no longer have great utilitarian value, arrows especially are valued for aesthetic and emotional reasons. The complex and brightly colored arrows, it seems, are no longer made. The bulk of arrows observed form a part of various transactions in Akuna and between Akuna and other villages, but these did not include 'awemi','ate?na', or 'eparati?i'. These latter kinds of arrows are to be found in the possession of many older males and those which I was able to acquire were delicately stored either in bamboo containers or wrapped in corn leaves and resting on the rafters in a hut out of the reach of children. In most cases the owner would explain that this was made by his father, or used "when we fought the men in Tompana".

But bows and arrows, though the most important weapons, were not the only ones used. There was a spear-like weapon, called 'wan'ka mi',[8] made from a hard black wood which looks like the wood used for bows, and tapering off to a sharp point at one end. It was nine feet long and a third of the way from its point a bunch of cassowary feathers spread around the shaft. These, it seems, had no aerodynamic functions for they faced the wrong direction, were situated medial rather than terminal, and according to informants the spear was hardly ever thrown. It was used primarily for in-fighting after an ambush, or while in pursuit.

For defense rather than attack there was a shield which was carried over the back and left shoulder and appropriately named the 'plank or wood which is carried on the back' (or 'yani upekunde?no?u'). This, it seems, accompanied the bow and arrows as standard equipment during warfare as a man could quickly turn his back when an arrow approached, or swing the shield over his arm. The shield was made from the wood of the 'mopu?uam' tree and was approximately one inch

---

7. Akunans traditionally started a fire by the sawing-thong method (Notes and Queries on Anthropology, 1954: 240). This was referred to as 'ikanami', a word constituted by the generic term for fire, namely 'ikai', suffixed by the word for string or rope, namely 'nami'. The bowstring is looped around the hardwood shaft ('ya?o') and some moss ('konku') placed over the place where they meet. The person now grips the two ends of the string and pulls first the left and then the right to cause friction where the string meets the hardwood.

8. Gadsup distinguishes by tone between various items of vocabulary. In this case 'wan'ka.mi' should be distinguished from 'wa.nkami' — the hole in the ground which is made by a rat.

thick.[9] It was fifteen inches wide and fifty-four inches long while decorated on top with cassowary feathers. About half-way down the length of the shield were two holes — one higher than the other and about twelve inches apart — through which a strap of pliable bark had been tied. This was the shoulder strap which allowed the bearer to hang the shield over and behind his shoulder leaving both hands free to shoot the bow and arrow.

While we were living in Akuna a man stopped by one day on his way to Kainantu. He had disarmed a fellow villager of his living in Kurangka of a new kind of weapon this man had made. This was a club with a head perhaps four inches in diameter into which had been driven a great number of six-inch steel nails so that all sides of the club-head had a sharp and irregular suface. Not quite satisfied with this product he had encircled this head repeatedly with barbed-wire. The product was as lethal a weapon as any that had ever been used in the Arona Valley. The Administrative Official in Kainantu was glad to confiscate this innovation.

The use of these weapons in traditional warfare, as has been suggested, was always legal or retaliatory. At least the way in which Akunans see it, they only acted to correct wrongs and injustices which had been committed against them.[10] These aggressive actions took mainly three forms: that of the surprise attack; that of ambush which was frequently associated with the first; and a retaliatory murder requiring stealth and speed frequently associated with sorcery.

After a period of quiet and recuperation when the community-life has returned to normal, or directly following the murder of a villager, the men would gather in the men's house and consider what steps of vengeance need to be taken. From this time, until after the fighting has been ceremonially concluded, there is a very strict taboo on sexual intercourse. It is said that if a man 'smells' the enemy arrows will find him quickly. After a ceremonial wash and the nose bleeding which follows the warrior is in a state of purity and neutrality.

They would invite clansmen living in neighboring villages or sometimes request a temporary alliance with a neighboring community with whom amicable relations existed. Those who were gathered in the men's house, and this would include only adult males of that or friendly village communities, would now blacken their faces with ash from the hearth and decorate themselves with feather-headdresses of

9. One of the shields which I inspected had been pierced by an arrow point — the latter breaking off. Upon removing it with a pocket knife the arrow point was one and one-quarter inches in length.
10. See also Berndt (1962: 408) who speaks of "to balance the account between districts".

cassowary feathers, and also bright cockatoo and other feathers. The
night before an attack is spent in preparing bows and arrows, checking for extra bowstrings and also boasting about previous encounters
in which they were victorious and in which their fighters and 'yikoyumi'
excelled. This it seems serves a dual purpose: first, it is a morale
booster and psychological stimulant to old fighters, while for the
young men it is an education and a blueprint for action. On this occasion, too, each man collects two pit-pit leaves and clenching them in
his hands the 'upaiyami'[11] is performed until the blood flows freely
from his nose. This, informants state, will clear his head and view
and make him strong as all obstruction in the system is removed.
The blood is now used for ceremonial decoration, and on the ash-stained faces two lines are drawn, one down the center of the forehead
to the tip of the nose, and the other horizontally across the face outlining the nostrils. The rest of the night war songs are sung.

Very early the next morning the 'yikoyumi' will address the men;
outlining their plan of attack and while speaking in the best oratorical
tradition, thumping his foot for effect, he will encourage them to
revenge the death of their kinsmen who had been murdered. Holding
their bows and arrows against their chests, the men will sing softly
while shuffling from one foot to the other. It does not seem as though
supernatural support was sought in any way. Omens, however, will
be heeded, and if the 'yikoyumi' or one of the older men dreams about
their defeat, this will be seen as a warning from the ancestral spirits
and the expedition would be postponed. There is also a little parakeet
called 'umpanda' which perches on high trees and surveys the surrounding country. Whenever the 'umpanda' whistles, it is a sign that
somebody is approaching and the men in the village will gather their
weapons and, dividing into two groups, approach that spot where the
'umpanda' was. Departing from their villages, the war party leaves
behind a village well stocked with food, water, and firewood for while
the warriors are away, the other villagers will remain inside the
stockade which surrounds the village. Their defense, if such is needed,
will depend upon the old and wounded men who were left behind and
who would guard against surprise attacks by watching through holes in
the stockade.

The attackers move as quickly and quietly as possible to the village
where they are to do battle. Apparently there was no system of scouts
as they moved in a body till fairly close to the village and then divided
into two groups, one going to the left, the other to the right of the
village entrance. Each man has on his left shoulder a wooden shield,
over his right shoulder a bag of arrows, and in his hands his bow and

11. Vide the discussion of male initiation in Chapter 7.

arrow ready for action. The warrior's face is stained and ceremonially painted, and clenched between his teeth or passed through the septum is the curved boar's tusk. Their 'yanka' bands around their arms contain leaves and also their backs are camouflaged with grass and leaves. Blending thus with the grass and brush, they form two long lines on either side of the village entrance. Three or four athletic young men who had not joined these two lines now enter the village gate, or at least go up to the entrance, and while shouting would shoot arrows at the first person they see. Aroused by this infringement of village privacy, the males with weapons in hand chase after these three men who turn and run. As they leave the village behind, while the village males are in close pursuit and shooting at them, their own fellows rise — first those nearest the village entrance to cut off the route of retreat — and the pursuers are caught in an ambush from which escape is only possible by the retreat or defeat of some of the attackers.

Another example of the tactics[12] used in fighting was an account of Akuna's attack on Tompena. Here the attackers arrived in a body to confront their enemy at the village entrance. As the defenders attack and press on, the attackers gradually lose ground, falling further and further away from the village entrance. Now once again a group of strong fighters who had moved up the sides, will attempt to intercept the village entrance, thus preventing a retreat from the defenders and simultaneously allowing them to lay waste to the village. In this the small group of fighters will simply form a wall behind their shields and attempt to keep the village entrance.

When they have counted the dead among the enemy an arrow is broken for each victim and these carried with them to the graves of their own soldiers. They also collect their own wounded and return to their village. As they retreat they destroy gardens and fences, but they keep a sharp eye for any enemy soldiers who might creep up on them from behind. When they are still a long way off, female informants explained, they would know what the outcome was and even have a fair idea of the number or names of the wounded. The latter are carried back to their village and surgery which may be needed is performed there. The broken arrows which were brought back from the battle are now taken to the place where earlier victims on their side were buried. For each person previously killed a broken arrow is stuck in the ground on the grave thus marking the balance which had

---

12. Both of these tactics are referred to by Flierl with reference to their early contacts with Gadsup. The ambush (op cit:18) was merely suspected, but this second form of entering the village and shooting did in fact take place on March 18, 1926, when at least thirteen women and children were killed (op. cit.: 21).

been restored. Important persons, such as a 'yikoyumi', might however require two or three enemy dead to restore the balance, and in this way both sides may never be satisfied at the same time. One side will always be in debt and for every community this means a debt to nearly every other community.

Following their ceremony to inform the dead that 'we have revenged your death', the dance and feast of victory takes place. Allies, either individuals such as kinsmen or villages who assisted or remained neutral, are also frequently invited as a show of appreciation in terms of the reciprocal relationship. This is the time when victory songs are sung, and when a man dances with one long feather on his head which swings to and fro accentuating the rhythmic shuffle of the dancers feet.

Another form of attack involved more stealth and preferably no fighting. In this case the enemy creep up on a village during the night or more frequently the very early morning, tying the door to the men's house and setting fire to it.[13] Each men's house, however, had a secret door, and if the enemy did not know about this the battle was on. According to tradition, the total male population of Akuna was killed one night during an attack by the men from Maropa?i. One man only, Ururato, escaped and fled to Tompena where he took up residence. While in the field I spoke to his son, an old man named Menu whose daughter has returned and married a man currently living in Akuna.

Even after the Australian Administration had taken over control of the Arona Valley, fighting continued for some time. The Akuna people recall how they awoke one morning to find Tompena attacking. Ari, a man of about forty-eight woke very early that morning and went outside to urinate when he noticed some men near his garden fence. Later when it was light enough for him to see he approached the garden which is very close to the village and was shot with a number of arrows in the lumbar region of the back. The attack did not spread as all the men from Akuna rushed to his assistance. He took a long time to recover but still has the marks. This same form of stealth was used to kill people as they sat by their fires or emerged from a house, and before a counter reaction developed among community members the attacker would be gone and unknown. Now the drawn out process would begin in which sorcery would be employed to ascertain which village and which clan — and possibly which person — was responsible for the hostile action.

13. Flierl in his discussion of their early missionary contacts with the eastern border of the Gadsup, mentions an occasion on which this technique was to be employed (op. cit.:12). He in fact calls it a Gadsup custom.

Material gains through war were mainly in the category of pigs, but women and children were frequently captured.[14] If the inhabitants of a village are killed and scattered, the conquering villagers will have no fear of settling on the land. It has been reported from other peoples that fear of the original ancestors and fear of sorcery spells which might have been left behind would prevent settlement of conquered territory. This is not the case among the villages neighboring Akuna. It is, however, of great importance to keep in mind that no hostile action was taken and no battle conducted with the aim of gaining in wealth, in members, or in territory. Battle was done to correct wrongs and to re-establish a balance in social order, and if gains were made this was accidental and beside the point. Nor was there a public or private dividing of the booty between the main village community and those who assisted. The latter were allies because of reasons more important than any potential gain.

Allies might be seen as including three primary categories of persons: those belonging to the same village as the family who is taking revenge; clansmen of this family who live in other villages; and friendly villages who might be closely tied to the first village or be in debt to it in terms of the reciprocal relationship. When a man has been killed, his brothers will start to mourn him. They allow their beards and other facial hair to grow and later they will blacken their faces with ash from the hearth in the men's house. Others looking on — and these will include firstly the men who live with him and share residence in the men's house and secondly clansmen who live elsewhere and visit the mourner — will feel compassion for them. They will say: "Poor fellows, look, they are mourning their brother who was killed by so-and-so" (naming the person and/or the village). From this moment a decision has been taken and preparation will be made. Occasionally when such a war (and this really implies no more than a single battle which is concluded within a few hours) took place, one village may be joined by another in defence, but usually in attack. This then led gradually to allies developing, and on the other hand to traditional enemies or 'namuko'. One such an enemy of Akuna's is Kundana across the Arona Valley. But such traditional enemies were not always members of opposite regional groups in the valley. It has already been pointed out that the district was not the warring unit, but villages of one district (such as Akuna and Amamunta) often allied themselves for an attack on another, or remained neutral when they had ties with both villages involved in the fighting. An example of each of these follows:

14. In a later chapter a number of examples will be given of both children and wives who were carried away and incorporated into the new community.

1. When, some years ago, the people from Tompena, the 'namuko' of Akuna, killed a man named Wopuma living in the latter village, his brother Yerai (the current tultul) left for Ikana and solicited their assistance for a retaliatory attack on Tompena. The brother who was mourning and initiated vengeance did not have clansmen in Ikana, nor was the deceased married. This clearly is a case of allies.

2. On another occasion Akuna attacked Onamuna, but Maropa?i (which recently relocated their village and are known now as Wopepa) remained neutral. The reason is that Maropa?i had very close ties to Onamuna — who in fact contributed members again when the village of Wopepa was formed — but possibly even stronger ties with Akuna. Most of the clansmen of Yerai, the man mentioned above, for instance, now live in Wopepa.

3. A more complicated case occurred between a number of villages in which Akuna was not directly involved. Some years ago a man named I?u wanted to visit his wife Yemano who, however, was in the women's house of seclusion. This was in the village of Ikana. I?u now climbed a tree which was close to the women's house and was seen by another woman as she emerged. She did not let on that she had observed him but immediately left to inform her husband Intato. The latter returned with his arrows and his bow and aiming at the man in the tree called out for the man to identify himself (so as not to kill a clansman or fellow villager). I?u must have been shocked by being discovered for he remained silent and was shot in the chest. As he fell to the ground he called, 'Intato, temi tironami!' — Intato, you have shot me! — but he was mortally wounded. Scared by what he had done, for one does not kill a fellow villager, Intato fled into the bush and roamed the bush from the Tairora border to the Markham river. When he was found at the gardens of Inunti and Waka — two Akunans — they fed him for he was about to starve. As his strength returned he decided to live in Wopepa because this latter community had very little contact with Ikana. One night while sleeping sorcery was worked on him by Ikama a clan member of I?u who had called in the assistance of Ampo?ima — the current 'luluai' of Aiyura. To make quite sure that their treatment would be lethal they also shot him with arrows. A clansman of Intato learned about this and threw himself upon them asking to die with his 'brother' but Ikama explained that they had merely revenged the death of I?u. The corpse was now removed for burial in Ikana. Yemano, the widow of I?u moved to Akuna and is now living there with her second husband Apaka and two sons by him.

It will be emphasized below that one never killed a fellow clansman and hardly ever a fellow villager. In most cases that were recorded punishment was meted out but one did not punish with the intention of causing death. Here again we are dealing with clanbrothers uniting, but we are not dealing with fellow villagers for Ikama was living in Amamunta with one wife — Uwandano who comes from Akuna — but her father and Ikama's second wife lived in Wopepa. When he and Ampo?ima united to punish Intato they were not bothered by conflicting loyalties. They were dealing with a stranger and not a fellow villager.

Our second example above points at a very important fact namely that a person never fights a village where clansmen of his live. In this way Maropa?i would not attack Onamuna, and if the action were reversed it is likely that those people who have clansmen in Akuna would remain neutral. Frequently an individual or a few men would not participate in an attack because of kinship ties, but since they were the exception the body of village males would take action. In exceptional cases a clansman may actually warn another village that action is to be taken, but this was not usually the pattern since loyalty to members of the community was usually as strong as kinship loyalties.

But there must be cases where members of the same clan, in the same or different villages, wanted to take action against each other, I thought. And informants agreed that this was true, but they would never fight to kill, only to punish. The killing of a clansman is as grave a matter as the killing of a community member, but in neither case will a wrongdoer be killed for his act. It is a well known fact in Akuna, though not discussed, that years ago Daweyaka who lives in Upper Akuna killed Te?u, the oldest brother of Yerai the tultul, who lives in Lower Akuna. In this case, however, Yerai did not take the same kind of revenge as when his other brother was killed by men from Tompena. In both cases, killing a clansman and killing a fellow villager, the culprit is made to stand in the village square holding his plank shield over his shoulder in self-defence. His mourning clan-brothers, or the clansmen of his fellow villager, will shoot their arrows at him. They are quite clearly not trying to kill or wound him for then he would not be allowed to hold the shield, but they are expressing their grief and sorrow.[15] But if this is true on the individual level, it also holds for community-wide action. Akuna would not attack Amamunta, Ikana, or Wopepa since certain clans represented in Akuna are linked to clans in these villages. But when action was obviously needed a stick battle would follow in which clansmen beat each other but do not attempt to kill.

From information gathered in the field it seems unlikely that cannibalism was practiced by Akunans, or the Gadsup in general. Leon-

15. There might also be a trace of symbolism in this action which is related to the action taken by mourners after death caused by sorcery. A case is discussed in our next chapter which is similar in certain respects.

hard Flierl, on the contrary, explains how they entered the extreme eastern section of the highlands and Gadsup territory in 1920. He continues:

"For several hours they wandered down comparatively good paths into the valley. It struck the black attendants that a few branches of fern lay in the path. They were not certain of the personality of the Gazub, of which they said nothing to the whites, but they whispered their fears among themselves. Later we learned that the Gazub make it understood to their enemies through fern bundles that they have the intention of striking them dead and of eating them if they should dare enter their territory. That is to say, ferns are cooked with human flesh in order to give it a spicy flavor" (op.cit.1932:9).

This particular statement reached me after we had left Akuna and it was therefore not possible to check its accuracy. It is well known that ferns, and a variety of greens are cooked in pork, and it is possible that the warning was symbolic or referred to a form of symbolic or institutionalized cannibalism.[16] The reason for feeling that Akuna did not practice cannibalism is based partially on informants' statements, but also on a culture hero tale which is told.

A long time ago there was a young boy. He lived with his mother and always walked about collecting edible plants and insects. One day he climbed an 'ondani' tree and found that its leaves were ready to be picked and cooked. He took a few of these.

He went down again and went to an old man, asking him whether he would like any 'ondani' leaves to cook, but the old man did not want them. He then went to another old man who was sitting in the sun and inquired whether he wished to cook any of the leaves. "No", said the old man, "but try someone else who might like to cook them." Finally, after visiting a number of old men, he arrived at an old man called Mani?i, and after showing him the leaves, asked if he wanted them to cook for himself. "Yes, I will cook and eat them, but will you go and collect some more, for I am too old to climb the 'ondani' tree", was his reply.

So the boy returned to the tree while Mani?i accompanied him carrying a 'wankami'. When the boy had picked all the leaves and started to climb down the tree, Mani?i took his 'wankami' and killed him, taking him to the house where the old man started to cook him.

When the sun had set and it was growing dark, the boy's mother became worried and went looking for her son. She came to the first

16. See in this regard the excellent analysis of the Marind in southwest New Guinea by van Baal (1947:139-144) and even more detailed in terms of this topic, the paper by Verschueren (1948).

old man and asked him whether he had seen her son. "Yes", said the
old man, "he was here this afternoon with some 'ondani' leaves
which he offered to me, but he left to give them to another man".
The mother continued from one man to another and finally reached
Mani?i. "Have you not seen my son with a number of 'ondani' leaves?"
she inquired. Mani?i replied, "Oh yes, he walked by here this afternoon with his age mates on their way to go and shoot birds in the bush".
So the mother kept looking and after a while she returned, saying that
she had not found him. "Well", Mani?i told her, "he came back but left
with his age mates to go and set 'kinku?yi' traps in the pit-pit for rats".
Again the mother left him to look for the boy, but for a third time she
returned, complaining that she could not find her son. Mani?i told
her: "He left a little while before sunset with his age mates to pick
'puyami' nuts to eat around the fire". The mother visited all the age
mates of her son, but could not find him.

When she finally returned to Mani?i she noticed her son's foot
protruding from a bamboo container which was on the fire. She became
very angry and departed to call her brothers who returned with her.
They killed Mani?i and burned his house.

Such is done to any person who eats the meat of a man.

This tale then explains what action is to be taken against any person
who practices cannibalism.

OWNERSHIP AND PROPERTY

A subject which is at the center of social order is that regarding the
rights which are recognized for each member of the community. It is
necessary for every person to know what belongs to whom and how
strict these rights are enforced. Related to this is the question whether
others have the right to tresspass and the right of use of this property.
I would suggest that no coercive forces can be brought to bear and no
judiciary machinery operate unless every member of the community
is informed regarding these rights and obligations. This in fact is one
function of the socialization process whereby a next generation is
familiarized with such rights and duties, thus assuring the necessary
conditions for social order within the community.

The rights to land which surrounds each village belong to particular
clans and in each includes 'kunai' and pit-pit stretches, garden areas,
and bush. The first of these is situated in the valley floor and serves
primarily as areas where pigs can forage. The second is a transitional
zone and includes land proximate to the village settlement and on the
verge of the forests where shifting cultivation is practiced. The real

forest or bush had primarily hunting and collecting value in the past, but with the arrival of the White man and the call for timber it has entered more squarely into the economy.

Within these areas — but more specifically in the first two — particular families have rights of use, and such rights endure while they are maintained. When a person clears a tract of land, erecting a fence of split poles and logs, he has established full rights over that piece of land. The fact that he might be assisted by a member of his interest group does not give them any joint rights for they will be remunerated in some feast and the owner always asks his wife to share the first produce from the garden with such persons who assisted. These gardens areas are recognized by all and it is the responsibility of a pig owner to keep his pigs from breaking into a garden. If, however, a person clears a garden and erects a fence in the pig area of the valley proper, he does so at his own risk. When Yifoka cleared a garden in the valley and erected his fence, old people shook their heads. When the pigs broke through the fence and uprooted every single tuber and Yifoka killed a pig, he was forced to make amends and repay Amonko for his loss. The older men explained that he should have known the valley proper is set aside for pig foraging — there are other areas where people make their gardens.

The garden zone is divided into numerous gardens, each surrounded by a split pole fence and these posts are all neatly and strongly tied with tree bark. The land thus designated as a garden remains the property of the family that erected the fences for as long as the fences are maintained. The test is not whether the garden is planted — though these will logically coincide — but whether fences are maintained. But while the fences and thus the garden area as such fall basically in the male realm, the same is not true of the garden proper. It is the woman who breaks the earth, who plants, who weeds, and who harvests. The garden, therefore, is hers and she may treat it any way she likes.[17] A good wife maintains the garden in order that she may have food to serve her family. A widow will ask her brothers or grown sons to maintain the fences and she will continue to plant her crops. Upon her death, her children may inherit these rights even though they might have migrated to another village, provided they maintain the fences, thus justifying their claims. Land rights, then, do not require village residence and hardly community membership, but obviously a person will maintain cordial relations with the people where his wife cultivates her gardens.

A number of such cases will be briefly mentioned to show the variation which occurs in such rights of ownership.

17. The case of Oturo will be discussed below (page 95).

1. Nokame, the 'luluai' of Amamunta inherited land from his mother who came from Akuna. Though neither one of his three wives came from Akuna, one of them walks to Akuna every day to work the garden and thus maintains Nokame's land claims close to Akuna.

2. Many years ago, Yata?uri of Akuna, the father of Duikama (himself now an old man of about sixty years of age), married Nana?o of Wopepa and went to live with her family while maintaining rights to his land in his natal village. When Duikama was old enough, he married Amapino of Wopepa and returned to Akuna, taking up his father's land claims but at the same time retaining his own by continuing to work the gardens in Wopepa to enable his children or grandchildren, should they wish, to return and claim rights on their mother's side or on their father's mother's side. This is one of the primary laws regarding land rights, namely that the fences must be kept in repair, or ownership lapses.

3. In the discussion of village economy mention will be made of Yapananda and Tana?o of Akuna who gave up all rights and left the community to settle in Ikana. After a few years both parents and a daughter died. At the latter's funeral one of the elders from Akuna told me: "O?e and Apopine (two unmarried sons of the deceased parents) can very easily come back to Akuna. Many members of their parents' clans are still alive and can help them to get settled. They don't have any land in Akuna because there are no fenced gardens. If Yapananda had kept on working the gardens in Akuna, the boys could quite easily have walked across and established residence. In this case they cannot but clan-members will be able and willing to help them establish themselves. After all, they belong there!"

4. Many years ago Awanano married a man Ya?e and left Akuna to live virilocally in Wopepa. She, however, retained her land rights which she had inherited from her father Kewama in Akuna. The old couple had three children and then died. The eldest married a man from Wopepa and used their land. The second went to Amamunta where they married and used the land rights of his wife. He became 'luluai' and is now deceased. The youngest son of Awanano and Ya?e namely Apayu decided "that the land was used enough for a while" and decided to return to Akuna claiming land his mother had inherited and retained. He still retains land in Wopepa and works a garden to justify this claim, but he states that his son will go back and claim it.

An owner of such a garden may in fact extend the right of use to another, but he cannot alienate it completely. If somebody were to use his land without requesting permission, hostilities would develop on an inter-kin or inter-village level.

While rights of ownership over land must be maintained and can therefore expire, the same is not necessarily true of all the contents of such a garden. On a number of occasions it was learned that men owned betelnut or pandanus palms in the garden of another. It was explained that these dated back to the time when that land was worked by another family who planted the trees or palms. Kaua claims a tree which was once in his garden, but the garden has since been abandoned and overgrown by bush. When Duka made a new garden in this same area while we were in Akuna, the rights to this fruit bearing pandanus had to be recognized since the palm still belonged to Kaua. When the garden is abandoned, rights over the trees or palms are maintained. To pick the fruit belonging to another is taboo. This in fact is not only a legal infraction, but also is seen as a taboo and as in so many other cases this taboo can be traced to an actual event: Napiwa recounted how his paternal grandfather. Pi?uma explained the origin of this taboo. One day Pi?uma's brother was returning from the bush and walked by another man's garden. He noticed a bunch of ripe betelnuts and picked a few to chew on his way to Akuna. Before he reached the village he was killed by some men from Tompena.[18] This was ample proof that he had done wrong, "and now we have a taboo on picking the betelnut of another man unless he is there or has sent you. I may walk through the bush and pass another man's betelnut palms, but I will not touch them; I don't want to die", he concluded.

These same rights which apply to trees in garden areas are also extended to the bush. Ownership claims are indicated by placing a small bundle of grass in a tree and the members of the community are then informed that such a tree has been 'marked'. This applies to any kind of tree and would include such as would be used in house construction, or which contains a special tree fungus which is eaten. This individual ownership also applies to bamboo clumps which are found in or near the village settlements, or further into the bush. On a number of occasions I spent a day with one of the men, for instance Napiwa, visiting all his trees. This entails visits to the gardens of other people, but especially hours of following one footpath after the other through the bush, climbing a palm to test the betelnuts for ripeness and size, and then again to the pandanus palms or some other tree. On one of these excursions we were accompanied by Kama, the son-in-law of Napiwa. It should be explained that Napiwa has only one daughter, Kankua, by a former wife who left him. His second wife has not given him any children, and so his inheritance will primarily go to Kankua and Kama, even though the latter has been trained as a teacher by the Lutheran church and teaches school among the Fore

18. Notice too that the name of the Namuko appears and is repeated in numerous contexts.

speakers, southwest of the Arona Valley. On this occasion Napiwa showed Kama which trees, palms, bamboo clumps, etc. belonged to him and which Kankua and he would inherit. Simultaneously he informed Kama which of his trees he had given to two other men in Akuna for assistance they had given him. This would still not be final because while we were in Akuna, Napiwa and his wife, because they have no children of their own to assist them in the daily household activities (and also because they are getting old and have no grand-children on whom to depend in their old age) adopted Apipeno, the twelve year old daughter of Ako and Urake who had seven children. Since adoption — and feeding a child — establishes the same relation-ship as actual kinship ties, it also gives the child full rights of inher-itance.

The house a family occupies belongs to the man since he was the one to construct it. In a case of the termination of marriage, then, it is the wife who leaves, or when she is widowed she will inherit the house and remain in it. Since the house is not permanent and the roof at least needs to be thatched again after eight or ten years, she will ask her brothers or her deceased husband's brothers to assist her.

All movables are privately and individually owned. The owner has the right to share, dispose of, or destroy any property which belongs exclusively to him. Any other member of the 'ankumi' — and it seems that in terms of our discussion of hospitality and assistance this was generally extended to all other in the community — has the right of use. This right of use may be practiced with or without prior permis-sion on condition that the owner or somebody close to him be informed as soon as possible after the use of the article. If, for instance, a man needs to cut down a tree and his axe breaks while he is working, he may use an axe belonging to a neighbor but should inform somebody about taking it or tell the owner as soon as he sees him. Should he wait until he is confronted by the owner who says, "I hear you took my axe", and he replies, 'O yes, I forgot to tell you" it is tantamount to theft. The same holds true for a person who enters the garden of another because he needs some food or moisture. He may pick a cucumber or some tuber since he needs it to stay alive, but the same is not true of betelnut as was seen above.

Also children own their personal belongings and these by and large will be sex linked as are the movables owned by their elders. A man owns his bows and arrows, his ceremonial decorations, his tools and his clothing, while a woman has little more than her clothing, per-sonal decorations, a needle of pig femur or similar objects. Nowadays Akuna houses contain an increasing number of enamel basins, spoons, and steel knives which the woman of the house owns, while her hus-band has his own knife, a steel axe, and in one or two cases trade store tools such as a hammer and pliers. Also children own an increasing number of such items, which they swap, trade or lose to an opponent in a card game or in other agreements or disagreements.

Pigs are owned individually and again children seem to have full rights of ownership where an adult extends these to a child. To at least distinguish between the pigs belonging to the members of different clans, a series of markings have been developed. These, it seems, might be post-contact in origin for the missionary people have their own marks. These marks are made by cutting the ears (and occasionally the tails) of the pigs. Thus the luluai's clan cuts off the tip of the ears of a sow while a boar's ears are split down the center. As will be discussed below, pigs are hardly ever raised by the owner and this spreads the rights and duties regarding them to what amounts to joint ownership.

Akunans are not familiar with rent in the sense that we think of it. There are a number of cases of people occupying the property of others. Just north of Akuna there is a large coffee garden and house belonging to the tultul. The house is occupied by E?arua and his family, and by Apaka and his family as well as Tinuyi, the aged brother of Apaka's wife. Neither of these two families 'pay' in any way for occupying the house, nor does old Tinuyi pay in any way for the privilege of living with and being cared for by his sister and Tiko?i. In Akuna (house No. 22) lives the owner and builder of the house, but it is divided into three parts. The persons who occupy the other parts do not 'pay' him in any way, nor do the two old women and one girl who live with his family in their part. While there is no regular rent, we are again dealing with the principle of reciprocity which entails that they might care for the owner's pigs or assist him when need arises.

PROPERTY INHERITANCE

By and large property is inherited bilaterally. Sons and daughters, ideally at least, receive equal shares of their parents' property, but certain categories must be excluded or are influenced by marriage.

Land rights may not be alienated by an individual, as they belong to the family and each one has rights to it, but usufruct may be extended on an indefinite basis. Upon the death of the user it reverts to the family which originally owned it.

There is no formal transmission of land, but the owner will usually inform his relatives of the heir to whom he is bequeathing the rights of his land. When land is inherited within the family, i.e. by natural heirs, these rights are not earned in any way but follow the kinship or affinal relations with the owner. Thus it is that children or a spouse have first rights to land and usually inherit it, but others may also inherit. Thus a man might find that some young person is especially friendly with him, assisting in work, or in some other way being of value to the family, and in such cases the person may actually earn or establish rights to inheritance. The same holds true for sociological children as in adoption. Ideally land is inherited equally by both sexes, but in fact it seems that boys get the first choice and the lion's share. Women do have land of their own and when widowed a wife retains all rights to her husband's land, while her brothers aid her in maintaining the fences. Should she die before the children are old enough to assume ownership, her siblings will justify the claim by working the land. Should a couple die without leaving legitimate heirs, the land will revert to the clan from which it was inherited in the first place. Should it be a new garden it will usually be claimed by the husband's clan.

A further illumination of joint claims to land appears in the case to be described briefly. A number of years ago Oturo married a young man, Arawe, but after he had beaten her in the course of a domestic disagreement, she was taken to the hospital in Kainantu for two months. Arawe was taken to jail for three months. Oturo left Kainantu while he was still in jail and upon returning to Akuna laid waste to the new garden they had recently cleared and planted, uprooting plants and even pulling down the fence. Arawe returned and married a divorcee, named Wio, while Oturo left for Aiyura where she married and subsequently divorced a man and then went to live in Omaura. One morning during our stay in Akuna, Oturo was found in the garden of Arawe busily taking out taro; and when he reprimanded her for her actions, she replied that she had made the garden and had full rights to its produce. Even though Arawe had mended the fences, and Wio had replanted some of the vegetables, Oturo still claimed joint ownership as it was still the same garden.

Trees and clumps of bamboo fall in the male realm of interest and activity and are therefore inherited by males. This is especially true of such trees as pandanus (both nut and fruit-bearing varieties), banana, betel nut, bamboo, and sugar cane, but it is not limited to them. Shortly after our arrival in the Arona Valley a number of men from Akuna cut a large tree which grew just across a stream called Tiwampa. One afternoon Adodi, an old man from Amamunta, came by and saw that the tree had been cut. He immediately went into mourning, started to cry, and covered his face and chest with mud and white clay. He wailed that the oldest tree which had been planted by the old people and his ancestors ('tinaputikaka')[19] had been cut down

19. For this composite kinship term see our discussion in Chapter 10 below.

to be used for timber. "If the young men keep this up", he said, "we will soon be without any trees for shade". Adodi demanded payment for the tree and his claims were backed by the Amamunta males. This shows that a regular forest tree which grows about 200 yards from Akuna can be owned individually by a person, in this case living in Amamunta. Trees which grow in other people's gardens may also be privately owned. Trees, however, may be given to other persons who are not related, thus altering the expected line of inheritance.

A person may also extend the right of use over land or trees, for the lifetime of the user. It is then clearly understood that the latter may not destroy it and that it reverts to the owner or his heirs upon the death of the owner. This also applies to the houses, which, however, generally last only about ten years. During that period the use may be extended to any person without expecting any form of remuneration. Houses belong to and are inherited by males, and this becomes clear from the terms applied to divorce, since it is said that the woman either 'leaves' the man, or is 'expelled' by him — the man never leaves his house. When a man dies, however, his widow may inherit his house just as she may inherit gardens. Though both of these are seen as a form of stewardship, she has full rights and not even her deceased husband's brother may dispute these or override her decisions regarding the use of land or house.

Pigs are owned and inherited individually, and while joint ownership develops either following marriage, or between the owner and the person who raises the pig, there is always one person who can make the decisions and at the most must give a small payment to the others for their loss. Most pig killings are followed by a court case where payment is made to the owner, who stands back before collecting it, thus allowing the person who raised the pig to select his share. When an individual has only one or two pigs he will not bother to mark them but simply keeps a close watch on them. Persons who own many pigs mark them by a particular notching in the ear or by severing the tail, and each pattern of notching has a proper name. These names do not have any particular meaning, and informants were not able to translate each particular word denoting a pattern of cuts in the ears. These cuts are made when a pig is about six weeks to two months old. The 'mother' of the pig (the use of kinship terms in this respect is a direct translation of the native use) sits holding it and talking to it while delousing it. In the meantime, the 'father', her husband, approaches with a sharp bamboo knife and makes the required notchings. The 'mother' now spits saliva (no other medication or other substance is added) into each ear and into the squealing pig's mouth. The explanation is that now the pig will hear her voice when she calls it and she will be able to recognize the pig's voice when it calls her, should it be hungry or sick. As the pigs will be inherited by the next generation, so the ear-crops will pass down within a clan.

At this stage a part of the litter may be placed in the bush and left to go wild by not being fed at the village, while others are retained as 'maponi'. They in fact do become 'house pigs', following the 'mother' to the garden and back and heeding her gutteral call of 'ma, ma' (here, here). Young pigs sleep in the house with the 'mother' and are fed at mealtime by the family. This is a carry-over from earlier days when the traditional Gadsup house had the front part set aside for the domestic pigs.

Pigs which are privately owned are never raised by the owner, but are soon given to some trusted relatives or friends to care for. Two reasons were given for this practice: Should some disease break out among the pigs (such an epidemic swept the Arona Valley in 1948), it is better not to have all your pigs in one place, for contagion will affect them all. The other reason, and one which differs only in degree, is the fact that the pigs that grow up together also herd together. If they enter another person's garden and are shot, it is desirable to have only one killed and not all at the same time. Though this was not mentioned as a reason, it seemed from the actions of the informants that persons were placed in special relations to others by taking care of and assuming the responsibility for their pigs. It was nearly as though they were tied by kinship bonds, for, as we shall see, there is a symbolic connection on this level. While all persons were able to take care of pigs, and herd the property of others, certain people had a special ability in dealing with pigs and they were the persons preferred when the young had to be farmed out.

Pigs that enter the garden of another person may be shot. It is said that ideally the owner should be given two warnings, but very often a pig is killed the first time it is found in trespass. The garden owner will then inform all that he has killed a pig which has entered and foraged in his garden, and that the owner should come and identify it. In such instances there are no court cases, for the legal agreement among Akunans accepts the fact that once a pig has foraged in a garden it will return time and time again to the same place in the fence and force its way into the garden. During our sojourn in the field one case was taken to court, but abandoned. In this case Yifuto made a case against Amosima (both from Amamunta) not because the latter had shot the pig, but because he felt that he had been abused by the fact that he had not been properly warned. Amosima had not adhered to Gadsup etiquette.

Some interesting rules govern the eating of pigs, and it is proposed that we might here recognize an analogy of the difference between biological parenthood and sociological parenthood association with the adoption of children. If person A owns a pig and has fed it, neither he, his spouse, nor their respective parents may eat of its flesh, irrespective of whether they had killed it or whether it was shot by another. As happens in nearly all cases, A might have farmed out this particular pig to B who fed and raised it. Thereafter B and his wife also become known as its parents, the relationship being expressed in Neo-Melanesian as 'small papa' and 'small mama'. They are not allowed to eat of the flesh of their charge, either, nor may their parents. Their children, on the other hand, may. As was discussed above, a person may not marry an individual to whom he can trace relatedness either through the mother or the father. Should such a person be adopted by another family, these same rules extend to the bilateral relatives of the adoptive parents. It is said that once they have fed a child there is an unbreakable bond between them and the child they have fed. The pig, then, has to be eaten by a third party, just as a child should seek a spouse from a third kin group. It seems possible here that the analogy points at the fact of social alignment and the establishment of a network of ties of friendship. The pig is shared as comestible with an outside family, which places them under the obligation to reciprocate at a later date, just as a child must seek a spouse from a third kingroup to establish new social ties and bind even another family into the web of rights and duties which accompany such ties.

It should be stated, too, that Akunans have guarded against the possibility of persons shooting pigs for meat. When a pig is shot after destroying a person's garden, it is claimed by the owner, of the person who cared for it, and under normal conditions a share of the meat will be given to the garden owner in compensation for his loss in garden products. If, however, he has not acted in accordance with the rules of etiquette described above, he will not receive a share. Under any condition, however, he may not eat the meat himself, but must give it to some relative.

Pigs are owned individually and a person may give away or sell this property as he wishes. Tapari bought a pig from one of the White planters in the Arona Valley, but the pig was killed in Auko's garden. Instead of giving the meat away, Tapari organized a 'market' by cutting up the meat and selling the pieces to compensate for his loss. He also offered native tobacco and goat's meat and netted a good deal of money, after paying two men who had originally helped him to buy the pig and paying the 'small papa' who had looked after the pig. His wife took the rest of the money, explaining it would be safer with her and he could not spend it so soon.

Personal belongings are inherited equally among boys and girls, but each follows the sex of the original owner. Thus bows and arrows are inherited by boys, while women's string bags and pandanus mats or objects of decoration are inherited by girls. These latter two categories of property — pigs and personal belongings — are the only two which include ownership per se, namely the right to alienate or destroy, while the other categories involve joint ownership and require consultation with the other persons involved. We may then distinguish in inheritance between the acquisition of full rights and privileges, and the acquisition of limited rights and privileges. This latter category implies that more than one person, such as husband and wife, parent and child, or siblings hold joint rights and privileges to the property. Every person then has certain duties and liabilities which influence his use of the property involved.

RIGHTS AND WRONGS

Generally stated, punishment among Akunans is the result of social sanctions in which a dual purpose operates, namely the element of compensation and the element of punishment. It seems, however, that the latter was of secondary importance in that it was a corrective action — an attempt to integrate rather than separate the members of a community. The sanctions which operate are mainly social, but certain supernatural sanctions were recorded. Once again it seems as if the latter have a social aspect to them as the supernatural punishment might affect not only the wrongdoer but also his family or the community to which he belongs. While social sanctions are more positive in their effect, supernatural sanctions are in the form of negative affects (du Toit 1965(b)).

In the category of supernatural sanctions we will mention a variety of negative regulations regarding the eating of pork, eel, soft food and similar preparations during that time when persons are especially susceptible to contamination. There are a variety of taboos on sexual relations at certain times, and while some of these will affect the parties involved, others might have a negative effect on a son who is being initiated — thus making it a joint social and supernatural taboo. There is also a general taboo on sexual intercourse with a Tairora (the 'namuko' par excellence of Akunans). Should a person commit such an act, his eyes would grow weak and in no time, with swollen sockets, he would be blind. It might also result in deformities developing in the bone or muscular structure of the body and a number of cases are quoted as proof.

Social sanctions differed in reality between members of the community and clan and those which applied to somebody not related in such a way. Since we lack any central legal or political authority, the

99

emphasis was placed on social action and community punishment, what Radcliffe-Brown has called "the embryonic form of criminal law" (1958:XV). It may then appear in the form of public disapproval, of a system of fines and reimbursement, or of vengeance when punitive action is taken.

Murder (except in the case of sorcery in which form it usually appears) and homicide are seen as the same act since consequences are what count rather than motives. The taking of a life is punished by taking a life in return and if the culprit cannot be reached a member of his community will be substituted. This then is the basis of vengeance and warfare. Should a man accidentally kill his wife during a beating, her clan members or if she was not residing in her natal village, her village community is sure to take vengeance against the husband and his group for it is said: "When a woman is married, her husband acquires all rights over her body, but her 'ankumi' members retain her arms, hands, and head". She may reproduce for the man and his group, but she does not belong to him as an object of property. Should a person in an act of fury kill a clan brother, the members of his clan will congregate and say: "One of our brothers is dead; let us not all die". They will then go to him and ask him to put on his shield. Then he will appear outside his house and his clan brothers, standing a little distance from him, will proceed to shoot their arrows into the wooden shield. Once this act has been performed the subject is not discussed again. Sorcery when detected is always reacted to by a penalty of death, whether the culprit is clansmen or stranger.

By far the greatest percentage of arguments, acts of aggression, and court cases between members of the village community, centered around two things: women and pigs. A large number of these arguments are amicably settled between the members of the two families involved, or in other cases between delegates representing the two clans involved. When the matter is too grave to be dismissed so lightly or one party does not wish to listen to reason, the village court is called into session. On these occasions the 'yikoyumi' would preside by sitting in the village square, chewing his betel nut or smoking a cigar from his long bamboo container. Around him there will be a number of the village elders and surrounding these any other person who wishes to attend. Court sessions in Akuna were not restricted to males or adults. Mixed with the audience would be a motley collection of witnesses for both sides. When the court session entailed intra-community issues, the plaintiff and accused and their friends were not separated and frequently might even sit together. In inter-community cases, where there was great possibility that the case might lead to open hostile action, it was more likely that two groups would be formed.

The court session would open on an informal note when the 'yikoyumi' commented briefly on the issue and the reason all were present. When the case had been stated by both sides and argued back and forth by representatives of both parties, the 'yikoyumi' or some other orator will start negotiating. With great eloquence he would state and restate the issue, bringing the parties closer together and then suggest action acceptable to both. This was the important point. The court did not impose a decision; it arrived at a decision, and since the court was the people, they had to agree and go along with the decision that was reached. This is why a court session might take all day, for nobody must feel slighted. Even the party that is reprimanded, or who has to pay a fine to the other party must feel that this really is the way it should be. It is for this reason that the 'yikoyumi' must be such a flamboyant orator, thumping his foot with great gusto as his voice rises and his arms move, and in the very next moment he will gaze far across the Arona Valley to the clouds rising out of the Markham and in calm almost inaudible voice remark on the sombrous conditions which caused this meeting. One moment he will seem close to tears, and then he will flare up anew, embroidering his statements and repeating points raised earlier in such a way that they now cast a new light on the situation.

By this I am not suggesting that all or even a large percentage of disputes end in a court case. Most disagreements are discussed, argued, and evaluated between the persons concerned, supported by members of their interest group. It is only when discussion and deliberation proves of no avail, and both parties are convinced that they are right that the matter is thrown open for discussion in court. One could actually conclude that many of the cases which appear before a court session are those issues which are not quite clear, which are not clearly predictable, and which therefore could frustrate the achievement of social order.

It was impossible, in covering all aspects of daily community life in Akuna and neighboring villages, to attend all and every court case. For that reason it will be impossible to suggest what percentages of all cases belong to each of the following categories of cases. It will also become obvious that some cases have elements of more than one category at stake. By giving a sample of the cases which arose, it will be seen how complex some of the issues are.

Sexual offences

Case 1. On Sunday Wowaya, the daughter of Kampuma and Didino went to Tiwampa, the little stream where a fountain supplies water for household use by Akunans. She is a young girl of about twelve years old who is just starting to mature physically.

While she was filling the bamboo water container for her mother, a young fellow came up and caught her by the arm suggesting that they slip into the bush just west of the stream. He was Duyawa, the son of Nomiwa, who lives in Amamunta and is about nineteen years of age. Wowaya refused his offer and called out hoping to scare him off. He took his knife — these young fellows always carry a knife but not the pocket knife folding kind — and just nipped through the cord which was the belt of her grass skirt and then disappeared into the bush. This obviously was complicating matters for it would be a great embarrassment to a woman to be seen without her grass skirt.

The following day there was much disgust at this act and a court case was threatening, partially because his father had already bethrothed him to Wana (the daughter of Nauna), but also because the emigration of a couple to Amamunta had complicated relations between the two communities.[20] It was finally decided that the case would not be pursued.

When older people in Akuna were asked what would have happened in the traditional set-up, it was explained that matters almost certainly would have come to open hostility. In those conditions Kampuma would have taken his bow and some arrows, and gone out to confront the boy who had 'shamed' his daughter. He would have been supported by the girl's mother's brothers and, if the need arose, by his clansmen and fellow villagers. No member of a community would allow a 'pumara' of nineteen to be punished by somebody from outside which would bring Akuna to open confrontation with Amamunta. But now things are different, and while some of the old men and some of the young hot heads occasionally wish that it was like 'time before', most of them agree with Opura. He expressed relief that one could now walk about in safety and that his wives and daughters need not fear going to their gardens.

Case 2. One day during February we were sitting in the sun outside Munowi's house talking with him and a number of men including Urinanda. The latter's wife came up and sat down close to him while calmly telling him something. His expression became strained, and taking up an axe that he always carried (he is one of the first 'line'

---

20. Informants stated that those people in Amamunta are really disgusting. They grow strong and increase the size of their village by drawing settlers from neighboring communities. This, of course, is the way it has always been except that now Akuna was giving rather than receiving. See also the discussion in Chapter 10 below.

workers cutting trees in the bush), he stalked off to Apapayu who was just returning from the bush and dealt him three tremendous blows over the back with the axe handle. A number of young men appeared and carried Urinanda away to his house where they pacified him.

It soon became clear in a court session that Tanano, the wife of Urinanda, had left for her garden that morning, and while she was taking out sweet potatoes for her pigs, Apapayu casually greeted her and suggested that they have sexual relations. She laughed him off and left for the 'kunai' to feed her pigs. He again appeared and repeated his suggestion, at which she gave him a blow over the back with the blunt edge of her bush knife and then returned to tell her husband. Following his beating by the husband, Urinanda, he was scolded by his own wife and criticized by all the people who were around, but the matter was soon forgotten. The fact that this adulterous attempt was unsuccessful is beside the point. Of importance is the fact that it was initiated by a married man of around forty-five or fifty years of age. The latter's wife was very disgusted with him and ignored him for a day, as did his household members, but tranquility soon returned.

Case 3. The following incident involves a single man and a married woman. In this case Apoya, a young man of Amamunta, had relations with Payanano, a young woman married to the older Dupo?ima; but to safeguard himself he paid her ten shillings before she returned home. She, however, decided that Apoya was preferable as a husband to Dupo?ima and attached herself to him. When he shunned her, she went to the 'kiap'[21] and explained that the young man had had relations with her and should marry her. When her husband appeared in court, he told Apoya that he could have the woman if he gave him (her legal husband) "a small payment". Apoya replied that he had enjoyed the woman and had paid her and wanted no further part of her, but that even if he wanted to marry her he would not give any payment to Dupo?ima as he had not yet transferred bride wealth for her. That is where the case was left. Nobody had won or lost, but each had played the game according to the rules his culture had taught him. These are the organizational features which delineate the attitudes and behaviors of the people and the choices in behavior which are open to them in a particular situation and at a particular time. In this case the woman initiated the relations.

In the traditional setting, these instances would not have been

---

21. This is something which absolutely infuriates the village officials, for it casts doubt on their ability to maintain law and order among community members. A person who acts like this on her own initiative and is sent back to discuss the case in the village does not receive much sympathy.

resolved so harmoniously. If a man had discovered that his wife had committed adultery, he would take an arrow and shoot her in the thigh or calf and take the same action against the man with whom she had intercourse. When both parties resided in the same village, the husband never shot to kill, but often a person's kinsmen would come to his aid, resulting in fighting. When this involved a person from a different village, or if the relations were forced on the woman (as in the case of rape), there would be a fight, the persons concerned being aided by co-villagers and kinsmen.

Case 4. About two years ago Dupopima left on indentured labor, but just before leaving he married Ma?e and placed her in his parents' home. Some time after his departure she had relations with his brother, and then left the house of her parents-in-law and returned to her mother. From here she started visiting Amamunta and was soon reported to be the constant partner in the bush of Apoya — the same young man mentioned in the previous case.

Early one morning a court case was held, and the tultul, acting as public prosecutor rather than as kinsman, laid charges against Wa?iyo, our interpreter, but these were later dropped. Ma?e claimed that they had had relations, but Yayo, a patrilateral relative of Wa?iyo came to his assistance and the charges were dropped as unfounded. By this time, however, Ma?e had left for Kainantu to see the 'kiap', an act that infuriated the luluai and the tultul (who as kinsman of her did not show it so clearly) as it suggested that they could not handle matters in the village. The evening after Ma?e had returned she was summoned to a court to discuss the matter. She claimed that men followed her, but two 'monkeys', those ever-present 'eyes' and 'ears' of all village happenings, explained that they had noticed her 'making eyes' at Apoya. Ma?e now admitted that he had been meeting her secretly, and in fact had had intercourse with her five times. With this, it seemed the onus was placed on him, for the tultul explained that they would "decide later whether to go to court (and he be jailed) or whether he would have to pay a fine to Dupopima (the husband)". This, of course, does not negate the possibility that she may leave on her own accord and join the man of her choice. She, however, promised to stay home and await her husband. Three weeks later she was back at Amamunta and was not so secretive about her friendship with Apoya.

A second court case followed the first in which the village leaders jointly prosecuted; and with a native councilman present, Ma?e promised to give up her past ways of visiting Amamunta. She was forced to make the rounds and shake the hands of all present, namely the luluai, the tultul, the council member, and the ethnographer. Two days later she was back at Amamunta, and on the day her husband returned from the coast she was away visiting yet another village. Great shock was expressed by all, not primarily because she had revisited the other villages ("some people like sex and companionship more than others do" was the tultul's rationalization), but because she had broken an oath.

The major court session followed more or less as a continuation of the previous one. Dupopima refused to take her back, explaining that when he left he had just married her and given her a place in his parents' house, but she had 'opened the door' and returned to her mother. Everybody agreed that he should have the right to decide as he preferred. She kept on saying, "He is my real husband, but he does not wish me to live with him". It was decided that she would go to Apoya, and the bride wealth the latter would pay to Dupopima was fixed at fifty pounds (approximately $140), which is the amount Dupopima had originally paid for her.

In this case, then, repeated adultery and unfaithfulness led to a decision for divorce. The fact that it was not carried through can only be explained as the result of the skillful oration of Apoma?i, the tultul of Amamunta, who in fifteen minutes swung the whole case around and adjourned court.

Co-wives

Case 5. It is important for every Gadsup adult man to be a father. Only then has he really become a 'wa.nta', a man, for then he has proven his virility and become the head of a family.

Many years ago Domandi married a woman called Dauwandino who gave birth to a daughter. The infant died even before she could be named and a while after this the mother died too. Domandi, however, had married Tea who comes from Tompena but in all the years they have been married she has never given birth. While this has frustrated the husband, it has also grieved the wife for she realizes that she has not been able to give the aging Domandi what he most wants — a child.

A little less than a year before our arrival in Akuna a young widow, Tero?ano (wife of the late Tu?a), became pregnant by Domandi. This caused some embarrassment and potential trouble with her mother's brothers, the luluai Nori and his brother Naniyu, all the more so since Domandi and his first wife shared the house of the luluai. The case was amicably settled when Domandi transferred some gifts to the brothers and the couple were accepted as married.

Shortly after arriving in Akuna I heard a ruckus one morning and saw two women waving their hands at each other. Men smiled and children continued their games. Enquiry brought to light that one of these women, Tea, was blaming and reprimanding the second woman, Tero (as Tero?ano's name is commonly abbreviated), because their husband, Domandi, had taken lime to eat with his betelnut from his main house to the house where Tero lived, and the latter was obviously trying to have him all to herself. This clearly was a case of sexual jealousy, for Tero was clearly pregnant and thus able to give Domandi what Tea could not give him.

On August 23, 1961, Tero gave birth to a daughter, and Tea was very jealous. During October Tea was working in her garden south-west of Akuna, and sometime during the morning Naniyu (younger brother of luluai Nori) was in the vicinity while taking care of his pigs. One of the other women (possibly Tero) reported Naniyu's presence to Domandi, who blamed Naniyu for committing adultery with Tea. This charge had delicate consequences as Domandi and Tea live in the house of luluai Nori, brother of Naniyu. Naniyu reacted by telling Domandi that he had two wives of his own, and thus he had enough avenues for sexual satisfaction without visiting the latter's wife.

To protect himself in the eyes of the White Man's law, Naniyu left for Kainantu early the next morning explaining his case to the kiap who suggested that all persons concerned come to court in two days. The luluai of Akuna and Amamunta had a joint court session attempting to clarify the case, as it is a recommendation for a luluai if he has conciliatory abilities and solves local village quarrels. The court case continued for a full day with much talking, smoking, and betel nut chewing, and it seemed that as the day wore on they became less likely to compromise — the case was finally resolved out of court with nobody winning or losing, but the matter simply being dropped.

Three days later this writer and Wa?iyo, the interpreter, were in the vicinity of the gardens, checking on some question regarding planting and horticulture. We happened to pass through the lower end

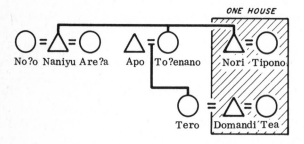

Diagram 4. Genealogy of adultery accusation

of a garden belonging to Tea, who called out as we passed by, "You guys had better not roam the gardens too much or you will be blamed of adultery and fought by the men". (It should be mentioned that this was one of the traditional places for adultery and also the reaction of the man and his kinsmen.)

From the cases discussed above it is clear that not every case of adultery leads to the dissolution of marriage, but every case of adultery contains within it the elements that may lead to divorce. On the whole it can be stated that marriage will not be dissolved on the basis of one or two cases of extramarital relations, but continuous infidelity undermines the very cornerstone of marriage, namely the principle of legitimacy.

Not all co-wives produce such problems for not in all cases are there such delicate problems as those faced by Tea. This might be the primary reason why every wife of a polygamous husband is housed in a separate structure and it might have some causal relationship with the fact that in many cases wives are allowed to live in different villages ostensibly to justify land claims and work gardens.

The case was already mentioned where A?utu and Nona?a, the co-wives of Eyo lived in one house while their husband was at the coast. When I expressed surprise that they could live together and not fight, they quickly added: "... while Eyo is away". But both have their own children and some women in Akuna in fact suggested that they welcome a co-wife, especially if the latter is much younger than she.

Pigs

The explanation which Akunans give as justification for killing a pig which has broken into a garden, is that the pig's memory is better than the proverbial elephant. It will return again and again, smelling, nudging, burrowing as it works its way along the fence, and at the same place it broke in earlier it will again break through. A sow would return and show her litter where to get in and there was nothing the garden owner could do about it except kill the pig. Yifuto, a man from Amamunta, was very upset when an Akunan named Amosima shot his pig without warning. He could not get the village officials to side with him and so he left for Kainantu to lay the case before the kiap. He was accompanied by villagers from Amamunta while Amosima went along in case he should defend his action, and he was accompanied by a number of Akunans. The kiap would not even listen to the case saying that Amosima was within his rights in killing the pig that wrecked his gardens.

Later informants explained that Yifuto had made the case against

107

Amosima because he had been abused, not because his pig had been killed. The pig was in the garden. That was sufficient reason to kill it; but etiquette required that a garden owner first warn the pig owner to allow the latter to place his pig in a pen or send it away. Amosima's actions were legal, but they were pretty bad manners they felt.

Case 6. Auko, the government appointed 'doktaboy' or medical tultul as the administration originally designated these people, lives just north of Akuna at the medical aid post. He acquired a bicycle with which he could travel to Kainantu when needed and around the Arona Valley the rest of the time. To lift the bicycle over the fence and climb over the stile himself was exhausting and not becoming a medical orderly. Also he was hoping to impress the doctor from Kainantu when the periodic visit occured. The result was that he removed part of the fence and made a beautiful strong plank gate which swung from hinges. That night a pig entered his garden and uprooted half his tubers. The following night the same thing happened and the pig uprooted the rest of his tubors before he killed it.

This was a problematic situation. According to legal standards accepted in Akuna, Auko could kill a pig that broke into his garden, but this was not the same for there was no fence — the pig had forced the gate open. Legally then Auko could simply go ahead and kill the pigs which are found in his garden or the owners would have to construct sties for their pigs ... and what about the bush pigs that came out of the bush at night to forage? The men in Akuna were in a quandry about what to do. One night at ten o'clock shortly afterwards Nori, his son Yajima, and Anupinkawa were seen going to the medical aid post. They removed the gate, standing it neatly against the fence and proceeded to plant and construct one of the strongest traditional split pole fences to be seen in the vicinity. Their explanation: "If he wants the pigs out of his garden, he must construct a strong fence; if he prefers to have a nice modern swinging gate, he has to face the consequences."

Case 7. An interesting family quarrel developed around the rights of ownership of pigs during our stay in Akuna. A number of years ago Wono who lives in Amamunta with his first wife, married Umeno of Akuna where she remained. During his absence she committed adultery with Kampuma of Wayopa, whose second wife also lives in Akuna, but was discovered and a fight developed between Wono and Kampuma. Instead of going to court, as both were at fault under the White Man's law, it was decided that Wono would pay for the injuries he had inflicted, while Kampuma would pay for the rights he had infringed upon. The result was that a pig was given to Wono, and payment — to which

Ako (eight shillings), Napiwa (two shillings), and Yerai (five shillings) contributed — was given Kampuma. These three men are not related to either of the persons involved here.

The pig which Wono received had piglets, one being given to Yerai and two to Ako in accordance with their share of the payment. Some time later the pig had four piglets, of which one was given to Ako, one died, and two were retained by Wono. Again, a litter of four was born of which two females were set wild in the bush to breed and of the two males Wono gave one to each of his two wives.

During January, Waropi of Amamunta shot two pigs which were in his brother Noka's garden at Akuna. One of these belonged to Umeno and returning to Amamunta, he told Wono that he had shot the pig. Wono immediately informed Umeno's relatives in Wopepa that they should fetch the pig as he and his wife could not eat it. In the meantime, Umeno had heard that the pig had been killed and informed her brother, Unanata, and Ko?ana in whose house she lived, to prepare to eat it. They set out for water, 'empomi' leaves, and firewood but returned at dusk to inform her that the Wopepa people had removed the pig. By the time Wono arrived, his wife was loudly proclaiming to all that her husband had no right to decide on the eating of her pig, and she threatened him with a burning log. In all of this she was sobbing and saying 'tentikai'. When the ethnographer inquired into the ownership, she did not point at herself, but used the same kinship term meaning 'my child'. Later she explained that she was not crying because the pig had been killed, but because her ownership and right to judge what should be done with it had been disputed. Her relations with her brother and friends had been endangered. In compensation Wono gave her one of his pigs.

Case 8. A good example of the value and complexities produced by pig ownership is found in the following case.

Some time before we arrived in Akuna, the adult males had decided that they needed a good boar to breed larger pigs. They approached a local White man and paid him fifty pounds (about $140) for a boar. All the adults contributed, and the boar was allowed to roam the valley below Akuna. During our stay among the Gadsup it was learned that the people of Apomakapa — a small daughter village of Kundana on the eastern border of the Arona valley — had killed the pig and eaten it.[22] When the news reached Akuna, the males set out in a body, armed with bows and arrows, to lay claim to their pig. Having arrived at

22. These villagers were apparently famous for this kind of "mistake", and at this time their tally stood at: 2 pigs belonging to Wopepa; a pig from Kundana; two dogs from Onanika; and now a pig from Akuna.

Apomakapa time passed as nothing could be done until the luluai returned from the bush. By late afternoon discussions and arguments were conducted and finally all admitted that it was in fact the Akunans' pig, but it "certainly looked just like one of ours" they stated. They decided then to pay the Akuna men in cash for the loss they had suffered and to replace the boar which had already been consumed. Twenty-three pounds in cash was collected from the men in Apomakapa, a large pig was shot to replace the one they ate, and with this the Akunans returned home that evening. They were still hurt that their prize boar no longer roamed the valley, but all clouds have a silver lining and the following day they would have a village feast where nobody would be subject to any meat taboos since the pig came from another part. The following day the earth ovens had scarcely been covered over and the people settled back to await the feast when two men arrived from Kundana. "Where is our pig?" they asked as they squatted down and started to smoke. It was then explained that the men from Apomakapa had simply shot the first pig they saw and presented it to Akuna, but that particular pig happened to belong to them. After this was confirmed, representatives from all three villages departed for Kainantu where the Apomakapa men came off third best ... they were forced to pay the difference in price to Akuna for the pig they had shot, and also to replace the pig belonging to the Kundana men.

Once again matters might have taken a different turn in traditional times, for the kiap would not have been there as a justice of the peace. Apomakapa had wronged both the other villages but Akuna and Kundana were not too friendly; whereas Kundana considered Apomakapa a daughter village, had close kinship ties with them, and in fact shared ceremonial activities.

Case 9. Claims are, however, not always paid. On one occasion Akuna men were joined by the Tompena males in a pig hunt. Why this happened in the first place is not clear for hostilities flared up repeatedly during our year in the field between these two traditional 'namuko'. On this hunt the whole day was spent in the bush and finally by mid-afternoon a group of about twenty-five adult males, exhausted and hungry, killed a pig at Yondonanomi, a stream which cuts a tremendous gap in the earth and is about halfway between these two villages. Since everybody had had a tiring day, the pig was cut up and eaten right there.

Upon arriving back in Akuna the Lutheran mission evangelist welcomed the men back, informing them they had just feasted on his pig. Once again the semi-official court huddle occured where their tracks were retraced, their actions relived, and much betel nut chewed before it was finally conceded that it was indeed that particular pig belonging to the mission. The men now apologized, saying that we all know how such a thing can happen and it surely looked like a wild pig, and finally agreed to collect money to reimburse the mission. The subsequent contributions netted a total of eighty-five shillings (about twelve US dollars) and a 'lap-lap'. Tompena refused to contribute, maintaining that they couldn't be taken in so easily since they knew it to have been a wild pig. When the Akunans presented their collections to the evangelist, he refused to accept it and claimed that the pig was worth at least ten pounds (twenty-eight US dollars).

Thus they waited, each maintaining that he could prove his innocence when the kiap came by. Long before that, however, the evangelist had decided that since he was an immigrant among the Akunans and needed their goodwill to crown his endeavors, he should drop the claim. And so harmony returned to Akuna ... for a while.

Case 10. During the week of September 4-9, Munka, who lives in Wayopa, stayed at the medical aid post due to ill health. While he was thus confined, he received a complaint from his fellow villagers that one of his pigs was causing trouble around the fences of Datipa's gardens. Not feeling in a mood for these worries, he suggested that they kill it. We should mention that Munka had five pigs which were not marked on the ears or tail, and that one pig belonging to Napiwa and raised by Anokara, also unmarked, foraged in this part of the valley. That same evening Datipa and Utima went out and killed the pig with the latter doing the actual shooting. The next morning Napiwa arrived in Wayopa and told the people that they had killed his pig. Present, too, at this time was Anokara, the sister's husband of Napiwa, who confirmed that it was the latter's pig. The owner, however, pointed out to the people that the pig was dead and that they might as well go ahead and eat it. Not waiting for a second invitation a feast was prepared.

On September 14, Napiwa walked to Wayopa accompanied by a large number of his fellow villagers. At about noon Napiwa stepped into the center of the gathering — his mouth filled to capacity with red betel nut juices — and explained that his pig had been killed, but had not been replaced. Should he not receive a young pig, he might consider going to the kiap in Kainantu. (Informants explained that if the case was taken up in the kiap's court, the pig would have to be paid for at a price somewhere between two and eight pounds depending on the size of the pig.) Utima responded that it was really pretty hard to identify the pig because it had not been marked, but Anokara responded that this was not true because anybody who knew pigs could have identified this one. While this stating and arguing of legal principles was going on, a number of the young men had already started killing a large pig which was kept in a pen adjoining the village fence. By about four o'clock the defence admitted that it had been in error, but in fact this had been admitted a few hours earlier, for now women started dumping large quantities of tubers, bananas and other valuables in the center of the gathering.[23]

Other cases

The owner is not really held responsible for the actions of his animals, but since he has more 'savvy' than they have, he can at least influence their actions by restraining them. That is why the owner of a pig is warned once or twice before his pig is shot, or the owner of a dog is warned that his dog is causing havoc among the chickens or pigs. In the last analysis the animal is beaten or shot, and the owner is not really involved. In the same way the child is responsible for his own actions. This seems to be even more so if the child is old enough to know better but does not consider others.

Case 11. Years ago Ko?wa resided at Ya?onankapa close to where the village of Wopepa is now built. A young boy of about twelve years named Tankumopiri started a fire and in no time the house of Ko?wa was burned down with all his arrows and ceremonial objects in it. As the flames started Ko?wa rushed to the house but could do nothing about it and learning that the boy had started the fire, he took hold of Tankumopiri by his neck and flung him into the flaming grass and bamboo structure. The charred body was removed later by Ya?e, the deceased boy's father, diced with pork fat and buried in banana leaves. A short while later Ko?wa left Ya?onankapa and moved to Akuna where his son Munowi still lives.

But this, informants state, was true in 'time before', since the kiap has explained to them that a child does not have 'savvy'. Only when a child is old enough can he be treated like a human being. Before that the parent must act on behalf of the child.

Case 12. To emphasize this change informants recounted the events when the four year old Mindapo, son of E?ananda, burr lown the house belonging to Padauno in Wayopa. Once again the structure went up in flames and in no time the grass and bamboo was a heap of smoking ash. The owner looked on as the last flames flickered and died, then taking

---

23. For a list of the items and the economic aspects of this case, see the discussion which follows in the chapter dealing with the village economy.

112

hold of the young culprit he dragged him to Akuna and claimed that
E?ananda pay for his loss. If this were not done, he would take the
case to the kiap, Padauno threatened.

Generally the payment is made in such cases to prevent attitudes
of hostility from developing and to keep the case away from the kiap.
In this case Padauno was claiming ten pounds (about twenty-eight US
dollars) and to assure friendly relations, the claim was paid.

EXTERNAL FORCES

All through the previous pages and in fact throughout this study, the
presence of the White Australian Administration is writ large. Nevertheless it is necessary at this juncture to outline briefly the specific
changes they have made in the coercive and judiciary forces, and the
ways in which they have brought about a social order — different, no
doubt, in certain ways from that which existed prior to their arrival.
When the Australians took over the administration of the eastern
highlands in the 'thirties, warfare and open hostility was a way of
life. The 'yikoyumi' was the driving force in every community and
was the most important person to be reckoned with by administrators.
He was the leader in warfare and by pacifying him or keeping him
under control the rest of the villagers would be more apt to adhere
to administrative orders. This does not imply that peace was immediately established, and in the early years of contact a police post was
necessary and frequent outbreaks of inter-village warfare had to be
quelled. Even after the Whites had established their peace and withdrawn the police from the area, intermittent attacks and counter-
activities occured. The new White administration told the people in
the Arona Valley not to attack each other and not to burn villages.
And then during the Second World War the Japanese got to Arona and
were supplied with food by Gadsup according to their tradition of hospitality. When the White administration, in the form of Allied Troops,
learned about this, they burnt houses and villages at Arona to punish[24]
the Gadsup for assisting the Japanese. This must have left a big question mark in the minds of these simple people.

The administration that came in attempted to recognize traditional
patterns and yet to establish a system which would best suit their aim
of improving living conditions. In the Official Handbook of the Territory of New Guinea (1937: 303), we read the following statement
about the appointment of native officials:

24. Patrol Report, Number K. 4 of 1948-49 by D. Whitforde.

"The 'luluai' is nominated by the people, and is generally an influential man in the village. The Administrator is advised by the District Officer in making an appointment. In actual practice, the District Officer picks out after careful inquiry the man who appears to be the natural leader. If he be incompetent or unworthy, the man who would, according to native custom, succeed him on his death will be selected if qualified.

"An election is not held, as it gives an opportunity to natives with a good knowledge of 'pidgin' to secure a dumb following in a show of hands.

"An active, intelligent man, with personality and a knowledge of 'pidgin' is selected as 'tul-tul'. Medical 'tul-tuls' are selected, trained and recommended to the District Officer by a medical officer or medical assistant before becoming eligible for appointment.

"A District Officer may recommend a native for appointment as 'paramount 'luluai', and, when appointed, he controls a group of 'luluais' and the villages under their jurisdiction.

"Paramount 'luluais' are paid £3 per annum. None of the other officials receives salary, but all are exempted from head tax. If any native Government official receives a bribe or extorts blackmail, he is liable to a considerable term of imprisonment.

"All instructions to a community are transmitted through a 'luluai', who is responsible for their enforcement. The 'luluai' is the mouthpiece of the District Officer; assisted by his medical 'tul-tul', he is responsible for the sanitation and cleanliness of his village, the health of his people, the removal of the sick to hospital, the keeping of the peace and the arrest of offenders.

"Paramount 'luluai', 'luluai', 'tultul', and medical 'tul-tul' are distinguished by peaked caps, which are issued to them as tokens of their appointment."

In much the same form the Australian administration affected the Arona valley. For each village which was recognized as an independent community a 'luluai' was appointed in terms of native values of leadership, and in the Arona valley these seem to have coincided very closely. Such a community would also have a 'tul-tul' appointed to act as liaison officer for the administrative officials touring the district. The decision of when a village becomes a separate community seems to have been a problem they never clarified. Thus Ikana had more than one 'luluai' and 'tul-tul', while Wayopa, Manunampi, and Apomakapa did not have an independent status though separated in most aspects from the mother village. The coming of a new system also allowed misuse of power and it seems that in some cases at least

constables at police posts set themselves up as 'kiaps',[25] exacting
tribute and swaying power beyond that which their office justified.

The same patrol report for the mid-forties comments on the exceptional abilities and great influence of Pumpua of Akuna. He was the
old strong man, the 'yikoyumi', who under the new system was
recognized and appointed 'luluai' of Akuna. But the Official Handbook
quoted above also mentions the appointment of a 'paramount luluai'.
This is something relatively unique in New Guinea, but it was attempted with no great success. Pumpua then was appointed 'paramount
luluai' for a large part of the western Gadsup, and while his name is
still revered in Akuna the experiment did not work. The only way a
'yikoyumi' could attain that position was by a gradual process of
allies and reciprocity, not by appointment through an external force.
Pumpua was held in such high esteem that he could not be admitted to
just die a natural death and when in 1948 he passed away, sorcery
accusations were made against the Tairora speakers neighboring the
Arona Valley on the south.

Another official is perhaps of nearly as great an importance in the
changes that have affected the Gadsup, namely the medical 'tul-tul',
or 'doktaboy' as he is commonly referred to. This is occasionally a
local man, but in many cases he comes from outside where he is not
caught up in the intricacies of clan membership, community identification and loyalties. His station is a 'haus sick' which adjoins but is
not part of a village. Here he has his own home where his family lives,
a garden and a neat two- or three-roomed structure. This includes a
place for people, males or females, to sleep if they are detained or
must be taken to the 'haus sick' in Kainantu, as well as a dispensary.
Here one finds a variety of disinfectants, aspirins, bandanges and
other medication for non-serious ailments or abrasions. The 'dokta
boy', with his white 'lap-lap' on which a red cross is stitched, also
has primary responsibility for the newly born. He must report these
recent births and any abnormality in health must be treated as soon
as possible or taken to Kainantu. In cases where such are not reported, the parents are blamed for neglect and may very well find themselves in jail. This action is an attempt to improve health in general
and cut down on the infant mortality which was traditionally very high
and was coupled with infanticide.

During our stay in Akuna a number of cases occured where the
conditions surrounding a death were investigated. We already discussed the birth of a daughter to Tero, but the child died after six
weeks and the 'dokta boy' was present to know why the child had not
been brought to him for examination. Some months later a girl of

25. Patrol Report, Number 32 of 1944-45 by Lieut. D.R. Blyton.

about seven died and this was worse for a police officer came by
some days later to ask what the conditions were which surrounded
the child's death. One other case can be mentioned since it resulted
in jailing for the father. It is taken from the field notes:

"About three months ago Amonko's wife, Dopaya, was getting close
to giving birth so she retreated to the bamboo bush southeast of Akuna
and remained there for two nights in the company of some mid-wives.
Rain was threatening and the party made a small hut of 'kunai' and
here Dopaya gave birth. The next day, Saturday, the party moved half-
way to the village but did not climb the fence, and they spent the night
in a small grass windbreak. On Sunday morning the midwives left to
get tubers from their gardens and while they were gone, Dopaya died
with the infant at her breast. One of the midwives was Didino, a clan-
sister of Amonko the widower, and upon discovering the death she
took the child and said that she would adopt it. Didino is the second
wife of Kampuma who lives in Wayopa, but she works a garden near
Akuna. She has given birth three times and only one child lived (this
is the same girl mentioned in Case 1, above). After about three weeks
in Akuna, Didino took the infant to Wayopa to visit with Kampuma.
Shortly after settling in Wayopa, the infant died and 'luluai' Nori had
to report the death. The 'dokta boy' inquired and upon being told the
history of the case, asked why the infant, when her mother died, had
not been brought to the 'haus sick'. Partly as punishment but partly,
I feel, as a warning to the other people, Kampuma was jailed for
three months. As head of the household he should have been responsible to report it or to insist that his wife take the adopted infant to
the medical aid post.

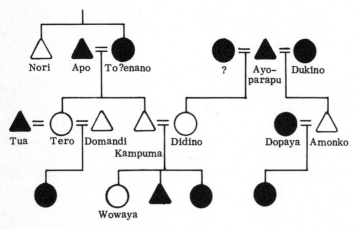

Diagram 5. Genealogy of infant death

This introduced the 'kiap' as court official and as a power that can enforce a new kind of sentence, namely to 'kalabus' a person. In such cases the prisoner is removed to Kainantu, but as in the case of persons taken to hospital there, the relatives and community members are expected to send them food. We have also mentioned a number of cases above in which the accused will run off to Kainantu asking the 'kiap' to hold court, or awaiting the regular visit of a patrol officer when cases can be heard. It seems that the 'kiap's' court is different from the traditional one where a decision is always unanimous and a verdict is agreed upon rather than a sentence passed. It would then depend upon whether a person is defending or accusing whether he would prefer this new court. Over and beyond this court of the 'kiap', certain cases may be heard by the district court and the 'kiap' alone decides which are weighty enough for this special consideration.

While we were in Akuna there was much talk of a survey to be conducted in the Arona valley to appoint a Gadsup Council. There was also some talk of a New Guinea Legislative Council elected by universal franchise.

"The ultimate goal, according to Territories Minister Hasluck, will be a Legislative Council elected on an equal and universal franchise, on a common roll which means, of course, a council elected by a native majority. Even Mr. Hasluck has some doubts whether the votes of native electors placed on a common roll at present might not be 'won too easily by extravagant promises or doubtful persuasions from the poorer types of candidate'." (The Bulletin: September 28, 1960)

Since leaving the field the first national election was held for the House of Assembly. It was based on universal adult suffrage on a common roll and therefore included the people we have been discussing in this study. An interesting analysis of this experience for Akunans and the Gadsup has been given by Leininger. She explains the effects of this election on the small community — resulting in a modified pattern of daily activities caused by regular morning and afternoon discussions in the village square where village court cases were all but abandoned. Women would spend most of their mornings sitting there; and then as soon as they had collected enough tubers in the gardens, they would return and again they would sit, they would listen. And after the elections took place, for many months an epidemic of influenza swept the Arona Valley and most villages experienced the 'election sickness' which was caused by an 'evil spirit' that had entered their heads causing the influenzal headaches. But throughout this talking and during the election itself, the Gadsup placed a great value on unanimity. As they would do in a court case, so here again a point was discussed and analyzed and argued until all agreed and all opposition was removed. Leininger explains:

"As one may have surmised from the above discussion, a significant characteristic of Gadsup behaviour is their strong desire for group consensus. In the past, achievement of group consensus had high survival value when they continuously fought with other villages, and now it was still a most apparent feature of their behaviour in inter-village discussions as well as intra-village discussions." (1964:31)

This statement confirms the point we made above and which will be emphasized again, namely that freedom and latitude exist, but are always tempered by community needs. This equilibrium is at the very core of social order.

Chapter 5

MAN, NATURE, AND THE SUPERNATURAL

Man lives in a world filled with powers and spirits, impersonal and personal. These supernaturals influence and affect all aspects of nature and human existence. To bring about an ordered universe man must continually work at maintaining temperance in social relations and while manipulating the good, he must always aim at counteracting evil forces. It is this temperance and restraint which assure social order, but this is only possible by recognizing the forces in nature and beyond man.

Living in Akuna from day to day, the observer is struck by the simplicity and even subordination of religious conception and action when compared with inter-personal relations.[1] It should also be granted that the Lutheran mission evangelist has lived in a house adjoining the village proper for more than two decades and his teachings have affected traditional beliefs. Furthermore it is striking that Leonhard Flierl who first entered the area makes very few remarks about religion — and one would really expect a missionary to notice such aspects — and hardly any notice is given to religion in early patrol reports. Both of these historical sources do remark substantially regarding other cultural features. Also Watson (1964b:15) remarks on the little work done on religion in the highlands when compared with other aspects of culture — an exception being the study of supernaturalism and ritual by Newman (1962).

As in other aspects of culture a wide range of belief and action is found regarding the supernatural. I spoke to people in Akuna who represented different degrees of belief and acceptance; and while they might in fact believe differently from what they professed, the variation was not great. Nor did it necessarily agree with an age dichotomy, for the critics and the dissenters were found among the older as well as the young people.

1. I should point out that shortly before going to Akuna I had completed a study of the African ancestor cult among the Zulu, which quite clearly permeates every aspect of their lives, even though missionized for a century.

MAN'S DUALITY

Every living person has a body or 'tuʔi' consisting of bones and flesh, and a dualistic spiritual counterpart. This soul includes both the 'timami' (visual dimension) and the 'aumi' (manes dimension). Informants knew very little about the 'timami' counterpart of the body except that it is always in the physical being. While it is an exact replica of the living person, showing the same characteristics and features, it is not located in any particular part of the body. The 'timami' is also associated with the shadow a person casts, as well as with his reflection in the water. Since it denotes these visual dimensions of the body it is also associated with the 'kayim' or dream vision. Thus the 'kayim' refers not to the body, nor to the immortal counterpart of the body (namely 'aumi'), but to that 'timami' which is seen in various forms during the waking hours.

Unconsciousness is seen as a temporary departure of the 'timami' from the body, and as is the case during sleep, it can travel and visit various places. But when it returns consciousness returns to the person. During these wanderings the soul can be attracted by an 'oyi wanta' (sorcerer) who may temporarily steal it. When it departs permanently from the body, the person dies and this, too, is frequently caused by a sorcerer.[2] I never heard anybody refer to the 'timami' of an animal or plant, but these are in fact observed during dream visions.

When a person breathes on a cold morning, the Gadsup say that one can see his 'taumi' or breath. This breath is also a manifestation of the soul but in this case it represents the 'aumi'. At the time when death occurs, but before rigor mortis sets in, the 'timami' leaves the body and does not return, while the 'aumi' remains in the 'anda' (corpse). One would have to conclude then that the 'timami' is not only shadow, reflection and dream vision, but also the essence of life since its permanent departure marks death. During the period immediately following death, the 'anda' is still occupied by the 'aumi'. Only when burial takes place does the latter depart, becoming a ghost or ancestral spirit. While this manes dimension of man's spiritual part continues to exist and has some relevance for those who remain behind, the visual dimension or essence of life becomes completely inconsequential. No information was forthcoming on what became of it and informants pointed out that the person now was dead, the life essence having permanently departed.

2. This is not quite the same as in West Africa where the witch steals and "eats" the soul of a victim (Field 1932: 135; 141). It simply refers to the fact that the sorcerer causes the person's death.

THE DEATH OF THE BODY

Akunans recognize only two categories within which death takes place. The first is that of old people who have lost their strength, who have grey or white hair, and whose coordination of thought and action have been impaired by gradual regression. In such cases death is seen as inevitable and natural, and occasionally even desirable. The wailing and mourning which follows the death of such a person is sincere but brief and the observer gets the impression that to a certain degree it is a relief, both to the suffering senescent and to those who had to care for such a person. The second category is that of young and middle-aged people for whom there is no reason to die, and here death is never natural — always the result of sorcery. This suspicion of foul play is especially strong when the deceased came to a sudden end, but also the person within this category who still has a contribution to make and who grows ill and weak is seen as the victim of sorcery. While Akunans, and the Gadsup in general, still hold this belief and may even go through the motions of proving it, there is very little they can do. No longer can vengeance be taken and so counter-sorcery is employed, hoping thereby to avenge the death of the deceased.

The 'anda' is left in the house to lie in state for anything from two to six days depending on the complexities which surround the death. While lying in state, the corpse is dressed in his best ceremonial attire complete with objects of decoration. In their 'yanka' or arm bands fresh colored leaves are placed and a new grass skirt — for both sexes — is donned. During these days that the corpse lies in state, visitors and mourners from various surrounding villages come to pay their respect and 'to look' (for the last time) at the deceased. They will cry a little or may even participate in the all-night wailing, and will join kinsmen and acquaintances outside in eating and smoking. During this whole time, the closest kinsman and a bereaved spouse will remain close to the corpse, and the most important person on these occasions is the mother's brother. At most wakes attended in the Arona Valley, the mother's brother was near the head of the corpse, wailing at intervals and telling about the life of the deceased, repeatedly calling the name of the departed. Mourners were asked whether any kind of prayer is offered or whether the repetitious calling of the name was aimed at introducing the deceased to the 'wami' (ancestral spirits) of those who had gone before. The answer was negative. They called the name, they say, because they had known the person and that person now had left them.

On the final day of mourning the feast is prepared in which large quantities of food are cooked in the earth ovens. In attendance will be the kinsmen of the deceased and the kinsmen of the bereaved spouse.

121

Also those persons included in the interest group of the deceased who were not related such as neighbors and fellow villagers. While the group is gathered in the village square or near the earth ovens, a number of males will prepare to remove the corpse for burial. These will be brothers or other male kinsmen. All the persons, male and female, who have been close to the corpse or who have touched the deceased whiten their chests and faces with light grey ash and white clay. When the men remove the corpse for burial, the decorations are removed in most cases and left in the house to be inherited, while the corpse is wrapped in a blanket of pandanus leaves or tree bark. Those who are gathered hardly notice as the bearers remove the corpse and take it out of the village for internment. Informants suggested that traditionally there was a common burial ground which was taken over as a cemetery with the advent of European administration. In the case of Akuna, at least, Onampa has remained the common burial ground for many years, and a few of the graves which are not older than five years, are clearly discernible in some cases with a marker. The rest of the area is covered with pit-pit.

The grave in which internment takes place is large enough to allow the corpse to be placed in it in a sitting position often with the face turned slightly upwards. The total corpse was covered, not leaving the head above ground as has been reported for neighboring groups.[3] Fencing posts are placed upright around the corpse making in fact a small hut in the grave. The top is then covered with a pandunas leaf blanket and stakes laid over this before the ground is leveled over the grave. The grave is called 'wamati?i'. With fence posts a little circular protection is made around the grave and a platform constructed on which provisions are placed in the form of food and water as well as the carrying bag of the deceased with the contents of betelnut, lime, pepperleaf and tobacco. A couple of sweet potatoes and some water in a length of bamboo is left on the platform until signs of use are present and then replaced with fresh supplies. This was apparently kept up for about a month at which time it is clearly evident that the spirit has eaten[4] and departed for the 'wamima?i' — the spirit home. This is variously described as being 'far away' or in the bush, but no clear concept seems to exist regarding its locality nor about conditions there or the continued existence of the 'wami'.

3. Flierl (personal communication 1964) quotes a report by missionary Herrlingers, dated 1934, that on their way to the Purari River, they happened into such a cemetery. "Die Toten werden scheinbar sitzend begraben, so dass der Kopf frei uber der Erde zu liegen kommt."
4. The first Lutheran evangelists observed this and told the Akunans: "Can't you see that rats which live in the pit-pit come out when you're gone and eat the food?"

The average mourning period does not seem to exceed one month and very often a widow would remarry after this time. The mourning climax is reached when the corpse is prepared for burial. The internment occurs in private while the feast is in progress and the period which follows the burial is one of gradual reincorporation in community affairs. There is no mourning garb, nor could I ascertain any taboos during this post-internment period which would apply to either bereaved spouse or kinsmen. The white ash on face and chest which marks the closest relatives during the mourning and wake, is removed when the corpse has been taken to the grave. During this period of intense mourning a person may mutilate himself as a sign of sorrow. A woman would often cut off the first joint of her little finger for a favorite child, or even occasionally the third or second finger's first joint. For her husband she might sever the joint of the second or first finger, but never the thumb. A man would occasionally cut off the joint of a finger for a loved child — but it was not common to do this when a wife was deceased. In these cases the hand would be shaken to force blood into the fingers and cause bleeding.[5] This, too, was true when in some cases a woman cut off a small piece of her ear lobe. As the blood covers her breast, it is a sign of supreme sorrow and grief.

The feast which takes place during the act of internment seems to have a dual purpose. Partially it serves to reintegrate the mourners with the community and partially too it serves to balance debts. This is the time when the death payment is collected and the affinal group transfers valuables to the cognatic kinship group to reimburse them for the loss which was suffered. This subject is discussed with examples in our chapters dealing with economy and marriage.

To show the variation which appears in the wailing and funeral of persons representing different statuses, three brief examples are offered here.

Case 1. A young girl of about seven years old

On January 26 at twelve noon Numino, the daughter of Duka and Munokino died. She had been ill for only one day. The previous day Munokino had been to her gardens but returned during the hottest part of the day to rest in her house. The house is a modern kind with a padlock and lock to which Munokino has the key. After returning to the house, she locked it from the inside and lay down to sleep with a

5. Flierl (personal communication 1962) reports that during his first visits to the eastern Gadsup he observed a male and females in this condition.

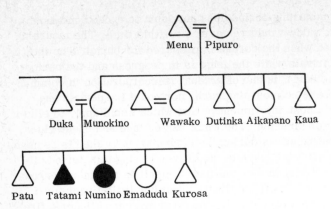

Diagram 6. Genealogy of child death

younger daughter Emadudu. When she awoke, the mother says, she felt something cold on her neck, just above the shoulder; and still lying down she put her hand to her neck and felt a piece of pork there. She immediately got up because she knew that they had no meat at the time and inspected the door. It was still locked. That night her daughter Numino became ill and by noon the next day she was dead. Both Duka, and later Munokino, stated that only the 'wami' can go in and out of closed doors, and they believed that it was the spirit of some ancestor — but they did not specify — who had visited their house, given the mother a piece of pork in exchange and called the 'aumi' of their daughter Numino.

We were just returning from the bush when word was received that she had died, and I immediately went to Duka's house, remaining there for the following twenty-four hours. At about sunset the wailing started. Word had been sent to Tompena where the child's maternal grandparents lived.[6] Present at the time were the parents Duka and Munokino; the mother's sisters, Wawako and Aikapano; the mother's mother, Pipuro; and the mother's brothers, Dutinka and Kaua. The two sisters of the mother were trying to console her, but she was visibly grieved. All the time neighbors and kinsmen were entering the house of about twelve by fifteen feet with a large hearth in the center. At this stage only the mother and close female kinsmen of the deceased were crying while the other people chewed betelnut and smoked.

6. This is one of the few cases where a girl from Tompena married outside the community and resides virilocally, but it may be explained partly by the fact that her father, Menu, is the son of the sole adult survivor when Akuna's men's house was burnt.

124

The grandmother, Pipuro, told a long tale about the fact that Duka and his widowed mother had come to settle in Akuna where he had no relatives, and when he married Munokino they came to Akuna; but Pipuro remained in Tompena and told them to come back and live there. But they did not heed her warning; and now she, the old woman, was still strong and healthy while this young girl had died. As the afternoon became evening, more and more people arrived and a large fire had been made outside. Persons who participated in the wake would periodically leave the room to sit and talk outside before returning to the room.

All this time the mother, Munokino, had been holding the little corpse cradled in her arms with the head against her cheek. Kinsmen who sat close by would stroke, touch, or caress the limbs or body of the dead girl. Thus emotionally charged, the mother would occasionally hold the corpse in her outstretched arms looking into the still eyes or stroking the pale cold face of her deceased daughter before breaking into loud sobs anew.

As it had already become dark and many people remained for the nightlong wake, food was prepared by members of the parents' interest groups and brought into the house to be distributed among those gathered there. Before it was passed around Duka made a short speech explaining that he had not expected his daughter to die and thus had not prepared large quantities of food, and especially tobacco and betelnut, which are consumed with great dedication during these emotional congregations. Everybody responded spontaneously with, what is best translated: "Sure, sure, after all we didn't come here to eat but to cry with you." Then somebody passed into the room one of the largest bunches of tobacco leaves I have ever seen, and bunches of betelnut, and enamel dishes with corn, taro, yams, sweet potato, and cucumbers and everybody relaxed and chewed and talked and smoked.

Night had descended. Those persons present would remain for the night because fellow villagers were close to their homes and visitors would not leave since Gadsup do not like to be outside a village at night. Now the wake started in earnest. The small room in which we were huddled had about fifty adults packed into it around a small hearth with glowing embers. The parents of the deceased child were still in the same positions. Closest to the mother were her sisters and brothers, and their mother, Pipuro. A primary part of the wailing and repetition of the deceased's name was done by the mother's brothers. As people finished chewing and picking their teeth, participation increased. Crying and name calling became louder with men and women alike weeping openly. They were also singing, and during the breathing spells between phrases the women would draw in their breath with a series of sharp glottal interruptions, and then with a loud sob, by a man or woman, the next phase would start. Such periodic wailing would usually last approximately fifty minutes to an hour and then during a fifteen or twenty minute breather people would sit around and talk or smoke, frequently smiling or laughing. Persons who needed to relieve themselves would go outside while somebody else took their places. Then, as if the theater lights had been dimmed, talking died down and the mother's brother started again, at first just calling in a distant high pitched voice, "Numino, Numino!!" Others would start shaking, crying audibly but not yet visually, but within ten minutes the mother's brothers' bodies would be shaking as tears stream down their cheeks. This was kept up all night long until just before six when, at the first light of day, the people, exhausted by their show of grief, settled down and dispersed to various houses for some sleep.

At about ten in the morning, January 27, the wailing started again, but the intensity, cause and result of group participation was lacking. The climax in terms of grief came at about noon when the corpse was placed in a roughly made box and covered with an old blanket. Now Munokino and her siblings and their mother stood around the coffining while the rest of the people sat. The singing and wailing became louder and the sobs and exclamations were made at regular intervals and loudly. Those standing around sounded as though they would not be able to bear the grief and they swayed back and forth, tears running down their cheeks. A number of old women sitting close by, both kinswomen and fellow villagers of the parents, held their hands outstretched with the palms turned upwards and rhythmically moved them up and down as though attempting, by the symbolic hand motions, to raise the corpse. This was kept up for more than an hour while Kaua and Dutinka, joined by the most emotional of them all, namely a cousin called Umato, finally closed the 'box' and nearly collapsed from sorrow. Now they walked outside and lit their cigars while talking and after a while also laughing. The wailing around the coffin, though subsided was kept up by the women. Most of those who were present by now had congregated about the earth ovens and as steaming food was passed around, they seemed to forget the reason for their congregation. Hardly anybody looked up as the mother's brothers quietly took the coffin from the house and departed from the scene. Duka was not allowed to participate, nor is the father — or bereaved husband — permitted to assist in preparing the grave. The mother's male kinsmen had taken the coffin to Onampa where it was placed in a grave and covered over.

They returned to Akuna to participate in the feast which was already in progress.

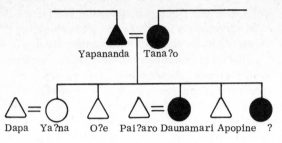

Diagram 7. Genealogy of adult death

Case 2. An adult woman who was married

Just before and after the Second World War, with the European Administration taking over and getting settled at Kainantu, it was still a novel and strange experience for the local people to be jailed. The result was that many of them left the villages which were closer to the administration's post and withdrew to one of the more inaccessible villages. These latter included Ikana. During these years of change a young man Yapananda, his wife Tana?o and two children lived in Akuna. They had been born there and were accepted members of the community.

Whether the husband had a brush with the law is not clear, but the family decided to withdraw from Akuna to Ikana. Here they had three more children of whom the eldest was a girl, Daunamari. Three years before the events discussed below the mother Tana?o died and a short while later her husband. The people of Akuna were upset by this and inquired from the Ikana community whether it knew how to take care of its 'visitors'. The Akunans clearly felt that the family in Ikana really belonged with their kinsmen and among traditional associates.

On Saturday, September 9 (1961), Daunamari, who in the meantime had married a man from Ikana, was taken to hospital in Kainantu with what sounded like toxemia from her right foot. After five days in hospital she died and was carried back to Ikana for burial. I arrived about förty-five minutes after the corpse had been placed in a house in Lower Ikana.[7] A large number of people had assembled and more were arriving while wailing and walking around the house or in some cases entering it. The name of the deceased would be called repeatedly in a high pitched monotone voice, "Dau?o, Dau?o!" (using the abbreviated form), and followed by wailing. In the foreground near the house

---

7. As in the case of Akuna and many other large villages, schism is already operating here in Ikana with the result that three separate hamlet concentrations are clearly visible — forming in effect sub-village communities.

sat kinsmen with their bows and arrows resting alongside their out-stretched legs, one or two of the closest relatives (e.g. the brothers of the deceased and the widower) were covered with white clay and would come out for short spells and then return to the corpse. Behind them was the congregation of visitors, chewing betelnut and smoking, while in the background a number of the young men played cards.

Inside the house were ten people including the widower, his father, the two brothers of the deceased and his sister's husband, while her sister and mother's sister with some other women sat next to the corpse which lay next to one side of the house. Most of the close rel-atives had their chests, shoulders and heads covered with a white clay and ash. Occasionally somebody would arrive and enter the house to look at the corpse and 'tikumikemi' — to commiserate.

For some time I had noticed the talking among the Akuna's men when they would glance up at the Ikana people. When one of my com-panions was asked about this, he explained that the Akunans were taking the Ikana people to task because they had allowed three mem-bers of one of the original Akuna families to die within three years. They were insisting now on the return of the two remaining brothers, O?e and Apopine, both of whom were single. The elder sister Ya?na could remain because she was already married (to the 'tul-tul' of Ikana) and well established. At this stage, as if he were conscious of the topic being discussed, Dapa (the 'tul-tul' of Ikana) arose.[8]

"Half acculturated as he should be since he is 'tul-tul', he is wear-ing a pair of old khaki shorts. His mouth is filled to capacity with betelnut juice, his left hand holds a cigarette rolled from newspaper, his right hand is tucked away in his trouser pocket, while his eyes roam the wooded mountains to the south that is Tairora country. He looks like a philosopher telling his audience: 'You know, it's strange, people are born and people die ...' What he in fact is talking about my interpreter says, is the displeasure of the Akunans. He says that one always helps those who need it and when a person dies his friends and relatives take care of his children and his pigs and see that they have enough to eat. This they have done for the family of Yapananda. And now the woman has died."

The men were chewing away peacefully at lengths of sugar cane which had been supplied, and toward evening many returned to their houses or villages. During the night the wailing started. One or two women at a time would keep up a high pitched monotonous nasalized whine for about forty-five to fifty minutes, calling the name of the deceased. Then it would be quiet, while people chewed and smoked and occasionally laughed. Then after an interval of ten to fifteen

8. This quotation is taken directly from notes I made at the time.

minutes it would be repeated. In the morning the visitors and other mourners arrived again. The earth ovens were cooking and people sat around the house in small groups while off to one side a pile of cloth, arrows, trade store items, and native materials was growing as people arrived.

About midmorning while the earth ovens were being opened and the attention was focused elsewhere, a number of men wrapped the corpse in a native leaf blanket, and carried it off to the grave on a bamboo carrier. Their departure was hardly recognized and those who did observe it paid little attention. As the feast was drawing to a close, the 'tul-tul' of Ikana, Dapa, started to distribute the goods which had been contributed by the people of Ikana community. Those who contributed were clansmen and interest group members of Pai?aro, the bereaved husband.[9] Those who received and shared were mostly kinsmen of the deceased.

Case 3. An old widow

Ever since we arrived in Akuna two very old women had seemed close to death. They could hardly move from the hearth and were extremely thin. As informants they were of little value since their memories had lapsed. In the morning a number of sweet potatoes would be given to them but there was little real concern. One of these women was Akatino — a person for whom any Gadsup would feel sorry. Akatino was in the unfortunate position of not only being widowed but having no grandchildren to care for her. She had three sons who were living, and she lived with Kaua and his wife. She must have been about eighty or more years old.

On February 22 at four thirty in the afternoon she died. The news spread from house to house much as if somebody had made the announcement that a tropical thunderstorm was approaching — everybody expected it, but they did not know quite when. I arrived about an hour later at the house of Kaua. By the hearth sat Kaua, Dapa?e, and Iyuno. They were talking and smoking while the corpse covered with a blanket lay outstretched next to Kaua. He greeted me and in response to my statement that I was crying for them, he said: "Yes, she has gone now. Look, she must have known her death was closeby; she only broke off half the sweet potato". In the ash of the hearth were three small roasted sweet potatoes — one had been broken in half and partly eaten.

9. This case is discussed again in the chapter on village economy under the heading of death payments (page 174).

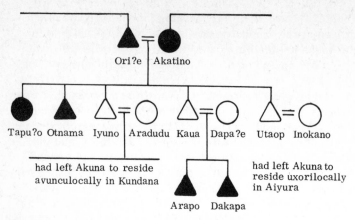

Diagram 8. Genealogy of the death of an old woman

During the evening and all of that night there was little crying, for
Kaua explained: "When a person is old like this, we are waiting for
her to die; we know it is closeby now and though our hearts are sore,
we do not cry as for a young person whose days are not yet lived and
whose strength is still in his bones!" With relatives and neighbors in
attendance only two wailing spells took place each lasting nearly an
hour. One was led by Kaua and participated in primarily by males;
the second was led by Tiyorono, the sister of Dapa?e. Both of these
occured about noon on the day following the death while the earth oven
was cooking large quantities of food.

While the women and most of the visitors remained around the earth
ovens, talking and eating, the corpse was taken to be interred. Here
both Kaua and Iyuno were present. Those who were participating in
the feast meal did not react in any way as the corpse was removed. It
seems that grief is expressed in wailing, but once this has been clim-
axed the mourners withdraw psychologically. Those who place the
corpse in the grave do not wail while doing it.

GHOSTS AND SPIRITS

The 'wami' of a deceased person remains near the grave for a period
of up to two months, while mourning is continued. This disembodied
ghost does not visit the village or haunt the houses of kinsmen or neighbors but remains there at the cemetery. During this transitory period
it is often seen there by adults — children are never allowed near the
cemetery since the ghost of a deceased person might narm them.

When an adult sees the 'wami' of a deceased person, he can readily recognize it since there are few changes. It does not decrease in size or take on any features or characteristics other than those it manifested as a living person. Now, however, due to the fact that it has been separated from the 'timami' and the 'anda', it is fleeting and will disappear from vision when a person approaches it too closely. Due to these ethereal qualities it is not possible to observe where it had been or to follow its tracks on the ground. The 'wami' of a deceased person seems to remain at the burial ground for as long as people mourn and actively remember him, after which it departs for the 'wami ma?i' — the spirit home.

The 'wami' visit descendants as they sleep to convey messages to them, but this communication is not actively sought by the living. I never discovered any service to the ancestors and feel that while we may speak of a recognition of ancestral spirits, there is no reason to suggest ancestor worship.[10] In fact, Akunan ancestral spirits seem to do the soliciting; they are not in turn approached by the living.

In addition to these ghosts which are human in origin, there are also spirits or forces in nature. They live in holes in the ground or in queer shaped trees and rocks. If a man has been in the bush and happened upon such a strange formation, he will return to the village and inform others about it. Should some ill luck now befall him, his kinsmen will blame it on this meeting and that place will be avoided in the future. These spirits or forces are called 'o?emi'.[11] This spirit, while primarily found in nature, is able to enter a person. Thus, when a person has obviously died, when his hands and feet are cold, when his eyes are closed and his breathing has stopped "but his heart is still beating" people say that an 'a?emi' is in him. The survivors will now place a reward in the form of food, arrows, and similar valuables in the roof of the house and scold the 'o?emi' which then leaves the deceased and returns to the bush with the spiritual part of the reward. This then confirms the clear statement of an informant that "only some people have 'o?emi', but all have 'wami'".

The 'o?emi' seems to enter some people during their lives, afflicting them with some kind of mental disorder. But informants were adamant that the 'o?emi' could not be brought on by sorcery — it simply entered a person. If somebody were to cut down a tree in which a nature spirit resided, even accidentally cutting the tree, the 'o?emi' will visit his house at night and haunt him. This takes on the form of

10. The features and prerequisites for an ancestral cult or worship of the ancestors has been quite clearly discussed by Fortes (1965) and Hsu (1948).

11. The relationship is not quite clear between 'o?emi' as a force or spirit, and the terms for magic and sorcery ('oyi') and thus a man who practices sorcery ('oyi wanta'). Informants were convinced, however, that the sorcerer does not employ or appeal to these 'o?emi'.

pains and cramps on the stomach and may cause death. Such a person or one who has epileptic seizures and fits or who is insane is referred to as 'o?emi memi' — loosely translated as "spirit possessed". It could be driven out by others as when a group of men beat the afflicted until the blood flows — luckily the latter is said not to feel a thing — and then put him in a small hut of banana leaves which is set afire. The skin burns and the 'o?emi' is driven out. A man of Akuna called Imunopi was afflicted with this spirit and he was "cooked in a house of banana leaves" and now is normal. In some cases it seems the spirit leaves on its own accord. One of the villages at Aiyura is called Anamunapa?i and in it lived a man called Yayinkama who originally came from Omaura. He was married to two women when an 'o?emi' entered him. For nearly ten years, informants state, he lived in the bush half wild. Then he returned to normal. His wives had been living with other men, but when he returned to Anamunapa?i they joined him again. He is said to be completely normal, in contrast to the years when he had the 'o?emi' in him.

The ghosts also cause a lot of trouble. Frequently one hears a person calling out in no particular direction, and you learn that she is addressing the ghosts of ancestors to return an axe they 'stole', or a knife they 'took'. When a little boy called Po?au was lost one night in Ikana, it was blamed on a ghost. The little boy had gone down to the fountain to get water and it was already dusk. He did not return. His parents explained that he knew the way and had walked it numerous times. Why would he travel in the wrong direction this time? Finally the mother of the lost boy, after people had walked and called in all directions, informed the spirit of an old woman who had recently died that if she did not free and return the boy, she — the mother — would burn the old woman's house. This belief that the ghost can steal a child is a common belief in neighboring villages. It is also only a variation on the belief that the ghost comes to fetch the spirit of a child as in the first case discussed above.

The anticipatory belief that the ancestors will return — a belief at the basis of cargo cults so common in Melanesia — is also found here. When the first whites arrived in the Arona Valley, they were thought to be returning ancestors. When the solar eclipse occured in 1962 (du Toit, 1969) there was the expectation that the ancestors would return. How and why these ghosts had undergone a change in color in their ethereal condition could not be explained.

While ghosts which are directly associated with the people who occupy a village seem to reside near it, they do not seem to remain if the people leave. Thus there is no fear of occupying the land or the village where others had lived. During traditional times the victors in a war would not fear settling the land of their enemy.

1. Sugarcane grown
to maximum length
for ceremonial
occasions
2. Yams presented at
the Orande ceremony

Death due to suspected sorcery (see pages 133-134)
1. Bereaved husband covered with yellow clay listens as senior kinsman instructs people to circle the house and lay hands on the corpse
2. Mourning husband shoots arrows into fence post which substitutes for the suspected sorcerer

DIVINATION

The primary way in which people learn about the future is through dream communications. These may be with the 'kayim' (dream man) of a living person, or with the 'wami' of somebody already deceased. Dreams are interpreted mainly as warnings. Thus if one dreams of a living person he is warned to take good care of himself because his 'kayim' had left him that day. Should a village elder dream about an enemy village he will inform the members of his community and the village will be guarded. Thus dream visions are translated directly into conditions of everyday life whether these affect pig hunting, personal relations or inter-village hostilities.

An important point is that this kind of revelation is not actively sought. The hunter does not go to sleep thinking about receiving some guidance regarding his future hunt. The 'yikoyumi' does not place an offering of food or tobacco at his side, hoping to receive some revelation about the impending war party. Yet both accept and value such communication when they receive it.

There are, furthermore, two ways in which people attempt to establish the identity of a culprit. Both are tied to cases of death in which sorcery is suspected:

Case 1. Flierl (1932:20) related that shortly after they entered the eastern extreme of the Gadsup area, two women died in close succession in Omisuan. The village leader was very upset about this as they were both female relatives of his and he attempted to learn the cause of their death. "At the grave of the deceased he pounded the ground with a bamboo cane and enumerated the sorcerers of the region. When the cane made a buzzing sound, he believed that the men whose names had just been called were the sorcerers."

In this case it turned out that two men in a neighboring hostile village, were the culprits. These men were lead into an ambush and while one was killed the other was hacked to pieces and tossed into the river.

Case 2. This second example occured during a visit I paid to Onamuna in May 1962. It would not have proceeded in this fashion during traditional times for now open hostility and acts of vengeance are forbidden, but sorcery still continues — it is said.

Many years ago a man Nanuwe married Amayano. They lived in Onamuna and about a year prior to the events to follow here, he died of a strange disease which people blamed on sorcery. The widow then married O?oma, the 'luluai' of Wopepa and since he had a wife in the latter village, Amayano remained among her interest group members and cultivated her gardens. A few months prior to these events she became ill and was taken to the Medical Aid Post and finally to Kainantu Native Hospital. She was told that nothing could be done for her, an informant explained, "except to go back to her village and die". She returned to Onamuna and died at noon on May 24th. Her kinsmen and the bereaved husband blamed the death on sorcery for it works slowly but surely, finally causing death.

The corpse was placed on a sleeping platform in the men's house and a number of close kinsmen and neighbors were in attendance. This group consisted primarily of females. The bereaved husband and clansmen of his and those of the deceased were covered with white ash from the hearth and with yellow clay. Their faces were drawn, the atmosphere tense as they squatted a small distance from the door chewing betelnut or silently smoking. Slightly further back a pile of food was forming as deposits were made primarily by Wopepa people as well as kinsmen of O?oma who resided elsewhere. But the valuables were few and the pile not very large. Near this pile the visitors and more distantly related mourners congregated. Then the widower and those close to him rose and started walking around and around the house. They did not talk but with painted bodies formed a single file circling the house while wailing and shaking. In their hands in front of their bodies they held drawn bows and occasionally shot an arrow into the ground. Off to the right there was a wide plank split from a tree trunk (and since they circled the structure anti-clockwise they would face it after every completed circle) placed upright in the ground. This was on the opposite side from the village and beyond it was shrubbery and bush. Nearly every time the men approached it they would call out the name of the deceased and let fly with their arrows piercing the plank.

This plank it seemed was symbolic of the sorcerer who had caused the death of Amayano. It was also a way in which pent up hostilities and feelings of aggression for an unknown murderer who had caused them grief and robbed them of a wife, and a kinsman, a childbearer and a food-getter, could be released. These arrows were not retrieved but as soon as the plank was covered with projectiles a man would approach it with an axe, cutting them all down in two or three sweeps thus leaving the face of the plank clear. There seems to have been something here of the disregard for personal belongings in showing how deeply grieved they were. Later the divining test was conducted — I did not witness this — in which everybody was made to enter the men's house circling the bloated corpse and in succession lay their hands on the corpse. When the corpse reacted, apparently due to gasses escaping from the abdomen, it was a sign in which village, clan or even family the bereaved clansmen should look to revenge the sorce

There is no aspect of community life which is safe against sorcery and every person must constantly keep a watchful eye for signs of sorcery. If somebody attempts to get hold of a person's excreta, of his carrying bag, his nail parings or betelnut spittle it is quite clear that he is planning to work sorcery. This calls for preventive action to be taken which frequently ended in hostile encounters. A simple case of this nature occured while we were in Akuna.

As a postcript to the Akuna-Apomakapa-Kundana disagreement discussed in the context of pig killing, the following should be included. At least a month after this matter was concluded and considered forgotten, Apama?no, the wife of Uriwa in Akuna died. Her mother had come from Kundana and upon her death kinsmen from Kundana came to Akuna on March 29 for the wake. The corpse was in the house and all the visitors joined by Akunans sat around the fire outside. As it became dark it was noticed that the Kundana people were very nervous. They did not enter the house as was customary but ate around the fire and smoked nervously while looking about them. The 'luluai' of Akuna suggested that they enter the house, but the response was very quick from their leader: "You men of Akuna are still very upset about what our clansmen (in Apomakapa) did to your pig; now you have caused the death of this sister of ours so that you can draw us all together here in your village and when we are all together you will work sorcery on us!" They steadfastly refused to enter the house. Later Munowi and his son-in-law, Tapari, suggested that the Akunans all leave the hut in which the corpse lay in state and leave the hut to the visitors from Kundana. The wailing continued through the night, but it was strictly one-sided as interest group members and affines were not present.

During December (1961) and January the Arona Valley was cut off from the Aiyura Valley due to heavy rains and a landslide in the mountain. The 'kiap' with his Melanesian policemen came across the mountain and organized work teams representing each village. On January 5 two women from Tompena walked through the stream and over Akuna mountain to offer a large bunch of betelnut to the 'luluai' of Wopepa as his people were on road duty. The younger of the two women Ta?apuro broke off one of the betelnuts and chewed it as they walked. As they crossed Akuna mountain she had had enough and making a small hollow with her spade next to the path she spat the betelnut juice into it, covering it with sand. As they started off again they saw two Akuna men, Torawa and Tipu?na on their way to locate a pig the village men had bought for breeding purposes. When they returned the same way late that afternoon Ta?apuro noticed that the ground had been disturbed

135

over the betelnut juice spittle. She pointed this out to her companion who was her mother Nontarino. They now made an act of neutralization over it which negates its value in cases of sorcery. It consisted of breaking a small bunch of pit-pit, spitting into it and then placing it over the spittle. Somewhat disturbed at this event they returned to Tompena.

The following day, Saturday, they informed the males in Tompena on what took place and charges were laid against Torawa and Tipu?na. In a court case held in Akuna that afternoon it was alleged that they had noticed the women spitting the juice there and had subsequently dug it up to work sorcery against Tompena. We should keep in mind that Tompena and Akuna were traditional namuko and that these sentiments are still present. The accused denied any knowledge of the act, and recounted how they had gone after the pigs which graze in the valley beyond Akuna mountain. A stalemate had been reached and the Tompenans returned home.

By Monday — the eighth — it was found that the Tompenans were growing increasingly hostile and restless. Torawa, who had been trained as a policeman in the Western Highlands before returning to settle with his family, suggested that the two villages get together and buy a ball, ask a policeman to act as referee, and play 'kick-bol' to settle the dispute. The Tompena men agreed, no doubt having in mind the only other time this had been tried in the Arona Valley when Onamuna played Wopepa. The first mentioned village lost the game but won the free-for-all with clubs and sticks which followed the game. Unfortunately this solution was never carried through.

On January 9 the total adult population from Tompena turned up to work on the road. There were only two things which were strikingly wrong:

a) Very early that morning while washing in 'yondonanomi', the stream which runs down the fissure separating Akunan territory from the Tompena land, Ya?pe had seen the Tompena men go by with bows and arrows. He kept very still and then watched as they hid them in the 'kunai' grass covering the side of Akuna mountain.

b) The men from Akuna and the policeman overseeing the road work were struck by the fact that all the Tompena men turned up at the road with knives and axes but no spades.

Ya?pe in the meantime had hurriedly returned to Akuna by way of the gardens and with a number of young men returned to Akuna mountain to collect all the weapons. Upon their return it was suggested that these bows and arrows be placed in the ethnographer's house for safekeeping. I calmly suggested that the 'tul-tul' as representative of the 'kiap' might be the person to take care of these captured weapons and was relieved when all agreed that this indeed was true.

136

Roadworkers usually quit at an early hour, allowing them to return to their villages and complete private or community projects which might be pressing. About an hour before the policeman and Akuna males left the road work, it was noticed that two Tompena men left and a few minutes later two more. They went to Akuna Mountain, but when all four returned to the road the 'luluai' welcomed them back with: "Yes, we collected them all and will give them to the policeman". All parties now retired for a court session which lasted a couple of hours. Akuna men charged that the Tompenans were preparing to attack them and that was their reason for coming to work with bows and arrows. The policeman admitted this to be a strange practice for people clearing a road. In a lengthy and eloquent oration the 'tul-tul' of Tompena, Aipako — whose father and his family had in fact come from Akuna — denied any such motivations. He explained that they had in fact been pig hunting all night long and brought their weapons since they wished to resume the search immediately following their workday. Arariwa of Akuna, however, had cut his foot some time before and had been at the Aid Post for treatment when he overheard the men pass by in the road that morning while agreeing on this alibi for carrying weapons. Again Aipako rose, disclaiming this and like a lawyer submitting conclusive evidence, he mentioned Era?u, son of the 'luluai' of Tompena, who had injured his foot during the pig hunt. This was the proof he said.

When mention was now made of the betel juice and the sorcery accusation which really caused all this strain, the policeman became completely confused. He adjourned the session instructing everybody to meet again the following day at the 'kiap house',[12] and he asked specifically that the women who had been party to the betel juice incident, should be there. He was going to retain the bows and arrows, he said (and had in fact had them all placed in the 'kiap house') and would send them to Kainantu where the kiap could decide what should be done with them ...

A man and woman arrived from Onamuna to report a domestic dis-agreement. While the policeman was listening to her accusations that the husband had beaten her, a young man from Tompena entered the 'kiap house' through a window and passed all the bows and arrows to a number of other males. Jumping the fence the weapons were dis-tributed and with the policeman in pursuit they withdrew to Akuna Mountain. In the semi-darkness he soon gave up, explaining that he would report the matter to the kiap rather than follow them and be killed.

12. These are government rest houses used by patrol officers on visits to the valley for tax — or other reasons. This one was at the Medical Aid Post about a mile from Akuna.

As we returned to Akuna the Tompenans were clearly audible singing as they returned to the village. The 'tul-tul' of Akuna stated: "You have been wanting to hear our songs, those the men of Tompena are singing are the songs of our forefathers." We listened and he named them: Kano?i and Yimana Kuku, the traditional Gadsup war songs. That night as the darkness descended and the singing continued, clearly audible across the fissure in this complete stillness, the 'luluai' appointed a number of males to do guard duty and they departed for Yondonanomi.

The Tompena men did not return to the policeman for a court discussion and two days later a man from Akuna, a good friend of Era?u went across to see him. The latter told him that he had hurt his foot while making a new garden fence, "and those bows and arrows were taken to kill you men in Akuna!"

The case was never followed through, but this is the kind of sorcery accusations made. It is also an example of the intensity of emotion and reation activated by such accusations. But sorcery may also be aimed more directly at persons or their property.

Prior to the Second World War inter-village warfare seems to have been frequent occurrences in the Arona valley. One primary aim of an attack was to kill the 'yikoyumi' — strong man — and so was the working of 'poison' — sorcery.

A person would attempt to get a hold of something which was in contact with a victim, such as excrement (people always defecated in the river, but now informants say they 'mess up' the bush and pit-pit close to the village), betel spittle, the butt of a cigar, finger nails, hair, 'clothing' or the band of a carrying bag ('unammi'), because it is always around the neck and shoulders. These are taken, placed in a bamboo and cooked on a big fire The 'poison' worked will cause the victim to get hot and cold fever, symptoms of 'ollesem malaria'. For women the same objects could be used, but in addition menstruation blood. Women in this period were not in their garden but in separation houses far from the village where fire was always burning. Often they were killed here by enemy by surprise attack.

It has been mentioned that the inhabitants of one village might use sorcery to kill the 'yikoyumi' of another village. He would be buried just as any other person is interred.

The corpse having been buried, gave occasion for the male villagers to prepare revenge. All the men would withdraw into the bush and then gather 'moku. mo' (red croton leaves) and sweet potato. If they suspect the murder of their 'yikoyumi' to dwell in village B, they will make a small representation of each man in village B. This consists of a small piece of sweet potato, rolled up in a piece of 'mo' leaf, and all of this finally rolled in 'moku' leaf. All these are now set aside,

138

each one being marked with a cut, or a tear in the leaf to represent man No. 1, 2, 3, and so forth. A fire is now made and all of these placed in a type of earth oven.

Supposing that man No. 2 in village B had actually practiced the sorcery to kill the leader, he has to do something in self protection now, something to counteract the test of the other villagers. At night he cuts a stick of about 5 feet in length and goes to the grave, he opens up as much as needed, and punctures the chest of the deceased with the stick, licking up the fat and blood that remains on the stick, this will make him strong and immune to the testing of the other men. He hurriedly covers the grave and returns to his village.[13]

Gathered around the fire, the men now start to test the representatives of the men. Those that are raw (uncooked) are put to one side, should there be more than one the test is repeated until by elimintation, one is left representing No. 2 man in village B. Bows and arrows are gathered, and the men start for village B and kill their victim. The victim's villagers when they realize he is gone, look for him and discover him, shot with an arrow in the chest and in the place where the arrow was cut out, the little bundle of leaves that represent him. They do not leave the matter there as justice, but retaliate in an attack and warfare.

A person who becomes known as the practitioner of sorcery always carries a leaf 'tankakosa', which when held between thumb and finger in front of the sorcerer makes him invisible. At such occasions he can approach others to steal some object he wishes or to make sorcery on them but nobody can see them. The general term used for a man generally known to practice sorcery is 'upuwanta' — i.e. a man with poison who walks about.

Beyond Kurangka toward Wampul, where the big river, Onanomi, drops into the Markham, there is a sheer drop of some 20 to 30 feet which is called by the Gadsup, 'nomademi' (place no good). One cannot walk down but has to go down on a rope, but once down at the bottom there are small holes in the rock with poisonous water which the Gadsup use in sorcery. This water is called 'koi?edan'. Should a man wish to revenge the death of a clansman, he will inform all the men in his men's house and have three pigs killed. In the men's house he will take the stomach fat ('poranoni') and smear it over his forehead and around his eyes. He also uses some of it to smear his hands and feet and then partakes in the feast that follows. There is a strict taboo on sexual intercourse during this whole operation. Either alone or in company of a 'brother' he sets out for Nomademi. At the top of

---

13. Somewhat similar is the guarding of a new grave by the Kuma (Reay: 1959:137).

the cliff he collects a strong piece of 'kanda' (a type of liana), he fastens it around his waist and lets himself down. At the bottom he cuts a small bamboo and fills it with 'koi?edan' — he may not touch or have physical contact with the water or it will lose its power. On his return journey he may not touch water and when he gets to a river must either cross it by walking on a log or must be carried through by another man.

During the night after his arrival he visits the village on which he wishes to work sorcery. He takes his container of 'koi?edan', opens the wrapping on a bunch of bananas, injecting the water into a few of the bananas by means of a sharp pointed stick, he also showers drops of it onto the 'moku' in the garden, or injects some into the sugarcane.

During this whole time, there is a strict taboo on, in addition to sexual intercourse, the eating of any delicacies — he is only permitted to eat roasted taro and sweet potato from an open fire. He does not see his family.

Following the operation he sits in the men's house awaiting a report on the deaths of his victims, often with other members of his clan. Only after this does he return to his family, washes and eats other foods. Now a feast follows during which he proudly announces: "I have taken revenge, I have turned back the death of my brother." This type of sorcery is never used on a village where clansmen live as it may strike them too.

A number of other ways exist in which Akunans practiced sorcery on other persons, not their gardens or animals. Perhaps the best known of these practices is 'uwa'.

This is individual sorcery, and like most forms of sorcery and magic among the Gadsup, could be practiced and employed by just about anyone. Usually, it seems, men use it. Should a person have reason for vengeance or feel that he has reason to kill someone else, he secretly, without the knowledge of his age mates in the men's house and without his wife knowing it, goes to the bush and collects 'warepami' leaves. He dries these and then cooks them in the bush on a stone. While it is cooking he hides in order that he may not be contaminated by the smoke — which seems to work like radioactive fallout, it is effective on an enemy but can backfire. When only the ash remains he picks it up with a leaf and places it in a small bamboo container. He is now ready to select his victim and to act against him. He now takes a long sapling or 'kunai' and dips it in the powder (ash) so that this sticks to it, and he hides the sapling in the grass near a footpath where he knows the person will pass. Should he walk by (and he will not notice it there) and the ash is brushed onto him, he will die that same night. Should a man approach along the path while the practitioner of sorcery is walking he simply holds up the bamboo container hidden away in a leaf of 'yipuyana' (brush). He will not be seen. If the ash on a sapling is discovered (this very rarely happens), or should a person see the ash on his leg, or should the carrier suffer ash on his own skin, all of these can be negated by submerging the affected limb or the sappling in clear water.

'Yandabitini' — this is one of the deadliest forms of sorcery practiced — it cannot be stopped once it is set in motion and cannot be averted by the victim.

The sorcerer — and again most any person can practice it — collects something that was closely associated with the intended victim, such as betel chewing, a cigar butt, the handle part of his string bag which goes around the neck and is greasy with dirt and perspiration from that person. The sorcerer also collects 'yanapa?i' (a variety of wild 'taro') and places all of these in a small hole in the ground which he covers over with dirt.

He makes a fire over this and stakes it so that it bakes well, but he must take good care that he does not sit in the smoke as this is dangerous. After it is baked well, he and an accomplice (usually a clan brother) remove it but don't touch it, and place it in the wild 'taro' leaves, wrapping it around well and then place it in an earth oven which is covered over. There it is left and the sorcerer is not allowed to glance back or stand around; he just returns to the village as soon as he can.

The victim to whom the betel chewing, or whatever, belonged will now be afflicted and gradually lose his strength. His death will follow surely.

'Wandim oyi' — in this type of sorcery one employs the 'wandi?mi' tree which gives out a type of gum along its bark which is very sticky (resin).

Should a person wish to work sorcery on a victim, he collects sweet potato peels handled by him/her, or betel nut pods (the outside of the betelnut is often chewed to add zest and flavor to the nut, lime and pepperleaf). You do not touch this but carefully collect it with a sharp stick or bamboo tweezer. This is placed in a fork of a tree where there is much of this gum and the peels or pod are stuck to it.

As time goes by and more sticky gum accumulates around this object, the victim will experience greater and greater pressure on his throat and lungs until finally he dies.

A number of years ago a man in Tompena wanted to have intercourse with or marry a woman called Ampinkano (originally from Akuna). She had a grown up son Matito and decided she was better off single as her son could help her with fences and clearing gardens. The man then practiced 'wandim oyi' on her and witnesses relate how she was stumbling around with outstretched hands, groping for breath until she died as a direct result of this sorcery.

A wealth of forms of sorcery are recognized and practiced by Akunans. Since there was not really a specialist, every person knows about these prescriptions and practices them where necessary. For an agricultural people, sorcery to protect gardens and products or conversely to harm those belonging to others, is of basic importance. This sorcery is called 'oyi' and is applied very widely.

When preparing for a new garden here, a number of men and women go into the bush and cut down small trees and underbrush in order that these may dry and the owner then can burn down the area he is clearing. After this he will build his fence around and start breaking the ground. Now if he heard that an enemy was making a new garden, he may decide to work sorcery on the ground and plants — but not the owner. He now goes into the bush and collects 'yayana' leaves. These are tied into a small bundle with any piece of rope or pit-pit leaf and placed in the garden just below the surface of the ground — just to make sure he may place two or three of these in various places. Now when the owner comes to burn down the dried trees, and bush these 'yayana' bundles burn too and cover the whole garden area under their layer of smoke. Products such as corn, taro, and tape will not grow in that garden.

While the previous kind of sorcery has no affect on sweet potato — and we should keep in mind that this is the staple crop, suggesting a specialized form of sorcery — there is a form which can be used. As soon as any person has planted his sweet potatoes, a person hostile to him or jealous of him may collect from the bush a liana-like plant with small, dark green leaves called 'dokatemi'. At night this sorcerer will go to the garden he wishes to affect and plants these leaves under the surface on each of the sweet potato mounds. The sweet potato leaves and shoots will grow luxuriantly on the surface but when he opens the mound to collect his garden produce, the garden owner will find only thin rope-like tubers.

There is a third kind of shrub which grows in the grass or in the valley. This is planted under the surface like the 'yayana' mentioned above. Unfortunately informants who could identify this shrub had no name for it.

A variety of kinds of sorcery apply to other valued products for Akunans. One is called 'yamaponi' ('yammi' is the generic term for taro and 'poni' seems to be used for sorcery as is the case with 'oyi').

A man who wishes to work evil on the taro garden of another, will go into the bush and look for a place where there is a small pool of stagnant water, black and slimy — this water is called 'puronomi'.

He now cuts a joint of bamboo, gets hold of some large leaf to act as receptable and collects this dirty water by means of the leaf, taking good care that his hands or body do not come into contact with it.

142

When the bamboo is filled, he secretly approaches his victim's garden, walks to the center of it and throws the water in all directions. "The taro will all die and stink". Nor can the gardener use any of the tubers for replanting a garden, and the ground where this garden was is contaminated. He is forced to get new tubers from a different garden, and must make a new garden, for this ground is no good until it has been overgrown and is cleared anew.

Sugarcane is primarily of ceremonial value to Akunans. It is tied up and supplied with supports while growing to lengths of more than thirty feet. While not of great importance as a dietary supplement, it is of great importance during the 'orande' and similar ceremonies. Here, too, there would be ample reasons for wishing another ill or wanting his product inferior. To reach this the person may work 'ya?aponi' or 'ya?oyi' (generic term for sugarcane is 'ya?i') on his enemies' garden. When a man wants to work sorcery on the sugarcane garden of another, this is what he does. It might be stated that here I found a fair regularity in the reason why a person wishes to work sorcery on another's garden — usually because the garden was made on ground claimed by the man who decides on sorcery.

Such a person sets a 'yawa' (trap) in the water and baits it with an 'umemi' (white grub caterpillar). When he returns and finds a 'wari' in the trap, he carefully opens it and kills the eel by banging it on the head. He does not look directly at it for fear that blood or moisture from it will splash into his eyes — it will be discussed in our next chapter that it is taboo for the catcher to eat an eel, as it will endanger his eyesight — but kills it while watching from the corner of his eye.

He now collects a number of 'empomi' (breadfruit) leaves, and makes sure that there are a number of young fresh leaves. While an earth oven is burning and heating the stones, he gets some 'wandinona' leaves (this is a wild variety of the 'moku' leaves which are planted in the garden) and prepares the food. Old 'empomi' leaves are used on the outside and younger ones inside. He breaks the 'wandinona' into small pieces and then places the eel on this without cutting it — forming a half circle on the leaves. This is tied up and cooked well in the earth oven. When well cooked the eel is gripped firmly and pulled so that the whole vertebral column comes out at once leaving the fatty meat behind.

He now places all the small young 'empomi' leaves on the ground and proceeds to divide the 'wandinona' and then the eel meat which he first twists in a leaf to break the meat and to free the oil — this meat seems to be very greasy. Tying up each bundle of food he will give them to his wife and children or if there are a surplus of packages he may invite brothers or his mother's brother to join in.

After the contents have been eaten he gets as many of the greasy
leaves as possible, and places them in the roof of his wife's house
till action is taken. This latter step consists mainly of breaking into
small pieces these leaves and tossing them over and into the sugarcane. "They will die and stink". Everybody knows too that it is very
dangerous to cook eel close to a sugarcane garden because the smoke
from the fire is filled with the 'wari' smell and will kill the sugarcane.

'Wari' or eel is also a potent means of working sorcery on the
banana palms of an enemy victim. Should a person plant his gardens
where you have claims, or if his bananas grow too well, you trap an
eel, extract its teeth and then at night go and scratch the bark of the
palm. It will split open and the inside will grow right out. The palm
will die and the bananas will not ripen.

Should a person catch a 'wari' and have enough to share with his
clan brothers but does not do it, they may want to take revenge. A
man will then get hold of the bones of that 'wari' and grind them to
powder between two stones. This powder he will carefully collect and
then either dust it onto the graves of the ancestors (no one specific)
without any 'prayer' or simply throw it to the four winds. According
to informants that 'wari' catcher has caught his last eel.

The fruit-bearing pandanus ('yommi') was introduced from Pundibasa and while it does have sorcery associated with it in that part
of the eastern Gadsup, informants claimed that these were not practiced in Akuna. Ritual protection, however, is and will be discussed
below. Akunans also know of sorcery which can be practiced against
the pandanus proper. They state that among the Tairora the leaves
are tied up, which prevents the nuts from forming, but that they themselves do not do this. "The leaves are used for blankets and raincoats
and the nuts are given to all and shared by all", they said. Yet it will
already be clear, and it is mentioned in a subsequent discussion that
pandanus is owned privately thus in the same category as some other
trees. It must be admitted that the nuts are shared fairly freely.

Since dogs' teeth were greatly valued, it is not strange to find
'iyani?oyi' or dog sorcery. In the bush there grows a pumpkin-like
creeper which bears a fair-sized fruit (six to eight inches long) and
is reddish when ripe. This 'pumpkin' is eaten by man, the inside being
scraped together and cooked in a bamboo after it has been rolled in the
new soft leaves of the 'empomi'. The peels are dangerous and must be
destroyed by fire or carried far from the village.

A man seeking dog's teeth will take your peels and hold them at
your dog's mouth or allow him to eat them and the dog will die.

An important variety of sorcery is 'pori oyi', or pig sorcery since
pigs were highly valued in Akunan economy and value system. The two
basic forms were:

'Kunawari'[14] — this is a large species of praying mantis with long legs which is to be found in the bush. Should a person for some reason like to kill the pigs of an enemy, he will get a few of these flying insects, break a sweet potato in half and place these inside. The sweet potato will be put together and placed at the entrance to the pig house where it will be eaten. It will affect all the pigs, which grow thin and skinny, weak in the legs and finally die.

The same effect can also be attained by catching an 'ikaki paya' (this is a beetle, something like the Rhinocerous beetle, only smaller, about a half inch long and black). They bake this in an open fire and then take the beetle, place it in a sweet potato and put it at the gate of the pig house. The fact that it has to be placed at the pig house door seems to assure the personal contact.

A number of ways exist in which people can safeguard their gardens and products. These will be discussed below as magic since they differ basically from the harmful intention involved in sorcery. In neither of these are there any spirits or supernatural powers which act. Magical acts differ, too, in their primarily positive contribution to social order in the community.

Akunans distinguish between sorcery which has an ill effect on the person, animal, or the garden and product it has been aimed at, and ways of counteracting these. They also distinguish from the former a number of ways to produce rain or win the attentions of a loved one. Another way to point at this distinction would be to see the former as representing black magic, and it contains elements of Frazer's homeopathic magic as well as his magic by contagion.[15] This has been designated sorcery. By magic all acts usually seen as white magic will be designated. While the knowledge and acts regarding sorcery are not restricted to specialists, so, too, magic is common knowledge; and while a leader or some specialist might act for a community, individuals may take action on their own behalf. There was, however, no magician specially designated by members of the community to act on their behalf.

When the ground is dry, the heavens blue and the plants in need of rain, a number of the village elders or an individual gather 'karomu' leaves, the soapy blue-green leaves of the 'dunki' plant (used by the

14. Upon recording these various types of sorcery it occurred to me that certain beliefs may be based on actual poisoning properties. A 'kunawari' was collected and sent to the world expert on the order of 'Mantide', Professor M. Beier at the Naturhistorisches Museum in Vienna, Austria. The specimen was a female of Tenodera fasciata blianchardi. I appreciate the assistance from Professor Beier.

15. Frazer (1954:11-12); and used in the same way in Notes and Queries on Anthropology (op.cit.:188).

ancestors as soap), leaves of the 'wari waro' which grow in the bush, and the leaves of the 'no?i' tree, of which the bark is used for making fibre for carrying bags. These leaves are all placed in a deep spot in the river and allowed to soak for a night. In the morning the men will gather, take out the leaves and beat them to a pulp with another stick. This is to cause rain to fall.

I had heard mention of old men taking an earthen pot, filling it with water and throwing this into the air near the gardens as this would bring rain, but a gathering of men denied this.

Regarding rain magic, I inquired from some old men whether there was the belief that they were influencing the 'wami' or a Nature Spirit to send rain, but they denied this and said that they beat the leaves, came home, and it rained.

Since spittle, excreta, and other things which had been part of or in close association with a person could be used in sorcery against him, there were ways to make these harmless. The point is that not every person neutralizes everything which he has been in contact with, or if he does, it is an afterthought after the harm has been done. It will be recalled that Ta?apuro had performed such a ritual act of neutralization over her betelnut spittle. Had she done it immediately there would have been no trouble, but her action came only after the ground had been disturbed.[16]

Magic is also used in drawing the attention of a member of the opposite sex. If a person is attracted to another — and it seems that this initiative may be taken only by a male of any age, irrespective of whether the woman is single or married — he acquires some 'watuye', or love potion. This is prepared by obtaining a leaf or piece of bark of the 'awewe yim' tree which is rolled into a small bundle · and placed in a fire. When it has been burned to ash, this is collected and an opportunity is sought when this can be used. The application can be made either by placing this ash in a leaf of tobacco the female will smoke, or by piercing a betelnut with a sharp object and injecting some of the ash into the betelnut. A man may also present this to his sister requesting that she dust it into the girl's grass skirt, her hair, or tobacco or betelnut she will use. Women had no magical potions to aid them, yet they very frequently initiated relations.

MYTH AND RITUAL

In an attempt to integrate Gadsup mythology with the ethnographic data, a variety of myths and culture hero tales have been included

---

16. With the 'luluai' I visited the scene after the inquest, and it was clear that foraging pigs had found the earth broken and inspected it.

in the text with the material to which they refer. There might have been good reasons to discuss sorcery and magic in the same fashion. We learn very little about Gadsup history or even the history of Akuna from their mythology. On the whole these tales must be seen, as is the case with taboos mentioned in the text, as explaining the origin of particular cultural practices. They nearly all refer to the culture hero, 'mani?i'; they nearly all outline expected patterns for action. While they do not prescribe roles, they suggest forms of behavior; while they do not enforce behavior, they set the limits of freedom of choice. These myths are directly related to ritual acts, but they are also directly related to the order which exists in the community. They justify the patterns of belief, regulate the behavior, and give antiquity, therefore value, and by extension stability to those institutions and customs which are practiced today.

The mythology is not only linked very closely with ritual but also with taboo. A number of taboos are in fact bolstered by these myths, but many more are found in various aspects'of every day living. Two of these taboos are in fact named.

1. 'Oyatani' — this is also the term used to describe a newly cleared garden which has not yet been planted.

The owner of such a new garden may not eat the eggs of an 'emo' (this is a turkey sized bird which keeps to the bush, but I never was able to see one). These eggs, which we tasted, have a distinctive taste, and should the person enter his garden after eating such an egg, his breath settles over the garden and will prevent plants from growing and developing. There is no ritual act for purification and the only way open is for him to sleep one night, after which the danger has left his breath.

2. 'Kayena' refers to the same kind of taboo and is the Gadsup term designating a variety of kangaroo. After eating the 'kayena' the owner of a garden may not enter his garden or his breath will stunt the growth of the plants. Here too purification cannot be achieved ritually but only by sleeping for a night.

A wide variety of other taboos existed and pertained in the first place to a sign which is set up to warn intruders.

The owner of a garden takes a piece of pit-pit or a branch, breaks it and spits on it, placing it in a footpath or at a fence of his garden where people regularly climb over. Should a person approach and trespass, he will develop a pain in the knee or ankle which can only be cured by getting the garden owner to spit on the paining joint.

Above we discussed the case of suspected sorcery when Amayano died. An old informant explained that the post which the mourners were using to shoot their arrows into could be used as such a taboo sign. It would be set up at the garden or bamboo bush which had belonged to the deceased person. This sign warned a visitor not to approach the property since the ghost of the deceased might endanger his visit.

A person with a 'namomi' kind of skin abscess was not permitted near a garden either. Native beans, 'ko?i', were very popular in traditional times and formed a part of many payments. A person who had a sore on his leg, and this does not apply to a cut that was festering, but to a sore which appeared from within on its own like a boil, may not go near the garden. Should such a person visit a bean garden, the bean pods will break out with warts.

Most of the taboos appear in the text where they apply. They include rules during menstruation and childbirth, regulations pertaining to the nose-piercing and 'orande' of males, to adults prior to warfare, before going into gardens and following sexual intercourse. Some of my informants suggested that mishaps which follow the breaking of one of these taboos was caused by a spirit in nature but others maintained that the 'wami' caused them to happen. There are two kinds of purification which apply generally. One is to cool down by sleeping a night; the other is to wash or bathe in clear running water. Both of these allow forces which have been activated or awakened to cool down. Such ritual protection could also be procured by the use of pig fat and blood which was smeared on various parts of the body. From the discussion earlier in this study it will be recalled that after the Akunans saw the first White man and observed the first aeroplane to fly over the Arona valley, they smeared their eyes and faces with pig's blood and pig's grease. Above it was pointed out that when a sorcerer prepares to leave his village to collect 'koi?edan' poisoned water, he ritually smeared his eyes and face as well as his hands and feet. This protected him. A similar ritual, combining the pig products and the cooling off is described by Flierl.

"Before us lay the territory of Afunakeno; we wanted to arrive today. To be certain the people of the village there, who were not a little excited about our unexpected appearance, acted strangely so that it appeared to us advisable to stay with them in order to calm them. We let ourselves be led into their village which lay at the edge of a high forest. A man met us in front of the village in order to greet us; he stroked us with a fan to which swine blood adhered. A little further lay a pig which we had to touch ... A group of women came to us with a bunch of Drazanen in their hands, which we had to touch. Evidently we were supposed to have been 'cooled off' as the natives take care of anxiety; that is, all harmful influences which went out from us had, according to their opinion, been rendered harmless" (op. cit. 1932: 49).

Ritually a garden owner may prepare to protect his plants or crops.

1, 2. Duikama in his normal village dress and during his incorporation as Lutheran church member (see page 156)

3. The table prepared for Holy Communion during the 'wash-wash' of the Lutheran Mission in Akuna (see page 155)

1. Breadfruit leaves used in cooking
2. Ti?**e** returning from her gardens
3. Pandanus proper; with
E?arua chewing betel nut
4. Fruitbearing pandanus;
Wa?ɪyo showing the ripened fruit

This will assure healthy growth and maturation, unless some person aims sorcery directly at them. These, then, are ritual means of assuring growth; they are not ritual or magical counter-actions. They apply specifically to:

'Yomi' — This small species of fruit bearing pandanus is planted by private owners, either in their gardens or at various places in the bush, and the palms are then watched and cared for. When a man now decides to plant a tree — which grows from a shoot or cutting from a full grown tree — he first prepares the ground and then collects the leaves and small pink flowers of the 'kamope' plant, and also the seed of a variety of pit-pit called 'indumi'. The hole for the cutting of the tree is made in wet ground or close to a spring if possible. He then places a mixture of the flowers, leaves and seed, which have all been ground up, in the bottom of the hole and the 'yomi' branch or cutting on top, and covers it. He then makes a small fence around it to protect it from pigs but also collects the seed and leaves of 'inumande' (a variety of tall grass) which he grinds between his hands. With this powder he regularly dusts the young tree and this ensures that it will grow up well and that insects will not eat it. He may step up his dusting if it does not quickly bear fruit.

'Yammi' — When a person wants to plant taro and ensure that he gets a good crop (unless somebody else works sorcery on it), he prepares the garden and completes the furrows, draining ditches, small mounds and the holes into which the taro shoots will go. He now gathers all the taro in one place in the garden and chews a piece of 'otnanakami' (i.e. bark ('akami') of the 'otnami' tree) which he spits over the taro and then quickly drops them into their holes. After this he may plant in between the taro such things as beans, cucumbers or greens and now with the coming of the Whites also peas, corn or tomatoes.

The ethnographic description of male initiation in Akuna will be discussed in a chronological consideration of socialization. There is much of it, though, which is primarily ritualistic and even religious in nature, for only by being initiated into this knowledge restricted to adults does a 'pumara' really become a 'wanta'. He is no longer a boy but a man. Eliade has said that "the puberty initiation represents above all the revelation of the sacred — and, for the primitive world, the sacred means not only everything that we now understand by religion, but also the whole body of the tribe's mythological and cultural traditions" (1958:3).

All his life the boy had been taught that the area around the men's house was sacred ground. That was the place for adult men, fearless fighters, people in close association with the unknown, the sacred, the supernatural. The boy was also taught that the sounds he heard during ceremonial periods were in fact the voices of the ancestors. At night, sitting in the semi-dark of his mother's hut, she would point out that 'tinapu', the grandfathers, had returned as the sounds circled the village.

When the time for his initiation came, this very strong maternal bond had to be loosened. The boy had to be removed psychologically from the mother's hearth where everything was done for him to an increasing independence. This increasing independence entailed, furthermore, the growing acceptance of institutionalized male status. This rite of passage starts when he is removed from the realm of his mother's protection, is continued when he spends the night with males and makes the psychological shift from a mother to a mother's brother — the realm of the female to the realm of the male — and reaches its climax when he is first introduced to the sacred objects. There before him are the sacred flutes, symbolic of the grandfathers, and those who have gone before, and in fact all of Gadsup tradition. But they are simply two flutes played by men. Behind the flutists is the man swinging a whip-like instrument and on the tip of the cord a piece of plank or bamboo about nine inches long. There was the bull-roarer[17] which causes the accompanying asthmatic breathing that he had always heard. And as a climax to this, there follows the treatment of his tongue and his glans penis. This opens the sacred and restricted world of males; it places within a few hours a vast distance between his boyhood and his status as an initiated youth. Once again the mother's brother, symbolic of the male realm comes to his rescue, and then as a finale as they approach the village, there follows the mock battle between the world of the uninitiated — the women — and that of the world into which he had just been admitted. He now had undergone a status change; he had been irrevocably removed from the world of the uninitiated.

"Thus, the initiation ceremony helps the boy learn his sex role by placing him, once he is physically ready, in the appropriate sector of the social structure. It is not that he learns during this short period any significant knowledge, or even attitudes, nor is it adequate to say that the boy internalizes his role by experiencing a shift in status from the team that is not initiated to the team that is. Rather, the ceremony gives the boy access to another symbolic world. The men allow him backstage to their 'show', and he is concomitantly barred from the doings of women. Now the real learning begins, and only a few significant experiences — such as hearing a man laugh at the women as he whirls the 'terrifying' bull-roarer — are necessary

17. Akunans refer to this as 'ireno?namu', but I was not able to get the etymology of this word.

150

before the boy understands a whole range of ideas that he only dimly perceived before. Then, as he emerges from the ceremony and its ritual seclusion, he immediately notices that the women and young boys treat him differently. He has previously been in their place and has implicitly learned the counterrole that he now takes as his own. In no time he adapts to the circumstances of being the 'person' and thinking of the uninitiated as the 'other'. Under such conditions learning is swift, sure and satisfying."

In these words Frank Young (1965: 32) summarizes the function of initiation for the individual. It is not merely a revelation of the secrets and rituals of manhood, but it occurs by way of ritual action and the ritual unveiling of the realm of men.

Ritual is involved when a girl first experiences menstruation and her procreative potential is awakened. It is involved when the mother's brother washes the newborn infant and ritualistically protects him from disease. A little ritual follows when siblings have been separated due to an argument and the sister presents her brother with a new carrying bag. But in none of these is there so great importance and symbolism attached to ritual as when the boy is incorporated into a new status.

THERAPEUTICS

Every person in Akuna has some basic knowledge of the human anatomy and also the types of treatment that can be made for various discomforts. Many of these require ritual acts to accompany treatment, some of them suggest the support of supernatural agents, none is restricted purely to some kind of specialist. Yet the 'uʔwata' was of great importance in the community, for as is the case with a herbalist in a folk society, or the folk remedy expert in an isolated rural setting, there is a combination of knowledge and distinct medical properties combined with belief and acceptance by the patient. When these two are combined, they are of great value. It should also be mentioned here that in Akuna there was for many years a female 'uʔwata'. This reflects the status of women, for it is a realm commonly restricted to males in the Eastern Highlands. Also during the period of our visit, Anapaiyu was the local doctor though somewhat hesitant since this status had been questioned by an administration that was appointing medical orderlies.

Much of this generalized knowledge for a people like the Gadsup, and perhaps even more so in communities depending on hunting to a greater extent, comes from the simple fact of observing animals. It is known that the organs function in particular ways and are situated with regularity in the same position. Adding to this is knowledge about men derived from frequent exposure to wounded persons. When a number of older people were questioned in this respect they explained in great detail — while pinching, poking, and in other ways showing — where certain organs were, and where it would be fatal to receive an arrow wound. This again would be explained with cases, as when a certain relative received a wound in a certain part and what the results were.

There does not seem to be much knowledge or concern with a germ theory of disease or with disease as such. If a person dies while he still has a contribution to make in any way, then obviously it was the working of sorcery. Above types of sorcery were mentioned which might kill slowly or more rapidly but both are certain killers. When a person gradually grew weaker medication was attempted, first by the family, as every family has some cures, but later by the u?wata. But then when the patient's feet and hands grew cold and the coldness spread towards the body, they believed that death was near and preparations were made in this regard.

In general Akunans did not utilize any great number or great variety of plants. This seems to be in keeping with the general picture we have for Melanesia as Hoogland reports for Australia too. The following applies as well to the Gadsup:

R.D. Hoogland writes from Australia that "Botanists in Australia have paid very little attention to ethnobotany. Apart from the enormous amount of work still needed in local taxonomy and the pressure on economic aspects, this can probably partly be accounted for by the small aboriginal population and the seemingly little use made of the native flora."

"The latter seems to be the case also in New Guinea. In comparison with the islands of Indonesia, where the native flora is extensively used in medicine, very little use is made by the New Guinea natives of their vast resources in this field. During my collecting trips in New Guinea, I have tried to collect material of native drugs for chemical testing and found it very difficult to get any at all. Others have noticed the same, e.g. D'Alberts (1880); Professor Baas Becking tells me that he has the impression that this applies to the whole of Melanesia." (Bank 1962: 21).

While a variety of medications were known, they also lacked the magical and ritual complexity in preparation and administration found elsewhere, as in Africa. The poultice might be made, a leaf crushed and rubbed into or placed over a wound, a piece of bark chewed but that was the limit in most cases. The 'u?wata', however, due to particular powers which entered her at the time of her calling, can also remove the cause of discomfort by extracting it with some visible object

Every village community has its own 'u?wata' and in some cases it seems more than one could be present but one would still be the senior specialist. At times a shortage of shamans may occur, or an older person may find the need for somebody to succeed her and then a special ceremonial was held. All the members of the village community would don their ceremonial attire and in the village square start dancing around a fire. One of the females in the gathering would get the call, upon which she will fall down or stumble about with bloodshot eyes, bobbing her head from side to side. She seems to go into a trance at this stage, for with eyes staring ahead, she stumbles about, even falling into the fire "and showing no effects of this". Those persons closest to her will pull her away from the fire, but she will get a hold of a bow and arrow while aiming at people and would have to be forcibly restrained. When persons act like this — and informants suggested that an 'u?wata' ceremony may produce more than one person going into this state — the rest of those who are present will say: "Oh, she has got the 'u?wata'!" It seems then that this 'u?wata' might be something which enters a person made susceptible by the ceremony; but if it is some kind of spiritual agent or supernatural power, informants denied that it was related in any way to the 'o?emi' which entered people.[18]

Such a novice would now receive advice on medication and technique from a senior 'u?wata' while also innovating her own forms of treatment. Payment would be made to a practitioner for services performed and these would include the various objects of value transferred in other cases of various payments. Once payment has been made it is not refundable if treatment is unsuccessful. It seems that we are here dealing with the embryonic form of specialization. While the 'u?wata' still worked her gardens and still got married, she was the closest Akuna came to having a traditional full-time specialist.

Diagnosis followed the question and answer method to locate the source and locality of physical discomfort. The u?wata in Akuna explained that a pig lives much like a person, and by noticing the breathing mechanism and other physiological features of a pig, she was able to notice with a fair degree of accuracy when the swelling of the chest was regular and easy or when irregularities were present here or in some other body part. Having decided on the body part which was causing the discomfort — partly by questioning and partly by shrewdness — she started to massage it. While allowing her hands to move over the body, she chews a piece of 'yatapu' bark and spits the juice onto the afflicted area. Continuing with her massaging she repeats the chewing and spitting of some 'opumi' bark onto the

18. Cases of this were discussed in the early part of this chapter.

painful part of the body. All her massaging and the systematic hand
movements are concentrated on bringing the discomfort to a central
point. The hand movements always leave the body at a particular
point where she finally places a bunch of 'onana' leaves — this is a
variety of pit-pit — which have been folded into a small bundle around
a tobacco leaf. The hands continue to move towards and around this
bunch of leaves which in turn attracts the pain. When the discomfort
and pain has been localized under the leaves, it is sucked out by the
'u?wata' in the form of an intrusive object such as a small piece of
wood or stone. This whole process is called 'airaupemi' which simply
means 'to extract' as an arrow is extracted from the wounded body it
protrudes from. Other practitioners are said to use different leaves,
e.g. banana or 'pomu', but the method of extracting the discomfort
in visible form is the same.

Another form of letting the cause of discomfort leave the body is
by blood letting. This, however, is a technique which everybody knows
and which need not be performed by a specialist. In this a man's
brothers or father may treat him for a painful or swollen joint by
'ko?ya'. The treatment involves a small bow of perhaps twenty inches
long called 'anankum' and miniature arrows. In traditional times these
arrows are said to have had points of stone flint but lately this has
been replaced by broken razor blade or a sliver of glass. This small
arrow is shot into the painful joint and it is believed that with the
blood, the cause of the discomfort leaves the body.

Everybody knows that for a cold they should chew a piece of 'wimi'
or native ginger and most gardens have ginger growing in them. Since
'wimi'[19] is also used in two important rituals — at a girl's first menstruation and during the initiation of a boy — informants were asked
whether this chewing had any ritual significance but they denied it.
The ginger and perspiration caused by it clears the head and respir-
atory channels, they stated. In some cases a broth was made and this
given to the patient.

The following is a brief list of plant medications used in Akuna:
1. 'Yatapu' — this is a tree which grows near the Markham divide
in the vicinity of Pundibasa. The bark is acquired from there and dried
in the house, where it dries very slowly. At the sign of a headache or
stomach discomfort — this suggests flu — a small piece must be eaten.
2. 'Na.nana akam' — this is the bark ('akam') of the 'na.nana'
tree which grows to a great height. It has seeds which turn red when
they ripen but the medical value, informants state, is in the bark.
Once again this is dried and used for colds or for a headache.

19. A specimen of each of these plants was submitted for botanical identification, but no identification has been received.

3. 'Karu?i' — this is a shrub which grows in the shade of forest trees. The flower has a long stem while the flower itself is red with white inside. The leaves are picked fresh and these are rubbed into painful joints.

4. When joints have been sprained or when a person's skin is cold with fever, informants suggested that the best treatment is 'kuman-kuma' mixed with 'dunki'. The leaves of these plants, both of which are small shrubs that grow along streams, are boiled in a bamboo and the warm pulp is then rubbed into the skin.

5. 'Tima.kukum' — this is a 'liana' type of creeper which grows in the bush while attaching itself to shrubs and small trees. The flower is a soft tan or beige color and bell shaped. Akunans collect the leaves of this 'liana' and either eat them fresh and green, or put them in an earth oven to cook them with pork. The value is placed in its properties as a tonic, for it is not only given to children who are thin and listless but also used by adults who feel rundown and tire quickly.

While Akunans traditionally had no way of affecting conception or causing abortion they have since learned about it, they say. Women scrape the bark of the 'yanapa?i' (alocasia macrorrhiza), or wild taro and eat this. In excessive quantities this 'a.yaye' is said to cause sterility.

THE NEW WAY

The coming of the Australian Administration is having an important change in this realm. Not only have the missionaries of the Lutheran church greatly influenced belief and ritual, but the Medical Aid Post is replacing the belief in traditional forms of medication and treatment. The one sphere which has been affected least is that of the belief in and fear of sorcery.

In Akuna the Lutheran Mission has established a strong foothold and a church has been erected within the village fence even though the evangelist and his entourage reside immediately south of the village in their own compound. At regular intervals, perhaps every nine or twelve months there is an elaborate baptismal ceremony accompanied with all the fanfare of a country fair. The local people have appropriately dubbed this the 'wash-wash'. For some time preceding it, people learn their catechism through the Lutheran lingua franca of Katé, but for those who are monolingual such as older converts Gadsup is sufficient.

The day of the actual ceremony is accompanied by a change in dress as each new convert must cast off the clothes which bind him to traditional beliefs. By donning a long white tunic men and women join those who are already members of the church. The long procession

is formed, which includes all the members and is led by the evangelist.
They walk through the village through a path hung with pit-pit and
palm leaves on either side. The vanguard of the procession is made
up by members of the evangelist's entourage, who have elaborately
constructed and decorated headdresses while they play their handdrums and dance. The steps which are executed are not Gadsup but
have been imported with the decorations and the music. They are
preceded by a number of women walking backwards while sweeping
the way with bunches of leaves for those going to church. The procession itself consists of all those being baptized. Here, too, is a new
ritual. Decked in white linen so that it is nearly impossible to recognize your fellow villagers of the previous day,[20] they must hold their
hands folded in front of their bodies while their faces take on the most
serious and repentant expression. I recall noticing a number of people
showing old Duikama, a mischievous old man who had adopted me as
his son, how to stand. There he was with white tunic, hands folded,
and in the place of his impish grin and fiery eyes a cold, faraway stare.

As this procession reached the entrance to the church the dancers
had reached a climax and were wildly jumping and swinging as they
executed their movements. The evangelist lit a kerosene lantern
which would light the way, and as he entered the church he unrolled
a scroll of Christ on the cross — visible to all the new converts.
Inside the church on this occasion and so too during regular Sunday
services, those who attend segregate themselves with females on the
left and males on the right — and if it is cold the churchgoer brings
his own fire and sets it on the dirt floor in the aisle.

The coming of Anutu — the Katé term for God — has changed belief
and action. Since there is opposition to polygymous unions by converts,
the whole social structure has been affected. Many persons who were
polygymists have been converted and sent their other wives home,
retaining only one. This has produced a number of unattached women
in every village community and also increased their potential as pros-
titutes and for adultery.

But the presence of an aid post where mercurochrome, sticking
plaster, or aspirins are available is important. So, too, is the pressure the administration is bringing to bear on parents if their children grow ill or die. The tendency is away from the 'uʔwata', for not
only are these new types of medication available, but Akunans are
starting to accept them and to believe in their value. With the practical substitution goes a psychological acceptance which is gradually
replacing traditional practices.

20. It was very interesting to note these converts throw on their new white
tunics over grass skirts and old lap-lap, and after the service to return to normal dress in no time.

Chapter 6

VILLAGE ECONOMY

While the family is the basic productive and consumptive unit, it
hardly ever happens that a family does not interact daily in joint
undertakings and sharing of products with other families. This, it
seems, is not so much a need on the material level as one of interaction and association. Since the daily activities are not very exacting and few enough to permit their performance at some leisure,
what joint activity and sharing takes place is not basically required.
Much of the work and daily living is and can be done by and within
the family unit, but the family is part of a village community and the
members work at giving meaning to their common membership of
the small group. This point will become clear in later parts of this
study. It is not suggested that people consciously aim at this association, but there are definite operative features which assure the
viability of the community.

The economic activities as they are discussed here apply to the
present time. Most of these aspects were observed and recorded,
and much of it reflects the traditional situation. People in Akuna
remember quite distinctly the conditions prior to the advent of steel
tools. Many of them made and can describe the production of stone
tools, while most females even today use bone needles, leaf blankets,
bark cord, and similar traditional products. In some cases reference
will be made to earlier practices and conditions but they will be quite
clearly contrasted with the present.

In outlining this section of our discussion an important problem
was faced, namely what to include under the heading of 'village economy'. It is now generally accepted by anthropologists, to use Firth's
excellent presentation, that "Economic organization is a type of
social action. It involves the combination of various kinds of human
services with one another and with non-human goods in such a way
that they serve given ends" (1951:123). Dalton even goes further.
"Transactions of material goods in primitive society are expressions
of social obligations which have neither mechanism nor meaning of
their own apart from the social ties and social situations they

express. In the Western meaning of the word, there is no 'economy' in primitive society, only socio-economic institutions and processes" (1961:21). Both of these definitions state or imply that we also look at the means which are available to serve these social ends. For this reason then, we will include subsistence here as it makes the initial goods available and is in fact at the basis of the social relations which form the topic of this study. Many of the subjects will be elaborated beyond perhaps the necessary minimum, but this is done in order that secondary features, the 'social actions' and 'social ties' referred to, may become clear. When focusing on the community and its daily life, the non-human goods are of importance only in so far as they add to our understanding of the villagers involved.

Akuna has a mixed economy, relying primarily on horticulture and animal husbandry, and in part on hunting, fishing, and the collection of edible materials. To be sure, hunting contributes relatively little to subsistence, perhaps because of the scarcity of game. The few wild animals and birds which are obtained serve primarily in a ceremonial context, e.g. at the feast celebrating a girl's first menstruation. Under modern influences even that part of collecting is diminishing and being replaced by the new agricultural products brought in by the White Man and various mission agencies. Domesticated pigs which have been allowed to go wild and which are hunted like game no longer constitute the primary source of meat, as 'seven day pigs' (goats) and 'bul me kau' (cattle) are taking their place. The former in particular is of great importance to the converts of the Seventh Day Adventist Mission in villages which border Akuna while Akunans occasionally use goat meat.

SUBSISTENCE ACTIVITIES

Under this general heading we will deal with all aspects of what traditionally was referred to as food gathering, hunting, and agriculture as well as the ways in which these goods are distributed. At the basis of this whole aspect of Akunan culture is the principle of reciprocity which was brought home to me a few days after arriving in Akuna. One of the families in Upper Akuna had killed a pig and cut off a piece of the meat which was presented to me, cradled in an 'empomi' leaf. It was accompanied by warmth of relations and the words that they "wanted to 'give' it" to me. My reaction was gratitude and showed appreciation but the following morning my interpreter Wai?yo suggested that "maybe you 'give' them a shilling" (about $0.15). So the process was set in motion by which I regularly was given a gift and they regularly received a counter gift, without anybody mentioning the direct relationship between the two 'gifts'.

158

The utilization of resources

The village of Akuna is bordered on one side by a heavily forested
mountain and on the north by the Arona Valley which extends for miles
without interruption by other villages. These are the two basic areas
where resources, which members of the community utilize, are available. East, down the hillside, and westward lie the gardens of villagers, but the forest is the important area for hunting and collecting
while the valleys are visited, mostly by children, to collect insects,
trap rats, or gather wild fruits. In all of these enterprises the division of labor according to sex is maintained. It is somewhat vague
during the early years of life, but becomes more pronounced and
more clearly defined as persons fit into the mold set by the elders
and as they identify to an increasing degree with their particular sex
role.

During the whole year women collect 'u?uanu', the immature tops
of and edible species of pit-pit (Setaria Palmifolia species) which in
form resembles very young ears of corn. These are baked in the
earth oven and when tender are extracted from the leafy covering and
eaten. Also 'woyami', a green resembling rhubarb, is collected and
eaten raw or baked in the earth oven; 'moku' (Chenopodium species)
and 'woru?ana' ('ana' — leaf, of the 'woru' which is a species of
squash) are greens, which are usually baked with pork, while 'meru',
a leafy shrub that is said to act as a tenderizer when cooked with pork,
is often eaten with it. Falling within this same category, but not seen
as food, is the small variety of the breadfruit tree which grows in
this part of the highlands. While the fruit itself is never eaten, the
very young leaves of this 'empomi' are used to wrap food in when it
is placed in the earth oven. The leafy covering is never removed but
enjoyed as part of the dish.

Nuts are very popular and form an important addition to a diet
which is poor in proteins. During September to November a large
yellow fruit, about one and one-half inches in diameter, ripens. This
'mi?yumi' is roasted in the fire and then cracked to obtain the kernel.
During this time, too, and into December, young boys and girls make
daily visits to the bush to return with 'puyami' (a variety of pandanus, but much smaller and with a very hard shell, commonly known
in Neo-Melanesian as 'wild karoka'). These are split open after successive hammerings on the stones found in all earth oven pits. The
most important nut is the 'imi' (Pandanus proper, or big seeded
pandanus), which ripens till as late as February. These palms are
privately owned, or if they grow in the bush, are claimed by certain
individuals. The fruit, a cluster of nuts, is split in two and roasted
on an open fire, which is said to add to the taste. The shells are soft

159

enough to be broken between the teeth, and while sitting around the hearth people visit with one another and gossip while cracking and eating 'imi'. Men often carry a number of separated kernels in their string bags to enjoy as a snack during the day. This is also the only form of food preservation, namely when nuts are saved in the house until they are needed, often for weeks at a time. During April and as late as July, a large pulpy fruit, called 'meka', ripens, having within it up to five compartments in which nuts are found.

Both men and women collect a variety of mushrooms and tree fungi which mature during the annual cycle. A collection of these was made and most of them identified and discussed by Professor Roger Heim, Director of the Museum National d'Histoire Naturelle in Paris (Heim 1964). The initial reason which prompted this collection was a report by Dr. Marie Reay that among the Kuma further west hallucinogenic effects were produced by such mushrooms (1960). This is not the case among the Gadsup and the identification did not produce the same varieties as are used by the Kuma. The two types of tree fungi which are most common and eaten by the people of Akuna are:

1. 'Onani' — (Polyporus Sulfereus forma Tropicalis). This is collected by women from the forest where it grows on dried rotting tree stumps during November to January. There are various colors of this fungus but none of them produce any side effects unless they are badly cleaned or undercooked. In such cases they produce vomiting. The identification of this variety shows it to belong to a variety which was very common in Europe and Madagascar.

2. 'Yamarana' — (Polyporus Frondosus forma Brevipes). This fungus ripens during November to March. The sample shipped out for identification was in fact collected on March 27, but informants said it was the last we would find. 'Yamarana' has soft brown 'branches' which get lighter as they grow older and larger. By the time the fungus is mature it resembles a large white spongy growth, perhaps twenty-four inches in diameter and the same height.

A number of varieties of mushrooms appear during the year and these are somewhat similar to the mushrooms usually found commercially. Two varieties which ripen between October and February, both growing in the forest, are 'Kantora' and 'Akaro'. It was impossible to get samples for identification. Of those which were sent, the 'Nanu?mi' and 'Impini' did not arrive in satisfactory condition for analysis, while the remaining four are: 'Kundara' (Pleurotus sp.), 'Wake?no' (Pleurotus sp.), 'Ku?ina' (Clitocybe sp.) and 'Namuni' (a small variety of Clitocybe). The collection of both tree fungi and mushrooms adds to the diet but the former is of greater importance since the person who discovers it may return again and again to pick pieces off the stem. Neither of these adds much in volume, but

160

supplies variety in a region where the staple may become monotonous.

This in fact is the same reason why children, and even adults spend time in collecting both plant- and insect- or rodent species which can be added to the basic diet. These then are the most important varieties of plant produce which the Akunans collect and which do not require a person to tend them. There are, however, other important forms of collection. Children, often in the company of adult women, may spend a whole afternoon catching 'yapumi' (grasshoppers), 'munani' (Melolonthidae Lepidiota vogelii — a beetle about an inch long that flies and is a copper brown color),[1] and 'kanki?i' (Brachytrypes Gryllidae — crickets). Children also spend much of their time catching 'wat?no' (a very large fleshy spider), 'watoya' (frogs), and lizards. For the children who collect these it is far removed from the economic activity that it would be for adults, as they have made it into a kind of sport. A special delicacy is a large white grub, called 'uma', which is found in rotten tree stumps and which induces even adults to chop and break at old trees. All of these are placed in bamboo containers and cooked, after which the mother will add them to her grated taro and sweet potato which she bakes in 'empomi' leaves in the earth oven.

When found, the 'panemi' (flying fox) is eaten, and boys spend much of their time shooting at birds, but for adult males this is a much more serious proposition. They never go to the bush without their bow and arrows and always carry a few of the three-pronged 'ayantakumi' arrows which are used especially for bird hunting. Nests are raided while young birds that are learning to fly are taken. All varieties of birds are hunted with the exception of the 'wuye' (cassowary) and 'emo' (turkey-like bird) which are trapped. All year around, too, whenever found, the eggs of the 'emo' — this large bird that resembles a turkey lays eggs about twice to three times the size of chicken eggs — are taken and especially enjoyed when the embryo has already started to mature. These birds are not very large nor found in sufficient numbers and therefore contribute only minimally to the native diet. Birds, and especially pigeons, are snared when they come through a hole in the fence and enter the gardens. This same type of snare ('wawami') is also used to catch rats, both in the gardens and in the bush.

A much more complex snare, 'kinku?yi', is usually made from

---

1. These 'munani' also form one of the pastimes of Akunan children. They catch the insect and tie a thin string fiber to its leg, thus allowing it to fly in circles around the owner but not able to escape. When the 'munani' grows tired and weak, the child flicks the string to simulate the flying and they look just like American children with a yo-yo.

bamboo for rats and small tree kangaroo. There are six named varieties of rats[2] that are caught, and they are identified partly by the size and the color of the fur and partly by the places they inhabit. The Gadsup never eat rats that are caught in or near the village irrespective of the species they belong to, because these rats, they say, are unclean and have fed on the faeces of humans. Two snares remain which are much larger: 'paro?uam', used to trap tree kangaroo, and the other, 'pom?uam', for wild pigs. All of these activities fall in the male realm, as do hunting and fishing.

Traditionally there was much greater ceremonial need for hunting. For example, meat of the tree kangaroo ('tapura') had to be eaten during the girl's first menstruation and during certain ceremonial occasions, such as the 'orande'.[3] A number of related men and friends would spend the night during the full moon in the bush, setting traps and waiting for the 'tapura' to travel its regular pathway. Also, 'kawena' (small variety of kangaroo or wallaby, according to picture identification by informants) was hunted, both of these with bow and arrow.

There are three varieties of aquatic vertebrates eaten by Akunans. Of lesser importance is the 'arauna' (tadpole) which is taken and added to the grated tubers much as insects are, and also 'kauya', a small fish which traditionally was caught in fish drives while people beat the water, or by tying an insect to a piece of string fiber and attempting to pull the fish from the water the moment it bit. At present, lines and hooks are often used. The third and most important kind is the 'wari' (eel). They seem to be more plentiful and make better eating but are also more dangerous physically and magically. The 'wari' are always caught in a trap ('yawa') made of hollowed-out tree stump with a number of small holes and a large entrance. Food is placed inside, and the moment the eel enters a small door is dislodged and falls into the opening. It is taboo for the man who catches a 'wari' to look at it or to eat of its meat.

While it was not discussed here we should state that men from Akuna very frequently participated in hunting parties. This sounds strange for a people living in a land so poorly supplied with anything more than marsupials, but it is of great importance. The prey might

2. 'Nantami' (small species) and 'A.kemi' (large species) are grey rats that live in the bush. 'Oyi' (lives in pit-pit), 'Uyammi' (lives in bush) and '?utonno' (lives in the ground) are all very light grey, nearly white, while the 'I?na.don' is a large, darker grey rat.

3. The 'orande' has fallen into disuse, but traditionally was one of the main rites de passage and took place around November or December. This ceremony is discussed in the following chapter.

include tree kangaroo or a variety of wallaby, which are found in the forests, but it more often involves pigs. These are domesticated pigs which are allowed to run wild, and every person is in fact responsible to assist the members of his community to restock the forests with pigs. While 'wild pigs' are perhaps of greater psychological than economic importance, it is really the domesticated 'village pig' which is important to the village economy.

Animal husbandry

Though pigs provide the only meat outside of an occasional tree kangaroo or the rats which are regularly hunted, we cannot even vaguely speak of 'pig complex' as has been done for the area further west (Reay op. cit. 1959:20). Akuna and the Gadsup in general lack the elaborate ceremonials surrounding pigs and also the exchanges which take place at these pig festivals. In addition they lack an elaborate set of terms for pigs with different colors, forms, and so forth. While the Bantu refer to the minute detail in color variation or horn shape in their cattle by a separate term, or the Camel Nomads distinguish the characteristics, durability, and physical condition of their camels, Akunans distinguish only three varieties of pigs, based on a color differentiation. The fact that they distinguish between only the white, red, and black pigs with no further breakdown suggests that pigs are not too important in their economy or ceremonial life (especially when this nomenclature is compared with that of various agricultural products), or conversely that pigs are of fairly recent introduction and have not yet received the terminological elaboration found in other aspects of the economy.

We have to distinguish, though, between two types of pigs, namely domestic or village pigs (called 'maponi', literally 'house pigs' from 'ma?i' or house and 'poni' or pig) and bush or wild pigs (called 'apatamponi'). Both types are owned privately by individuals, and while the former remain in or near the village the second are domesticated pigs which have been allowed to go wild or their offspring born in the bush. House pigs remain in the house where they are fed for a few months and are then removed from the village but remain close to it. While they forage on their own during the day, they sleep at the village fence and are fed there at least once a day. On this occasion the 'mother' who is feeding the pigs will call to them and while she has a long stick in one hand to keep filchers at bay, she will feed her own pigs from the other hand. They are quite tame and frequently she will stroke them or rest a piglet on her lap while delousing it.

The ownership of pigs as well as the use of their meat are subject

163

to strict regulations. The rules governing the eating of their meat have been discussed in the fourth chapter, but in general we can state here that the person who feeds a pig may not eat of its flesh whether it is slaughtered by the owner or shot by another party, e.g. after breaking into a garden. But the person who has slaughtered or shot the pig is also restricted from eating it. For various reasons we find that Akunans farm out their pigs to be cared for by others. This spreads the pigs in case of disease, but also ties persons together since it is somewhat like having a neighbor's child in the house. The following is an example and the justification given:

The pig of Torawa that was killed had two piglets. One was given to Ko?ana and one to Dupana. Asked why he had not given both to one person to care for and raise for him, he thought for a moment and then replied that it was dangerous if disease broke out to have all your pigs in one place. The most important reason was that pigs often grazed together and a person did not want his pigs to enter one garden and all get shot by the owner. Rather spread them out. He also stated that of the pig that was shot, neither he nor his wife, nor his parents nor her parents could eat the meat. His children could.

Ownership is identified by a complex combination of markings which involve the ears, tail, or both. Each clan has its own way of cutting the ears or tail of their pig, and this prevents chaos breaking loose due to the shooting of pigs; it, however, does not prevent trouble because each pig has only two ears and one tail.

The cutting of the ears is performed when the pig is about six weeks to two months old. The 'father' of the pig, i.e. the owner will take a piece of sharp bamboo to perform the operation. The 'mother' of the pig sits with the small pig in her arms delousing it until her husband is ready. He then cuts the specific marking, and while the blood dries from the cut ear and the piglet squeals, the 'mother' spits into each ear and into the mouth of the pig. This, she says, is in order that the pig may hear her call when she wants to feed it, and also that its voice may become known to her when it needs her or is sick.

Shortly before we got to the field, the men in Akuna had pooled their money and bought a boar from a European planter in the district. They paid the outrageous price of fifty pounds (or $140) for it only to have it hunted down and eaten as a wild pig by the men in Apomakapa at the other end of the valley. The culprits were immediately confronted with the fact that they had just enjoyed the most expensive meal of their lives and they had to reimburse the Akunans.[4] Individual con-

---

4. This case was discussed in the fourth chapter and under sorcery we mentioned the fear of the Kundana men that the Akunans would harm them when they were in the hut mourning the death of a relative.

tributions here had varied from five pounds each by Munowi and Tapari, and four pounds by 'luluai' Nori, to two shillings by Kipayowa and one shilling by Wira. This is then truly community property since even women contributed.

In addition to pigs but not nearly so important, the Akunans have chickens. These, informants explain, were traded in from the Markham Valley to Pundibasa in the Eastern Gadsup and thence throughout the Arona Valley. This was apparently of recent origin and no Gadsup term was recorded for chickens, as the Neo-Melanesian 'kakarok' has been taken over. Chickens are owned privately but hardly ever eaten, nor are their eggs commonly eaten, so that their economic importance is negligible. The only time that a chicken was killed it was included in a legal payment to the man on whose land the ethnologist was living, and without much ado was turned over to the ethnologist and his family. As with pigs, on each individual bird private ownership is indicated. In this case, however, a short piece of colored cloth is tied to the bird's wing where the wing meets the body, and should some disagreement develop regarding ownership the chicken is run down by the children and produced as concrete evidence. There are not many people who own chickens, but the few who do usually own from ten upwards.

Goats are slowly being introduced but are only rarely owned by the Akunans individually especially since we are dealing with a 'Lutheran village'. They were of course originally introduced into the Arona valley by the Seventh Day Adventist Mission because the people were not allowed to eat pork. In this context it is not strange to note that the Gadsup refer to these goats as 'seven-day-pigs'.

Many Akunans own dogs, and there are at least five to every village; but they are no longer of importance in the diet. At present these are especially prized in the hunting of tree kangaroo and other marsupials, or in chasing pigs when they are hunted in the bush. Traditionally dogs were eaten, and their canine teeth collected for necklaces which gave pride and prestige to the wearer, but since contact with the coastal natives the Akunans have discontinued the practice. According to informants, indentured laborers talked with coastal Papuans during their time at the coast and the latter told them that it was wrong to eat dog's meat. "Look", they said, "a dog goes around smelling the excreta of pigs and people,[5] and it retains the smell in its nose. A dog, also has 'savvy', just look into his eyes and you will see". This is what stopped people from eating dogs, but dog's teeth necklaces and the 'warunkara' — which are worn on a man's temple — are still very highly valued.

5. It will be recalled that this is the reason they don't eat 'village rats' either.

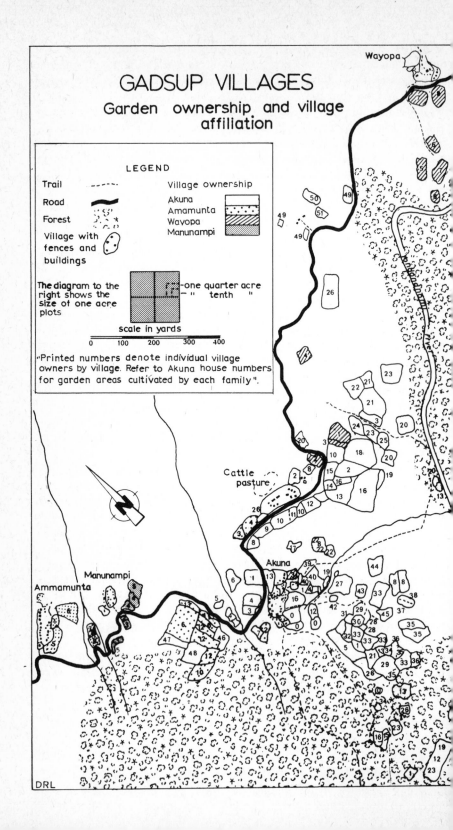

The agricultural base

In spite of the important variation produced by hunting and gathering, or even fishing, the basis of subsistence for these people is tubers. The production, care, and harvesting of these garden products employs a woman the greater part of each day, for food cannot be preserved and has to be collected in fresh supply nearly daily.

As has been stated above there is no shortage of land in the Arona Valley and there is enough garden land for whoever needs it. The only limiting factor in this respect is that caused by previous claims and ownership. This is primarily true of garden land, though it is becoming increasingly more so for tracts of bush where coffee gardens are planted. The bush can be divided between true bush and bush which has overgrown earlier gardens gone fallow. Within the latter there may be privately owned trees or even parts of gardens which are still recognized but in the true bush there will be no remnants of gardens. If trees are claimed it will be because they were planted there or claimed by hanging a bunch of grass in the fork of the tree and informing all of rights which have been established. While individual trees are owned by individuals, large tracts of the bush are owned by extended families or referred to as belonging to particular villages. This is no more than localizing ownership and the right to cut trees, just as the bush which surrounds the Arona Valley is distinguished from Tairora bush, and should some Tairora cut a Gadsup tree (as was done during the period of the field study) there will be serious repercussions.

All the land surrounding villages is privately owned by individuals who have prior claims to it. As long as an individual maintains his interest, or as long as it is maintained by his widow or children if he dies, the land belongs to the family and no other person has the right to start a garden.

In the bush any person may stake out claims and inform the others that he intends to make a garden. Privately owned trees which may be on the land remain the property of the owner, and his rights of free access must be honored. A number of years ago Kampo?i had started a new garden where the present village of Akuna is situated, but he did not honor a prior claim to a 'Pandanus' tree by Afita, father of Napiwa, and when he cut it down was very severely beaten by the owner. He could not bring legal action against Afita, as he had acted against the rights of ownership recognized in Akuna and in fact by all the Gadsup.

The Akunans' agricultural practices revolve around a cycle of clearing new land for gardens, while older gardens are allowed to grow wild and in time to turn to bush. With a population density of approximately fifty-three individuals per square mile, there is no shortage of land and even cultivable land is readily available to the person who wishes to clear it for use. While large areas are, of course, not suited for gardens, remaining under permanent forest or brush and are grazed by pigs, the simple fact that it is situated on a steep grade does not disqualify land from being used as a garden. On an average it seems that gardens are used for six to eight years, and then simply abandoned, either by taking the fence posts for a new garden plot, or simply by not repairing the fence once pigs have broken in. The climate is of such a nature that abandoned gardens soon show weeds and brush when not tended and very soon are completely covered by shrubs and small trees, or, at a lower altitude, by tall waving 'kunai'. After a fallow period of eight or more years, an old garden plot can be identified only by the relatively immature bush which covers it. The process of transforming it into a garden again takes from three to six weeks of cooperative labor by a nuclear family, nearly always assisted by consanguineal relatives and interest group members of both sexes from the village.[6]

There is no particular time of the year when gardens are started or specific seasons when they believe it to be most advantageous for them to plant their products. It is usually the older men, the keener observers of climatic changes, who will suggest to the women when gardens should be planted (about November), or when new gardens should be cleared and the brush burned. The work is started by the man, who inspects the land and decides where the fence should go, after which he sets fire to the bush and shrubbery, while his wife sees to it that the fire does not get out of bounds. He then starts to cut down the tree stumps which have been burned down, but very thick stumps are allowed to remain and should they sprout again they will offer welcome shade to the women working there during hot days. It very often happens that pandanus or banana palms and bamboo bushes grow in this new bush from a previous garden, and if so they are carefully guarded and cleared around before burning for they remain the property of the original owner, even though growing in another person's garden. Once a person has cleared the trees, he leaves for the bush to split logs for fence posts. This is often done before the trees are burned and the garden cleared so that the fence posts are already dry

6. This kind of agriculture is frequently referred to as slash-and-burn or swidden agriculture or shifting cultivation. As this last term suggests it is based on the rotation of fields rather than crops (Pelzer 1945:17) in which clearing is effected by fire and the ash forms the only fertilizer. Human labor depends on the use of the digging stick or bladed implements (Conklin 1961:27) and the period of use is short relative to the fallow period.

by this time. His wife, usually assisted by other females, will start breaking up the ground with a digging stick and now also with a metal spade bought at the European store. Roots and other organic materials are removed from the ground and burned, and so in time a layer of fairly clean top soil is established. The man in the meantime has started to plant the fence, digging down perhaps six inches and forcing the post about four feet high, into the soft soil. Between posts he then ties three horizontal saplings, one placed about six inches off the ground, the second near the center, and one near the top of the posts, using as cordage 'nani', anomi', or some other liana.

These days of cooperative labor often develop into pleasant social meetings where the owners will offer a meal to all who happen to be present and a long day is spent by a group of relatives and friends. This meal serves in a dual fashion, namely in partial payment for services rendered and also as an expression of that characteristic Gadsup practice of always offering food to every person present, and to assure that no person for whatever reason will be in want. Should some person assist who has not been asked to help, the man will tell his wife to remember that when the garden bears fruit a share should be given to the helper or to his wife.

Once the fence has been erected, the man's share is completed and he will return from time to time only to repair weak spots or places where pigs have broken through. The woman's work, however, has just started, and she will now begin the laying out of the garden and the areas to be planted with various food crops. Akunans do not segregate their crops or assign separate gardens to each type of garden product, but interplant a number while assigning the central or most important position to the staple, sweet potato. Almost every garden will contain sweet potato, which is planted on mounds situated about thirty inches apart. Between these are beans which grow on the ridges and a number of greens of which a mucilaginous spinach is the most popular. It should be kept in mind that this work traditionally was done with the use of a stone adze and a digging stick only. While the stone adze was replaced a number of years ago and the people no longer use it they all remember how these were made, and can re-enact the felling of a tree or the cutting of posts for gardens. This transition is not complete, for the women, many of whom still use the digging stick for garden work, and even men, were noticed to prefer this implement when digging holes for house or fence posts.

The products of greatest importance are all tubers or root crops, and among these we find the greatest percentage of the Akuna diet. The staple is the sweet potato ('kama?i', of which there are nineteen named varieties), which is always planted in a small mound. The great number of mounds all set in straight lines bearing bright green leaves on the top as the tubers sprout forth and vines grow lend an enchantment to such a garden. Also yams ('oba', of which there are nineteen named varieties) are planted in a mound with the tuber set at an angle so that it can grow down. In this case, however, the vines are not allowed to rest on the earth, but are tied to sticks.[7] There are sixteen named varieties of taro ('yammi') and also the slightly smaller variety called 'tape', which was introduced by the Lutheran evangelists from Finschhafen on the north coast and which is very popular in Akuna. In this same category is the 'irinapu', manioc or cassava, which is either grated or simply roasted in the open fire before eating.

Every garden will include sugar cane ('ya?i', of which the Akunans distinguish twenty-five named varieties), which is tied up against saplings planted next to the stalk. As the sugar cane increases in length the supporting sapling is replaced by a longer one, and in time the sugar cane stalk may grow as tall as thirty feet. Sugar cane is a product which is primarily handled by males (in many cases women are allowed to handle only broken lengths of sugar cane, but never the long stalks) and which functions prominently in ceremonial contexts. A stalk of this length will assure its owner much pride as he first displays it and then breaks off pieces to be distributed among those who are present. Frequently the garden has 'pamparimi', the pandanus palm which yields the edible 'imi' nut, and the 'yu?i' leaf used for mats and rain gear; and 'yomi', the fruit-bearing pandanus of which the red juice is extracted and added to grated taro. Both of these palms as well as the bamboo to be discussed below are very frequently planted in the open bush, in gardens, or continuously visited where gardens have been allowed to go wild, for they are always privately owned. An Amaranth species is planted in the gardens to provide a kind of native spinach, and other greens are also eaten. Native ginger ('wimi'), though not tended, is planted in all gardens, and people enjoy the sting when they eat it and especially the effect it has when they feel a cold starting. Also 'kopunayun', a fleshy quadrangular bean is planted and even though it is being replaced now by other varieties of beans and peas, it is still very popular and traditionally formed part of gifts and an important item in the Gadsup trading inventory. Banana palms ('e?i') are either saved if growing in old gardens or new palms are planted in newly cleared gardens. New varieties have been introduced by the agricultural officer at Aiyura, but the Gadsup boast seventeen named varieties. Many gardens also have clumps of

7. Among the Eastern Gadsup, it is said, yams are of much greater importance in the diet than among the Western Gadsup here discussed; and they have more varieties which have been borrowed from the Markham peoples.

bamboo ('onai': there are five named varieties) in or near them as these are used in house construction, for water containers, and similar purposes. In the absence of cooking containers the Akunans simply place the grated tubers in a length of green bamboo and place this in the fire, having closed off the opening with a stopper. This steam cooking produces a well-cooked mushy product even though the burned bamboo container must often-times be removed piece by piece before they can get at the food.

Introduced vegetables include beans and peas, tomatoes, Irish potatoes, and groundnuts ('kata'), the latter being quite popular as a substitute for other nuts. The seeds, of the groundnut are first dried in the sun and stored in bamboo containers. When well dried they are shelled and soaked in water in order that the seed may swell and start the germinating process. A few hours later they will be planted at distances of about eighteen inches apart. This same soaking treatment is given to corn, beans, and peas as it causes quicker growth. Corn ('dankuna') is also very popular and is roasted in the fire being eaten before the main meal, and frequently cold roasted corn is carried in the string bag as a snack. Perhaps the most popular of the introduced varieties of edible plants is the cucumber, for in a land where the sun is so hot people enjoy the juicy flesh while working in the gardens or while being engaged in some physically taxing work.

The layout and organization of these gardens immediately strikes the observer. The split pole fence which surrounds each garden sets it off from others and creates the impression of a large number of artistic patchworks, each bearing the individuality of its owner. Favorite plants and products may dominate the scene as is the case with the orchard of Napiwa. Frequently the gardens too, as is the case in most village settings, are bordered by decorative plants. These colorful edgings of Cordyline Terminalis and other ornamental shrubs and herbs make the village and gardens of Akuna not only neat but also attractive.[8]

Referring to the Gadsup of Aiyura village, the head of the Agricultural Experiental Station has stated that most families try to farm a plot of garden in the valley and one on the fringe of the forest (Schindler 1952:304). He calculates that the following acreages are available per village.

Forest	280 acres
Cultivable flats and slopes	1300 acres
Uncultivable slopes	800 acres
Swamps	300 acres

8. Flierl (op. cit. 1932:50) remarks on his observations of the popularity of sugar cane and a native kind of bean. He states that "these fields make a good impression because they are so surely divided and kept so clean".

Based on the average village populations given earlier in this discussion, it is clear that ample land is available both for gardening and pig grazing. We should mention here that large areas of forest are now being cleared and coffee plantations layed on. Schindler concentrated his observations on two families — he admits that this is not an accurate sample — and, based on their activities, figured the average size of land cultivated while keeping in mind the fact that about half the sweet potatoes gathered go to feed the family's pigs. He then experimented on the average-size garden in similar soils and obtained the following yields:

Table 10. Crop yields at Aiyura village

Crop	Yield per acre per year	Acreage per person	Yield per person
Sweet potato	8 tons	0.36	6200 lb.
Maize	40 bush.	0.05	112 lb.
Peanuts	1200 lb.	0.05	60 lb.
Sugar cane	—	0.10	unknown
Beans	—	0.05	unknown
Miscellaneous	—	0.05	unknown

If we take this to be an average which need not, and most likely does not, compare with any actual case, it is then possible to suggest that the food available for consumption compares well with other areas in the highlands. This means then that the average woman produces a minimum of 6372 pounds of food. The fact that people in the Arona Valley produce much less peanuts and very large quantities of taro and yams, need not enter here. The point is that if we accept these figures as rough estimates and keep in mind our demographical figures for Akuna which give an average of about 4.4 persons per family we find that for each person there is an amount of food which compares well with that calculated by Salisbury for the Siane (1962:80).

	Aiyura	Siane
Sweet potato	3.91 lb.	4.20 lb.
Maize	.69 lb.	.25 lb.
Peanuts	.36 lb.	—

To this we must add yams and taro which are grown in the Arona Valley as well as vegetable greens, cucumbers which people eat during the heat of the day, sugar cane, the protein-rich nuts, and occasionally me

9. A nutritional survey which was made on the coast found the diet to be sufficient in bulk, but deficient in calories, in protein, and in calcium (vide Hogbin 1951:301).

This, I repeat, does not propose to substitute for a detailed study of the subsistence in Akuna. It is, however, a guide to the amounts available since Aiyura is just across the mountain from Akuna. Furthermore, since I was not in a position to weigh amounts of food produced during my stay in Akuna we can use these averages as a rough guide to the subsistence level in Akuna.[10]

Distribution and consumption

Since every household produces more than its members can consume, even in times of relative scarcity, there is a constant process of redistribution which takes place. Part of this is in the form of prestations to persons who assisted in putting up the fence, in breaking the ground, or in constructing a house. Other occasions are when a man's wife is in confinement, or when some ceremony calls for interaction between members of a part of the whole village community. All of these occasions call for distribution of products among different households but it should also be kept in mind that hospitality requires an open hand and a warm heart at all times.

In his study of the economy of the Mt. Hagen tribes in the Western Highlands Gitlow (1947) distinguishes between three types of feasts. He speaks of religious, social, and economic feasts. For Akuna, and even the Gadsup in general, it would be very difficult to draw the line between these various types and to suggest the dominance of a certain orientation in the aim or outcome of a feast. To be sure, there are certain overriding characteristics in a death payment, or a wedding festivity, but the distinction is not always as clear.

In the cases to be discussed below, I have attempted to enumerate the contributors and their contributions as well as the receivers where at all possible.[11]

10. One of the best studies of carrying capacity as compared with nutritive value and energy expenditure has been done by Roy Rappaport (1967). Using Carneiro's formula for carrying capacity he applies it to pigs and humans (op. cit. 285-298). He also measures the nutritive value of the various plants consumed (278-284), and this can be compared with the energy expenditure in gardening (256-262). This refers then to the major source of nutrition as well as the major consumer of energy.

11. I would like to acknowledge valuable suggestions made by Dr. Cyril Belshaw during a discussion in Vancouver, B.C. in April, 1961. He emphasized the redistributing value of ceremonies and suggested that the goods and the time be counted. My only regret is that I have not done it nearly as completely as he suggested. See also Belshaw (1959:560).

173

Death payment — The full description of this case was given in our previous chapter when we discussed death and mourning. It will be sufficient to give a brief summary here.

A woman whose parents had originally come from Akuna died in a neighboring village. A large number of the Akunans, primarily but not exclusively males, attended the wailing and the feast which accompanied the redistribution of the death payment. This payment had been contributed to by members of the Ikana community consisting primarily of the interest group members of the widower and was shared by a large section of the Akuna community.

In the accompanying table I have attempted as far as possible to designate the relationship of contributors and receivers to the deceased.

In the original distribution of this payment shortly after the corpse had been removed from the village, the bulk of this payment as it appears in the left-hand column was presented to Amata and Detio. These two young fellows live in Amamunta and stood in the favored avuncular-nepotic relationship with the father of the deceased prior to his own death. During their initiation at the time of the 'upaiyami' he was the mother's brother who assisted them. Now when payment was made the Ikana people turned over the total prestation to them which caused great concern among the Akunans. Before leaving for Akuna that night they lodged their protests. On October 3, seventeen days after the first prestation the tultul of Ikana returned from Kainantu where he had gone to discuss the payment with the 'kiap', and requested the ethnographer to attend the redistribution of the payment. The right hand column represents this second distribution.

It is of importance to recognize that many of the receivers are in fact patrilineal clan members of the deceased woman, but the rest are not related in that way, belonging either to her father's 'ankumi' or even to his interest group. The deceased father, Yapananda, had belonged to a long line of Akunans but his sister, Atuno, the mother

Table 11. Daunamari death payment

Contributor (Relationship to deceased)	Item	Receiver (Relationship to deceased)
Tultul Ikana (SisHu)	35 arrows	Unanata (3 arrows)
O?upe	26 arrows	Tultul Akuna (1 arrow)
Mano?a	30 arrows	Uriba (3 arrows)
Yai?o	30 arrows	Iuwa (3 arrows)
Podiri	19 arrows	Amo?a (4 arrows)
Torana?o	25 arrows	Apuneo (3 arrows) (FaFaBroDaDaHu)

Contributor (Relationship to deceased)	Item'	Receiver (Relationship to deceased)
I?ampa	10 arrows	Amata (8 arrows) (FaSisSo)
Atu	10 arrows	Detio (6 arrows) (FaSisSo)
Yai?o	1 laplap	Amata (FaSisSo)
O?upe	1 laplap	Detio (FaSisSo)
Tuma?o	1 laplap	Detio (FaSisSo)
E?ino	1 laplap	Tultul Akuna
Duna?o	1 laplap	Nona?a
I?a?e	1 laplap	Dupana
Podidi	1 laplap	Itana (FaSisSo)
Paita?o?a	1 laplap	Itana (FaSisSo)
Kainu	1 laplap	
Purori (HuFa)	1 laplap	
Pai?aro (Hu)	1 sweater	Iripe
Ma?a	tobacco	
Ma?a	3 newspapers	
Apapa	5 newspapers	
Tuna	2 newspapers	
Puro	5 newspapers	
O?upe	1 soap	Inuntino
I?a?e	1 soap	Itana (FaSisSo)
I?a?e	1 body oil	Itana
I?ampa	1 spoon	
I?a?e	1 spoon	
Uromu	1 spoon	
O?upe	1 mirror	Uriwa
I?a?e	1 mirror	
I?a?e	2 boxes matches	
Duna?o (female)	1 box matches	
Tultul Ikana (SisHu)	1 pig	Eaten by all in a feast
All (everybody contributed from 3-5 a piece)	67 andimi (barkrope)	4 Nona?a (female) 4 Nuruno (female) 4 Yoy?a 4 Apaserano (female)
All (everybody gave one or more)	Yienni (28) (for grass skirts)	4 To?omo (female) 1 Inuntino (female) 6 Ropayowa 2 Iripe
	3 arrows	Inuntino (female)
	3 arrows	Dupana
	3 arrows	Ropayowa
	3 arrows	Duka
	5 arrows	Iripe
	3 arrows	Wana
	4 arrows	Ana?o
	1 native blanket (emkomi)	Ana?o

of Detio, had married in Amamunta and resided virilocally. While certain members of Yapananda's patriliny were included in the payment, it seems that they were there due to their loyalty relationship with him and with his daughter who had just deceased rather than due to their kinship. They would then be included in his 'ankumi' and his interest group, and that of the recently deceased daughter. Yapananda's father's brother's son's son, the Akunan Napiwa was not included among the receivers, while Yerai and Iripe, who belong to different kinship groups both were included. This distribution shows quite convincingly that: (a) males and females may both be contributors and receivers, and (b) in both cases the 'ankumi' and the interest group members are primarily involved.

Betrothal — During the past two decades an increasing number of young men have been leaving Akuna as indentured laborers. They go to the coast or even as far as Rabaul to work in copra raising. Since these men must be single, or in exceptional cases, very recently married, it also means that they are not available when many of the young women reach marriageable age. If a person marries and leaves his bride chances are she might be attracted to others,[12] but if he returns with money and know-how he might find all the attractive girls married off.

While we were in Akuna, Urinanda decided to 'mark' a very attractive girl for his son Yantape who was on the coast for two years. She was Tududu, the daughter of Opura (leader of Wayopa) and Monana. Table 12 details the price which he paid to the father of his son's fiancé. Needless to say members of Urinanda's interest group (relatives and friends) contributed to this payment.

As in the previous case there is a mixture here of trade store items and traditional objects. There is then a clear shift in values, as will also become clear from later examples.

Marriage — At the marriage of two persons we again find the transfer of valuables accompanied by large quantities of garden produce for a feast. As in these other cases we are discussing, the man's interest group contributes while the bride's interest group partakes of the prestations.

While ethnographic descriptions of marriages and ceremonies were collected none occured during our year's stay in Akuna. It is thus not possible to list comparable data under this heading.

---

12. In our discussion of court cases we discussed in some detail the case of Ma?e who was involved in adultery with a number of males while Dupopima was away on the coast.

Table 12. Urinanda's betrothal payment

$30 (or 11 Australian pounds) in cash which was paid to Monana, mother of the betrothed
1 pig which had been killed and was eaten by the people of Wayopa in the feast
15 new laplap
5 sugar cane, each about ten feet in length
2 large bunches of bananas
2 pieces of soap
2 boxes of matches
1 spoon from the trade store
1 file (to sharpen the axe)
2 enamel dishes
14 bunches of yienni (for grass skirts)
16 andimi (traditional barkrope)
4 bunches of adnumi (betelnut)
3 rolls of dried tobacco
4 sheets of newspaper

Court settlement — The discussion of this case was given above in the fourth chapter. It will be sufficient at this juncture to summarize briefly the fact that it dealt with a pig that had been killed. Many years ago a Patrol Officer suggested that most legal cases in New Guinea centered on either women or pigs — this is still basically true for Akuna.

On this occasion a pig had accidentally been shot by a young man named Utima who lives in Wayopa. It had been accidental in so far as it was a case of mis-identification since the pig really belonged to Napiwa, an important man in Akuna. A court case was held. By the time the first statements were being made in court other members of the defence were already killing a pig to offer in payment. This clearly was an admission of guilt, even though they argued for some time about the case.

Table 13 lists the items which were collected by Utima and paid to Napiwa. I have again shown the known relationships between persons. Included are also persons who were admittedly not related but simply lived in the same village.

As in earlier examples we find here a mixture of traditional objects and products of value which are matched by trade store items. As in most other cases we find bows and arrows; 'yienni' which has been scraped and dried and can be used for grass skirts; pandanus leaf blankets; and a variety of the usual foodstuffs as well as betelnut and tobacco. Added to these are the utility objects acquired from the trade store, the laplap which is wrapped around the waist, the enamel basin and spoon which is now important when you entertain a visitor to the

177

Table 13. Payment in Napiwa's court claim (The case of the mistaken pig)

Contributor	Relationship to Utima	Item
Opura	WiMo#2Hu	Pig (1)
Opura	WiMo#2Hu	Small pig (replace 1 killed)
Puna?o	none	Fowl (1)
Opura	WiMo#2Hu	Bows and arrows
Toma	none	(3 bows and 54 arrows)
Marepo	none	
Ipawa	none	
Datipa	MoFaSisDaSoWiBro	
Makada	MoFaSisSoWiBro	
Miwa	WiMo#2HuSo	
Utima	ego	
Opura	WiMo#2Hu	Bananas (3 bunches, 150)
Padauna	WiMoSi	Corn (30 ears)
Makada	MoFaSisSoWiBro	Sugar cane (26, 10ft. long)
Munka		Taro (15 large yammi, 50
Opura	(above)	smaller tape)
Puna?o		Betel (2 bunches)
Datipa		Fruit-bearing pandanus (5)
Opura	(above)	
Padauna		
Datipa		Tobacco (6 bundles)
Padauna		
Makada		
Didino		
Utima	ego	
E?ananda	WiMo#2HuFaBroSo	
Ipawa		Yienni (14 bundles and 1 skirt)
Didino		
Padauna		
Daitno	WiMo	
Data	Wi	
Utima	ego	Laplaps (8 new ones)
Makada		
Opura	(above)	
Padauna		
Ipawa		Money — cash ($5.88 or 42
Datipa		Australian shillings)
Opura	(above)	
E?ananda		
Amaro		
Puna?o		

Contributor	Relationship to Utima	Item
Tiwa		Enamel bowls (4 medium size)
Data	Wi	
Didino		
Amaro		
Na?o		
Awandi		Spoons (6 new ones)
Na?o		
Dume		
Dunduya	WiMo#2HuSoWi	
Nompamu		
Miwa		Carrying bag (1)
Data	Wi	Cowry shell (1)
Panauna		Soap (2 pieces)
Berora	WiMo#2HuSisSo	
Nai?a		Matches (6 boxes)
Makada	MoFaSisSoWiBro	
Opura	(above)	
Abida		
Tiwa		Knife (1)
Puna?o		Squash (1)
Awandi		Dried beans ($7\frac{1}{2}$ bamboo containers each 3 feet long)
Dunduya		
Nondoti		
Uararena		
Imuandudu		
Daitno	WiMo	
Daitno	WiMo	Peanuts (2 bunches)
Data	Wi	Pandanus leaf blankets (2)
Yodino		
E?ananda		Paper — smoke (6 sheets)
Miwa		
Opura	(above)	
Napa?o		
Utima	ego	

village and other less permanent products. A number of items here merit special mention. In a reference above we quoted Flierl's remark that a certain native bean was in great abundance in the Akuna gardens. These are the dried beans, 'kopunayun', which are included here and which formed such an important part in traditional trading relations between Akuna and neighboring villages. The inclusion of a cowry shell in this payment is an exception rather than a representative act. On

179

no other occasion was a cowry shell, or the gold lip shell, included in
any kind of prestation nor was either of these shells mentioned as
prized objects by Akunans. This then contradicts Read's suggestion
(op. cit. 1954: 9) that gold lip shells and the large white cowries are
valued throughout the Eastern Highlands. It also confirms, on the
other hand, observations by earlier patrol officers who remarked that
these varieties of shells were not valued (Stevenson No. K 7 of 1945-46
and that they are not in demand (Blyton No. 32 of 1944-45) in the Arona
Valley.

The most important observation to be made at this stage is that
those who contributed to this payment were all members of Utima's
interest group and also his 'ankumi'. They are people who lived together and developed loyalties due to this association rather than due
to any kinship links which bound them. They include in fact very few
clan members but mostly affines, matrilateral relatives belonging
to his 'ankumi', and the rest are persons who reside with him and
associate with him in daily activities. It also leaves the choice to the
individual concerned whether he wants to contribute or prefers not to.
Certain members of the Wayopa community did not contribute either,
to this payment, while certain members of his clan did not contribute,
due to conflicting loyalties. At various places throughout this study I
have remarked on the fact that residence is of greater importance than
kinship. It is therefore significant that Utima's own sister Inuntino
and her husband Ko?ana, who live in Akuna, did not contribute to this
payment. In other words. they were not assisting a kinsman against a
fellow villager.

The redistribution of this payment is illuminating. The food was
shared during a community feast, since there is no way in which foods
can be preserved in these climatic conditions. Before Napiwa started
the selection of his goods, he paid, from his own money, $ 3. 64 (or
26 Australian shillings) to Anokara who had cared for the pig.[13] Then
he selected from the payment two medium sized enamel bowls, one
spoon, and the one carrying bag. He also turned over to Anokara's
loving care the small pig which was to take the place of the one that
was killed.

While Napiwa's items of payment were being sorted and tied
together by his wife, he repeatedly made short speeches about the
value of pigs; the fact that people do not simply go around shooting the
pigs of others; the payment that was being distributed here was a just

13. It will be recalled from an earlier discussion that pigs are farmed out
to be fed and cared for by others. Anokara was especially esteemed as pig far-
mer since he knew pigs so well and frequently spent days in the valley at an old
house at his gardens. He also knew how to protect pigs against sorcery and
assure their health.

payment. While the last of these items were being put together and tied up for transporting to Akuna and distributing, Napiwa stepped forward. In a short speech he thanked everybody especially Opura for their contributions. He gave Opura one of the new laplap, a used steel axe and 30 Australian shillings (about $4.00).

During the feast that followed it was noticed that most of the Akunans were eating pork, including Napiwa. It was explained that had this been the original pig which was killed, the symbolic kinship bond would have prevented his eating. Since it was a pig which had been owned, raised, and fed in Wayopa all could eat of its flesh. One important remark, which is discussed elsewhere, needs to be made here. While sorting his loot when we had returned to Akuna, Napiwa was very upset when he discovered in one of the bundles, a 'kanna'. This is a very rough tree limb, totally covered with small thorns. The women use it to grate their gams, taro, and sweet potato; but in this context it was symbolic. It was explained that some person in Wayopa (in this case we learned later that it was Miwa) was not very satisfied with the outcome of this case and was sending this message to Napiwa. "As men shun this tree in the forest (due to its thorns) making a footpath around it, may they treat you in the same way." Napiwa did not allow his wife to use this as a grater, but cast it away.

Other occasions — These occasions for the transfer of valuables which have been mentioned are by no means the most important occasions or feasts during which redistribution takes place. We should mention the traditional 'orande', the modern 'sing-sing' which tends to substitute for it, and the novel 'wash-wash' which was brought in by the Lutheran mission — all of which we attended as well as the traditional war celebrations after an attack when allies are repaid for assistance and loyalty. These occasions are important from an economic point of view for they allow the redistribution of wealth and surplus. They also serve basically social ends of forming or confirming social ties.

This same aspect is tied very closely to the value Akunans place on hospitality. What a person has is to be shared and a request is never turned down. This does not mean that a person is likely to request food, for a person would be embarrassed by openly asking for food. When an informant was asked why he could not request food, he answered in fact 'tiyore?u', meaning 'I shame' or 'I am embarrassed to do it.' Since the one in need will not openly request food or assistance, the responsibility is placed on every person to be mindful of the needs of others. When a person enters a house, he will immediately be offered something to eat.[14] He will do no more than hint at certain

14. I never saw a person turn down proferred food and frequently a fellow would leave a house he had been visiting with a cold sweet potato in his hand.

needs when the host will consider this assistance. As in the case of the origin myth, the people of this community find reason and justification for this practice in their culture hero.

This story is about a man whose name was Mani?i. He lived long ago at a place called Maropa?i and always he would watch his betelnut palms and fruit trees while his wife gathered food and cooked for him.

One day his wife was visited by the moon and retired to the women's house. There was no food in their house and he was very hungry, so he went to his brothers. After entering the house he sat down by the fire and asked his brother whether he had any 'inami', for his own was used up. "Yes", replied his brother, taking from the roof a roll of 'inami' which he gave to Mani?i. Mani?i took this, returned to his own house, and sat down by the fire.

Soon, however, he felt the hunger pangs once again and left his house to visit another of his brothers. He entered the house and sat down by the fire. After a while he asked his brother whether he had any 'inami', for his 'itanda' was broken. His brother gave him some bow strings and after receiving them, he remained for a while and then returned home.

Again he left the house going to yet another brother. He repeated his request, this time asking for 'adnumi' and after receiving it, remained for a while and then went home. He was hungry, but there was no food.

Another time he visited a brother, asking for some 'puka'. Again it was given him but no mention was made of food and he returned to his house placing the tobacco in the wall.

Soon afterwards, the call of the 'yikoyumi' went out for all to attack the enemy village Abinapa?i, an Agarabe village. Every man in the village left for the attack but while walking through the bush, Mani?i grew faint and died in his tracks. His brothers were very sorry that he had died. They picked him up and carried him back to the village while every one cried. When they got to his house they took him inside and placed him on the sleeping platform where he would lie until the time of burial.

After putting him down they looked around and noticed that there was a very large heap of 'inami' alongside the fire and also plenty of betelnut and tobacco, but no food in the house. Then one of them said to the deceased man's wife, "What have we done? Our brother is dead because he was hungry, yet when he visited us we did not offer him any food."

The tale clearly enjoins the wife to make provision for her husband and family while she is in seclusion, both during menstruation and for the post-partum period. The tale is also related to give charter to bond between brothers. To the biblical question: "Am I my brother's keeper?" the answer here is definitely in the affirmative and 'brother' is broadly defined to include all members of the village community. This also extends to the right of use which we mentioned earlier.

The basic process of food consumption — and ceremonial or festive occasions will be discussed in context — can best be discussed in terms of the regular meal and the more elaborate meal around the earth ovens. The usual family meal takes place in the house around the family hearth. It usually consists of roasted sweet potato and corn which has been steamed over the coals by not removing the husks. Frequently yams or taro, especially the smaller 'tape', will be added. Very often some of these tubers will be grated into a fine mush which is then placed in a green bamboo container. A grass stopper prevents it from leaking out. This bamboo length is then placed in the coals and repeatedly turned. The effect is a steamed pudding which has become quite solid. It invariably happens that people get to talking and forget all about the container and it is burned through on one side. Needless to say these containers are only used once unless they are very care- fully handled. Since money has become more readily available due to lumber work and coffee sales, some households boast a pot from the trade store which is used to cook sweet potatoes with a little water. On the whole, however, the readiness of the glowing embers and the ease of roasting a tuber are important deciding factors. When this forms the evening meal, roasted tubers will remain in the coals and be enjoyed for the morning meal. While Akunans eat only two meals a day, they invariably have snacks during the hottest part of the day. A man usually has a handful of 'Pandanus' nuts, a cold roasted sweet potato or even an ear of corn in his carrying bag. When the sun is hot they are usually in the forest and will rest while they eat, but this is not a meal. Passing through a garden a person will pick a cold cucum- ber and eat its juicy contents with relish. Children also enjoy cucum- bers while they spend long hours collecting wild fruit and nuts, or roam the 'kunai' in the freedom and aimlessness which marks child- ren's play.

During our stay in Akuna, members of various households combined frequently for an 'onakareno', the more time-consuming earth oven meal. At least once a week some family would join neighboring house- holds in the arduous preparation of foods for the earth oven, but it should be admitted that this meal is much more varied in its offerings and more savory in its taste.

When an earth oven meal is to be prepared the women return from their gardens slightly earlier than usual, perhaps two-and-a-half to three hours before sunset. Most of them squat outside while peeling tubers and usually grating these on a 'kanna' stick. Every woman who is grating has by her side a number of small rolls of yellow material. These are the very young inside leaves of the banana palm which have not yet opened to the sun and become fibrous. The woman places a mature, dark green banana leaf on the earth and rolls onto this the soft yellow inside leaf. On this last mentioned receptacle the mush of gam, taro, or sweet potato is made into a smear of perhaps half-an-inch thick. Since salt is a delicacy which is expensive, the ash of burned banana or pandanus leaves is placed in the bract of a banana palm which has a number of small holes pierced in it. Water, which is transported and stored in bamboo lengths of up to six feet of which the internodes have been knocked out, is now added. The result is drop of dark brown salty water which is worked into the smear. One day an interesting innovation was observed. One of the males had been to the trade store in Kainantu and while passing the garbage dump of one of the European administrators, he saw an old rusted enamel wash basin. That evening his wife used it as a substitute[15] for the bract which gets soggy and has to be replaced after an evening's use.

Frequently manioc is added to a mush of this kind, but the most frequent additions are the red juice which is squeezed from the fruit of the fruit-bearing pandanus, greens or spinach, and a variety of meats. This may be in the form of rats or insects, and occasionally pork.[16] One occasionally nowadays sees a man sitting by an earth oven while opening a small tin of canned fish which he will eat with his staple while the rest of the group enjoy their pork.

The mush smear which has been prepared is now carefully enfolded in the banana leaves and tied into loaves by tufts of grass. In the meantime one of the other women has removed the rocks from the earth oven, kindled a fire, and replaced the rocks which by this time are well heated. They are handled by a "pair of pliers" made by bending a split of bamboo. This is the 'data' which was mentioned when the Akuna referent for the Markham people was discussed in the early part of this study. The rocks are now placed in such a way that the leaf-covered mush loaves can be placed in the hole on warm rocks. These foods are interspersed with hot rocks and are added to by a variety of peeled tubers, corn, the immature tops of pit-pit, greens, squash, and similar vegetables. In exceptional cases the wild banana is used in this form but usually it is baked separately in an open fire. The new varieties of banana which have been introduced are eaten as a fruit.

15. In the sense in which Barnett conceives of these (1953:181 et passim).
16. It should be emphasized as has already been pointed out that there are strict sanctions which regulate the eating of pork.

Should there have been a pig killing for the occasion, the bristles are scorched off in an open fire and the intestines removed. The carcass is now placed atop this pile of food, much the way a floor mat of goatshide will be spread over a couch. With great speed, while the rocks are still hot, the pile is covered over with leaves and grass and occasionally a piece of old sacking, and then covered by sand. The women then relax for approximately ninety minutes while the food is baking and then the pile is opened and the male's share in taken to him where he is discussing with the other men.

Water is the only liquid regularly drunk by Akunans. When it is carried back to the village in a length of bamboo, the container is called 'nonana', being constructed from the contracted form of 'nomi' (water) followed by 'nana' (bamboo). In most cases the water is collected from fountains but never from the center of streams or rivers, and where such is the case a slight hollow is made below the trickle of water, and a V-shaped leaf such as pandanus inserted. This acts as a duct allowing people to use clear water. In many cases the water is enjoyed right there and in these cases a leaf-cup, usually from the breadfruit tree, is used. Half way between Akuna and Aiyura, where the mountain streams run clear and cool, there is a small stream called 'ite dunapa?', but better known in Neo-Melanesian as 'Wara karoka' — water pandanus. An informant explained that the latter name was known from Chimbu in the west to the Markam Valley in the east and that weary travellers always looked forward to reaching this sweet water.

In addition to those varieties of plants which produce sustenance there are others which for these people are of nearly equal importance. Usually planted in special gardens near the village are a number of tobacco plants. While there are two named varieties, they go by the generic term of 'puka?i', which is derived from the verb 'to smoke'. These plants are not specially treated but are tended better than other garden plants and kept free of weeds. The leaves when picked are dried in the house and then in a semi-dried form smoked as cigarettes or cigars, and lately in pipes. When used in the former two cases, the tobacco is wrapped in a wild leaf, corn husks, or a young banana leaf, but all of these wraps are now being replaced by newspaper. Growing either in the gardens or protected in the bush are a few varieties of betelnut ('areca') notably 'kopemi' and 'yapapimi' which are native to the Arona Valley, but once again they go by the generic term of 'adnumi' which really refers to the large variety introduced from the coast. This is of the greatest importance to the Akunans and, as is the case with tobacco, is indulged in by males and females alike. The nut is of course chewed with 'kote' (lime) and 'opo?i' (pepper leaf) and produces a red juice which causes a stinging sensation in the inside of the mouth. While children of about puberty start smoking, young children are often encouraged to chew betel. Once while we were deep in the forest attending a full day's activity to prepare and burn a supply of lime, called 'nora' in its raw form, it was observed how an old grandfather was showing his granddaughter of about seven or eight years old how to prepare the lime and pepper to chew with her betel in such a fashion that it produced the greatest satisfaction. On a number of occasions old people were observed who had lost their teeth but not their taste for 'adnumi'. They were not able to bite the fibrous covering which surrounds the nut, nor able to chew the nut while adding lime and pepper leaf. They had taken a thick pig femur and broke it into two pieces leaving a tube about an inch or two above the joint. This was neatly cleaned and scraped and was now used as a container. They would place their betelnut in the hollow and grind it to a paste while adding lime and pepper leaf. The pestle is bone or any length of hardwood and with it the paste is placed in the mouth when it is ready to enjoy.

Value should, of course, not be mentioned in terms of products alone for even in these communities time is of importance. While it is quite easy to calculate man hours of labor, or the time which has been spent productively in our own culture this is not equally true for a Melanesian community where the very fact of inspecting his areca nut trees, or sitting smoking in the village square while a court is in session are productive and of great importance. On three different occasions four of the people in Akuna were followed for a whole day, studying their activities, timing their involvement with a particular project in an attempt to get some idea of the way a day is spent by male and female members of the Akuna community. The tabulations of these twelve days are given in Addendum B at the end of this study.

PRODUCTIVE ACTIVITIES

Every community has specialists, even if these are only distinguished by age and sex, who are directly involved with the production of durable items. Whether we are speaking about clothing or housing, each of these requires certain secondary products, certain tools, and certain persons who have the know-how and the time to spend in their creation, construction, or repair. In communities such as Akuna, perhaps more than in some of the more complex settlements, these basic products as well as the tools are found in their natural form. I am referring here to the use of things like shredded tree bark, rather than a sisal product in the form of yarn or string. Also the products produced by these activities would necessarily be classified in this discussion of productive activities.

186

One of the most basic materials is a variety of barks and cordages since these are used to attach other products or to produce secondary materials. There are two kins of cordage which are of great importance and which are planted and tended, though this might not be in the garden proper. The 'no?i' tree branches are cut while they are young and pliable, and from the center a woman segregates the bark which, with treatment, produces a fluffy white fiber. The women sit in the sun twining this into rope or string, and in time such a ball of string will be used to make a string bag. In addition to this, and closely related, is the tree bark called 'amunka', but it is usually used to sew together the leaves ('yu?i') of the 'Pandanus' tree for use as a sleeping mat or rain gear. In addition to the bark, there is the modern type of fiber which was introduced to the Arona Valley by the Lutheran evangelists from the north coast. This 'repa?i', as it is called, is a sisal-like plant of which the leaves are cut into lengths, and then scraped to remove the green watery fluid from the white fiber. The 'repa?i' fiber is then allowed to dry and between hand and palm rolled into cordage as mentioned for the 'no?i'. A variety of carrying bags are made, and often the cordage is colored to produce a pattern when the bag is completed. While traditional prepared coloring agents were used to dye the fiber the Akunans now acquire this from the store. The agents they used in the earlier years were the reddish bark of the 'o?i' tree, and the leaves of a small plant, called 'kuyandu', which grows in the valley among the grasses. In both cases the bark or leaf is beaten to a pulp and moistened after which ash is added to the paste. It produces a blue dye while the 'o?i' bark allows the user to color her fiber red. Since the original fiber was white, they make three colored carrying bags. The technique employed in producing these bags is a series of loops which are interconnected producing the cycloid technique.

This is women's work and a woman is expected to make these and present them to relatives and also friends. While each woman is primarily responsible to her husband, her brothers and other male kinsmen she will in some cases present an 'unammi' (carrying bag)[17] to a friend if she notices that his is old and torn. An old carrying bag is never abandoned but always carefully burned. It will be recalled that men wear their bags around the neck rather than around the head, like women, and the grease, grime, and sweat which is concentrated around the neck produces in a very short while a strand of fiber deeply

17. This is the generic term, but Akunans distinguish terminologically between 'kupanammi' which is used for personal paraphernalia such as a man's betelnut, lime, tobacco and the like, and the 'akoni unammi' which is a large bag which may be used to convey anything from garden produce to a baby sleeping on a folded pandanus leaf mattress.

embedded with personal filth. This would be the prime article sought by a person who planned sorcery against the owner. The sister will also use the presentation of a new carrying bag as a marker of reunion when there has been a rift in a family or hostilities between her and her brother. By presenting him with a new 'unammi' she is clearing the air, and reaffirming the basic sibling bond.

It was stated above that making the carrying bag was woman's work, but this does not prevent the man from catching a stitch if he notices that it is becoming undone. Many of the older carrying bags are dotted with knots. In making of the 'unammi', as is the case in producing the large sleeping mats and rain gear produced from treated pandunas leaves, the woman employs the 'u?uma', a needle of pig femur. This needle is about one-quarter of an inch thick and four to five inches long and made by grinding and polishing a piece of bone split off from the femur of a pig, since their yarn and bark is so thick it would be impossible to employ steel needles from the tradestore for these activities. On one occasion I was approached by my interpreter, Torawa, who requested that next time I open a can of fish he be given the key. I immediately presented him with one on condition that I be allowed to see what use he had planned for it. He walked about ten yards and squatted down to the first rocks he saw; the rounded grip of the key was removed and he proceeded to sharpen the end from which this rounded grip had been removed. In about five minutes he had produced a 'needle' which served as well as any 'u?uma' and had the added advantage of not cracking or breaking when the hole was drilled in the back since it was there already.

Sewing was of course needed in the various items of clothing, but especially in the traditional bark cloth, made from 'kamuna' bark. This 'enkomi' is slightly thicker than the paper-mulberry product used in Fiji, and is not decorated the way 'tapa' is. It was produced from strips of bark, and while narrow strips were used for the grass skirt, a number of wider pieces are beaten together with the 'yan-kosi' to make a large piece of cloth material. The latter is a small baton made from the wood of the 'wandimmi' tree.[18] Sizes vary, but an average of the 'yankosi' would be about twelve inches long and five inches in diameter at the widest end. The face of the baton is carved into a pattern of grooves of one quarter-inch deep which produces the rough surface to break and flatten the bark into a relatively thin fabric. In addition to being a blanket or body covering wound about the waist, the narrower strips of this bark fabric are often used to attach the

18. Informants explained that they used this type of wood because it was so hard, but Brass (1964: 168) identifies this particular tree, Araucaria Cunning-hamii, as a soft wood.

'yienni' to for grass skirts. This latter material is acquired from an onion-like grass, thick and fleshy, which is collected in the marshes or along streams in the valley. Frequently, when garden work is not pressing, one finds a large number of women, including girls from about their tenth year, sitting in the village and scraping with a bamboo knife the green pulp and moisture from the leaves. What is left behind, then, is a flattened damp grey strand perhaps one-quarter to one-half inch wide and up to twenty-four inches in length. After this 'yienni' has been allowed to dry, individual strands are tied to a length of 'amunka', 'no?i', or 'kamuna' bark, and hung around the waist as a grass skirt. All women, and even girls who are no more than three weeks old, wear these skirts and are never seen without them.

This, in effect, is the woman's only clothing. To be sure changes have taken place and many women in Akuna now have some kind of top or loose hanging dress acquired at the trade store, but these are used primarily for missionary activities like the Sunday service or the 'wash-wash' (the Neo-Melanesian term for the baptismal ceremonies). While the traditional clothing was scant, each woman had a variety of personal decorations. These ranged from the 'tilim.po?i', a nose stick extending from the front of the nose and usually being the wing bone of the flying fox or fruit bat. On ceremonial occasions they wear their necklaces of pig's teeth (known as 'iyunawa') or cowry-shells (known as 'puyataki'), though the latter were of secondary importance. They also used to make a very attractive necklace of beads which in fact are the seeds of a local plant. This is now replaced completely by multi-colored glass beads acquired from the trade store. Many of the women, like their men, had the septum of their nose pierced to receive the 'porawami', or semi-circular decoration produced by attaching two curved boar's tusks to each other at the base.

Males also used the 'porawami', and the dog's teeth necklace, but in addition the 'warunkara' were worn during ceremonies. These consisted of two small circular sets of dog's teeth which were attached to a man's temples. Men also allowed the 'porawami' to hang around their necks or during fighting they would frequently grip the circular form of boar's tusks between their teeth to make them look fierce. On festive occasions which followed a successful attack, the men would decorate themselves with a 'wuyena.ni' (the feathers of the 'wuyemi' or cassowary) headdress while they danced gripping their bow and arrows. On these ceremonial occasions they also pass through the nasal septum very attractive length of sandstone, ground to a pencil-form and decorated at the ends with colored bark (either yellow ('tawarimi') or black ('moni')). This is called 'para?wami' (nose stones) and must have formed the prototype for the New Guinean in the former Dutch part of the island who found a ballpoint pen less trouble to care for and brighter in color.[19]

In the discussion above, we remarked on the ways people employ the variety of barks and cordages to join things together. There are only two cases which come to mind where these are not tied together. Akunans make a wealth of varieties of arrows in which each of the points is separately carved out of hardwood, decorated, and then attached to the shaft. This occurs by true hafting in which the point, averaging from twelve to eighteen inches in length, is sunk into a shaft of about thirty-six inches. This lighter shaft is usually made of 'wiaka', a stem which resembles 'pit-pit' or bamboo but is smaller than the latter and stronger than the former. Before forcing the point into the shaft, it is covered with a black gluey substance which they get by cutting down the 'imanu' tree. A length of this wood is then covered with its own leaves and stored in the house. Whenever the glue is needed, the piece of log is placed in the fire which produces this black moisture and it is rubbed onto the surface of the arrow. In addition, however, the shaft is tightly lashed with a thin narrow strip of bark. This serves to strengthen the joints but also in a decorative sense.

The other case in which the cordage is done away with is in the making of the traditional hour-glass shaped handdrum. A log of the 'koyami' tree is cut and then shaped and hollowed out. Over the one end the skin of a 'tiwaru' or tree kangaroo is attached. The skin, which must be well dried in the sun, is soaked in water until it becomes pliable and stretches. It is stretched very tightly over the opening and with the help of glue from the 'akatan' tree, the skin, still moist, is fastened to the side of the drum. As a temporary device a piece of bark is wound around, but the drum is now placed in the sun to dry out and as soon as the skin and glue are dry this bark is removed. It is interesting to note that they are never satisfied with the sound of the drum at this stage. For this reason they attach 'eyes' of beeswax onto the face of the drum where they beat the skin covering. Four or five little 'eyes', each about one-quarter inch square, supply the resonance and sound. I asked them about this wax and in Neo-Melanesian my informant explained that it was "pek-pek belong fly" (lit. excreta of a fly) but this was not the regular fly, rather it was the "fly-talk-talk" — obviously a bee. They call this wax 'yari wayanara' ('yari' is the bee; 'waya' is a continuous talking, they

19. Brongersma and Venema (1962) have very clear photographs of a variety of nose plugs used by people they met on their way to the Sterren Gebergtes. One of these people (p. 157) had substituted the ballpoint pen for the traditional decorations.

explained, like when the 'luluai' is really fed up; 'N' seems to be an infix; and 'ara' is excreta' thus 'wantara' is excreta of a man, and 'porara' is the excreta of a pig). This little black stingless bee was identified as Trigona cincta by Mr. John Barret, entomologist at Aiyura Agricultural Research Station. I am also grateful to Dr. Charles D. Michener, Watkins Professor of Entomology at the University of Kansas and Dr. Hugh Cutler, Executive Director of the Missouri Botannical Gardens who in personal communications discussed this subject (1962). See also Michener (1961).

Of greater economic significance is the house structure. In traditional times these were round, but under pressure from the administration most of the houses today are square'. In many cases even composite structures are appearing which we could best denote as duplexes and even triplexes. The construction of the house in all cases is men's work, whether this was the men's house ('wadna.ma?i') or the smaller structures where a woman lived with her small children. The latter was called the 'ina.ma?i'.

Men's houses have fallen into disuse in Akuna since the men no longer have any military and ceremonial roles to perform. From recollections, since I did not record this in the field, the abandoning of the men's house may be related causally and temporally with the surprise attack by their enemy from Onamuna-Maropa?i area who burned the men's house and killed most of the men in the process. During this attack they tied the doors, both the entrance and the secret escape door which was one of the major structural features of the men's house.

In modern housebuilding the man goes out to decide on the site and prepares the posts which will form the major structure. Once these have been planted and cross-posts attached with either of two types of bark, 'amunka' or 'ano?mi' or even the liana 'nani' — which is also used to tie garden fence posts together — he goes to collect the bamboo. Bamboo clumps are privately owned, and if he or somebody who is indebted to him does not own such plants, he will have to buy them from somebody else, meaning that he is placed in the relation of indebtedness. The bamboos are cut to the same length as the wall, some for horizontal and some for vertical application. Each bamboo is now split down the center and each of the halves is flattened by beating the nodules and nodes so that the product is a flat strip perhaps four to six inches in width — depending on the size of the original bamboo. Informants explained that they preferred the hard, mature plants and these are readily identified by their size, texture, and yellowish color. Men use sticks, or the blunt end of a steel axe for this breaking and flattening process, but I never observed a 'yankosi' being used in this context.

Once enough lengths of bamboo have been prepared, a number of males will join together to plait the walls. The method which is used produces a square section of solid wall which is then attached to the frame. The simplest form, known as 'pani undakemi', is a simple form of twilled plaiting[20] by passing a horizontal piece of bamboo over three vertical pieces and under three vertical pieces and so forth. The next horizontal piece is started one space further and plaited in the same way. When a very tightly-knit product is required as, for instance, in the raised floor and sleeping platform, a two-two design will be used rather than the three-three just discussed. Most men know how to do this kind of plaiting and while it requires a number of people to cooperate no special knowledge is necessary. It is different, however, with the design twilled plaiting known as 'ko.ma?i'. Only half a dozen of the adult males are completely familiar with the requirements of this work, for while it requires counting and noticing the design, it also requires seniority, since the person directing the work must instruct the others when to raise and how to plait. The leader stands erect while a number of males squat while holding the bamboo; he now calls out for each in turn to raise or lower the strands he is responsible for. This kind of twilled plaiting produces a square design.

When the number of these wall sections that are needed have been completed, they are attached to the house frame by tying them to it. The roof frame is made of poles, and once again the structure is completed and tied with bark before the thin sticks or reeds are attached to the frame, but at right angles to it. In some cases the house builder may now lay down a layer of pit-pit to which the thatch is attached, while the traditional method was simply to fasten the thatch to this frame. The wife and daughters, frequently assisted by female relatives, will go into the valley to cut 'kunai' grass ('Imperata cylindrica') which they tie in bundles and deliver to the place of construction. In each case it was noted that the builder shook and spread each bundle of grass before he actually used it in thatch. When asked about this little ritual, one of the older men explained:

Long ago there was an old man whose name was Mani?i, and he wanted to build a new house for his wife and children at a place called Irupikimi. Though he had already started the house he did not have enough kunai for the roof. He sent his wife out into the valley to get some more 'kunewa' (kunai grass), but she returned stating that there

20. Vide Notes and Queries, 6th ed., 1954, p. 273 which mentions this method for basket weaving, but it is essentially the same process which is employed to produce these sections of the wall.

was none. He then told his wife: 'I will go up in this 'puyami' palm
and take its leaves to cover the roof of your house.' So he climbed up
the very high 'puyami' palm and collected a number of its leaves.
When he again stood on the ground, he shook and shuffled them, for
there were not many, but he wished them to spread and cover the
whole roof of the house.

That is why it is common practice for all Gadsup to shake the bundle
of kunai before starting to thatch the roof. It must be increased.

In attaching the kunai to the roof structure, the builder will pass
over every second horizontal lath while folding grass double over the
next lath and then passing the bottom half of the grass out to rest on
the lath that was skipped. From the inside, looking up, one can see
the thick timbers of the roof structure and also every second lath
while those in between have grass around them. It clearly gives a
step-effect, but on the outside looking up one sees a thick even-car-
peted effect. Since the thatching had been started from the lowest
point, every successive layer covers the former leaving only the
ridge of the roof where the grass is systematically woven from one
end to make it watertight.

An interesting case of culture change which has taken place is the
fact that the one woman's isolation house in Akuna was built by
Urinanda, who is male. In the traditional set-up males were never
allowed near the isolation houses, neither during nor after their construction. These structures, menstrual lodges as they have been called in anthropological literature, are used not only as places of retreat
during menstruation or of recuperation after childbirth, but also as
havens of escape. On a number of occasions while we lived in Akuna,
a woman would withdraw to the women's house for the night after an
argument with her husband. He could not follow and usually there were
one or more females to talk to while calming down.

The inside of the modern house in Akuna does not show much
uniformity upon first observation, but fairly soon a pattern emerges
in which the general organization of the interior is the same. Many
of the houses are built on poles which raise the whole floor about
eighteen inches to two feet off the ground. In others only the sleeping
platform which forms about one-half of the room, is raised. In both
cases the raised portions are made from tightly-woven bamboo. When
the house is built directly on the ground, the central point is the hearth,
lined by rocks in a pentahedral or sexagonal shape.[21] However, if the

21. This may in fact be something for prehistorians to keep in mind. Once
while we were hiking to another village we came upon a very neat organization
of rocks. It turned out to be the hearth of a house long since destroyed. In most

floor is raised, this same pattern is followed. Four heavy logs of approximately eight inches in diameter and thirty inches in length are placed in the center of the floor in such a way that they form a square container. This is filled with sand to prevent any heat from reaching the floor. This, then, is the hearth and a fire burns here constantly with food being roasted on it.[22] When the house is on the ground and a platform raised, this serves as sleeping area for the family and also as general storage space. In a number of these houses, an especially good example is Torawa's house since he has been out of the Arona valley for a spell, a small table is attached to the wall or platform and used for storage. In the other houses people simply curl up next to the fire and many a night while we were still talking, children would nestle close to us to get direct heat from the fire and be lost in dreams in no time.

These houses we are speaking about do not have any form of window opening. The result is that they are fairly dark inside and at night the only light is reflected from the glowing coals in the hearth. Men are starting to acquire a few kerosene lanterns and when visiting at night these will be used to light the way rather than for any illumination of rooms. While visiting, however, they never put them out but simply place the lantern to the one side while allowing it to burn. Upon their departure they simply pick it up and leave the room while still talking. Under modern conditions and with the prototype of Europeans who carry their keys with them, an increasing number of doors are being fitted with padlocks. It is a frequent sight to observe a woman going to her garden, digging stick in hand, grass skirt swaying from right to left while a thin chain around her neck holds the key to the house. This is a novel practice, one which has come together with increased metal tools and household utensils, but one which contradicts one of the basic tenets of life in Akuna, that of assistance, hospitality, and the right of use on condition that the owner be informed as soon as possible. But this also fallaciously suggests that the woman is the owner of her house since she carries the key.[23]

---

cases too, the shape of the house was dug into the ground in earlier days, something which might mark an earlier site with depressions.

22. In a recent paper (du Toit 1969), it was explained how we climbed limbed up into the luluai's house and sat down around the hearth. This is the situation described above.

23. The whole aspect of property and ownership has been discussed above in our fourth chapter.

TRADE

It seems that the residents of Akuna have always been in a trade relationship with other Gadsup village communities, as well as linguistic neighboring groups. This is, of course, essential when there are certain luxury items which are not found locally; when there are certain utilitarian items which facilitate life, such as pottery; or when products, such as colored clay, are used in aesthetic realms. It is generally accepted, however, that trading and other forms of intergroup contact were not all done in response to practical needs. Much of this occured in reaction to alliances, obligations, and similar relationships on a ceremonial and a socio-political level.[24]

The objects which Akunans traded for, and gave in exchange, can perhaps all be classified as "luxury" items. By using this term we have in mind here those commodities which are not basically required in the process of feeding and protecting the body. With such an approach we may in fact be approaching more closely Firth's definition given above. Here, in fact, are commodities which express relationships, which allow recognition, which form the basis of prestige.

An important requirement, as soon as trading with members of another language group is discussed, refers to the ability at least to understand. As a result of their geographical locality certain villages acted as trading partners with non-Gadsup. Tompena and Ikana traded with the Tairora of Batainabura and Noreikora; Wopepa with the Agarabe of Awinapa; Arona with the Markham Valley people at Wararais; and Pundibasa with the Markham Valley peoples at Onga. In each of these villages a group of persons were found who could understand and often speak the language of the neighboring group, but it was strictly for use in trade relations. This is why the people of Wopepa do not know Tairora, or the Tompena people Agarabe, while a centrally located community like Akuna had only two people who were bilingual (Neo-Melanesian and Katé not included).

One of the important utensils which the Akunans did not make was clay pots. In fact, none of the Gadsup villages produced these 'a.nati' but traded them from the Markham Valley villages. From Pundibasa these earthen pots were then traded through the rest of the Arona Valley and also directly to Akuna. The pots observed in Akuna and neigh-

24. I am thinking here of perhaps the best known example of trading relations in less complex societies of which Malinowski (1922:83) says: "The Kula is thus an extremely big and complex institution. Both in its geographical extent, and in the manifoldness of its component pursuits. It welds together a considerable number of tribes, and it embraces a vast complex of activities, interconnected, and playing into one another, so as to form one organic whole."

boring villages were all plain and lacked any kind of decoration. This
is a very interesting problem. By looking at the map of the valley, it
will be observed that Akuna is closer to Agarabe villages in the northwest than to Pundibasa in the northeast. We have also remarked on
the division which existed between eastern and western Gadsup. Akuna
thus was fairly close to Punano and related Agarabe villages where
pottery was made.[25] This is even more important when we keep in
mind that Akunans did trade with the Agarabe at Awinapa for 'tiwami',
a large wooden plate. While Akunans made their own wooden plates,
they claim that the wood which is used by the Agarabe just mentioned
and the Kamano at Tapopa was much better since it was a harder wood.
In return Akunans traded bows and arrows and a native bean called
'kopunayun'. This is a fleshy quadrangular pod about three to four
inches in length and a bluish-green color. These beans were dried in
much the same way now used to dry coffee,[26] and the bean then turned
a brown to copper-red hue. These are even today used in prestations
but do not have nearly the significance they used to. The Akunans, it
seems, also acted as agents in these trade relations for they frequently
exchanged the 'kawe?i' (earthen pot) which were traded from the
Markham via Pundibasa, to the villages further south and west. From
the west they now acquired a red clay and from the Tairora to the
south the same product (called 'kawani') which they used in coloring
and decorating their elaborately carved arrows. When this clay was
traded from the Tairora village of Batainabura just across the mountains to the south, Akunans would offer arrows and grass skirts in
exchange.

It might be correct to argue that since these valuables just discussed formed a part of all prestations, e.g. marriage, death, debt payments, and so forth, we should see each of these gift exchanges and
payments as informal trade between different village communities.
This is all the more true when we keep in mind that any relations and
interactions outside the social realm of the village were in fact foreign
relations.

With the coming of trade stores much of these earlier ties have
been disrupted. Enamel basins have appeared which can serve better
— in place of native pottery. The arrow has become a prestige symbol solely since its use has been curtailed by a strong administration.
An Akunan would never think of using one of his elaborately carved
arrows to kill a pig and so they are traded back and forth for peace,
for alliances, and for wives.

---

25. An interesting discussion of this male activity among the Agarabe has
been given by Virginia Watson (1955).
26. See discussion which follows below.

DIRECT EFFECTS OF EUROPEAN CONTACT

There has been very little which we have discussed thus far in which the presence of the European, and the changes brought by this contact, was not evident. While the earlier discussion may have represented a gradual and indirect effect, we are here dealing with such things as require continued contact and for that reason leave effects which go much deeper.

Shortly after the first contact by Australian administration officers, primarily Taylor and McCarthy, a base was established at Kainantu and a landing strip completed. For labor they employed people from the Markham, and later from Chimbu in the Western Highlands. These latter were also the people who came in to develop the Agricultural Experimental Station at Aiyura. The people in the Arona Valley, and among them the Akunans, were slow to respond to this new contact. By the early fifties, however, an increasing number were leaving the valley for work elsewhere. One avenue was the indentured labor system whereby young adult men could go to the coast and even as far as Rabaul or New Britain to work on plantations. These people returned with thrilling tales about the outside world; with accounts about the ocean; with money and European clothing; and usually with red hair produced by peroxide treatment. Shortly after our arriving in Akuna, two letters arrived from young fellows who were at Kokopo near Rabaul. One was written in Katé, the Lutheran mission lingua franca. In it Dupopima wrote to his young wife, Ma?e,[27] whom he had married just before departing for the coast. He relates very little about his experience except the fact that he had been ill, but informs her that he has sent two laplaps; one trousers, one towel, and two shirts, that his two year spell is drawing to a close, and then the rest of the letter discusses his pigs which she should care for and she should prevent from going through the fence belonging to another fellow for they will kill his pigs. These are the persons who are indirectly agents of change. The year before our arrival four or five men had returned, and some time before our departure four young men returned and four others left.

In the past, too, a number of young men have left for police training, among them Torawa who became a very good interpreter and in fact co-researcher. One man was still away and was a policeman near Goroka. During his leave he spent a week in Akuna and brought his bicycle which was the only one in the village. Young boys spent hours learning to ride and repeatedly falling.

27. We have already met this couple in a rather lengthy court case which occured and produced a somewhat less amicable atmosphere than the letter did.

Due to the Lutheran mission agent in Akuna a number of young boys have been encouraged to go to school first, then Raipinka, and two, in fact, have gone to 'high school' in the Chimbu area. Out of this group three became teachers who are attached to Lutheran missions in other areas. Only two young women who had completed the village school course had gone beyond it, by attaching themselves as servants to the Raipinka mission. They speak Katé fluently but not Neo-Melanesian. The establishment of Ukarumpa as headquarters of the Summer Institute of Linguistics has created the need for timber since all the buildings are produced from local timber which is processed at their own sawmills. To satisfy these needs of 'Amerika' (meaning the Americans) the men of Akuna have formed themselves into three 'lines' or teams, each with about ten men, headed respectively by Tapari, Urinanda, and Torawa. These work groups are not organized according to kinship ties but are based on free choice and association. In the line of Torawa is the luluai's own son, while the luluai's brother works in a completely different line organized in Amamunta. Torawa is the liaison officer between the Akuna lumbermen and the Ukarumpa tractor driver who fetches the logs and pays the men.

When a 'line' goes out to work, the men are not allowed to cut a tree before showing it to the leader who is responsible. They first have to make sure whether it is any of the privately owned species such as areca, pandanus, and similar varieties or whether such will be affected by its cutting. They also have to know on whose land it is growing since certain parts of the bush are owned by families. When the tree has been felled, it is cleaned of branches and carried to the road where it will be collected by tractor and trailer. Payment is based on the length and thickness of the log, in other words the number of board-feet of timber it will supply. Out of a payment of, say, $36.00, which is given to the leader, he will pay $6.00 to the owner of the land. This is called 'ars money' referring to the base of the tree. The rest of the money is then divided among the men of the work line.

While this is no problem as yet, it is a factor which the administration is going to have to keep in mind, namely the gradual removal of all the old strong trees. A number of tall 'wandimi' or Hoop Pines (Araucaria cunninghamii) are now growing in the valley or near the villages and since these seem to do well when they are kept free of the wood borer they might well be planted in reforestation.

Above we remarked on the occasion when Auko, the local 'dokta boy' killed a pig which had entered his garden. This pig had been the property of Tapari and had in fact been bought as a breeder in which transaction Tapari paid fourteen pounds for the pig to Kundana. When Tapari complained to Auko and asked for reimbursement he was told

that the pig had been in his garden and he was acting within his rights
in killing it. When Tapari could get no return for his loss he decided
to break with Akunan tradition in not distributing the meat among his
relatives and friends, but rather to have a 'market'. Word was sent
to the neighboring villages that the market was going to take place in
Akuna and by early afternoon Tapari had in fact rigged up a crude
table outside his house on which the products appeared. The meat had
been cut into smaller pieces which lay on 'empomi' leaves and were
priced from one shilling to ten shillings a piece. On the posts overhead hung betelnut, bunches of tobacco, and decorative leaves while
Tapari, Datipa (his father's second wife's son — thus a member of
his 'ankumi'), and Aika (his wife's younger brother) stood behind
the table. During the afternoon at least thirty people from neighboring villages attended and bought meat and tobacco but not betelnut.
Later in the afternoon Tapari killed a goat and placed the meat on the
table as well. The proceeds were as follows:

Pig — $14.00 (five Australian pounds)
Goat — $2.80 (one Australian pound)
Tobacco — $4.20 (thirty Australian shillings).

While these proceeds give some idea of the relative value of meat
versus tobacco, it was also important that all the money did not belong
to Tapari. Certain people had to be reimbursed:

Dowaka — five shillings
E?arua — eight shillings
Apaka — forty shillings.

The first two men had contributed to the original price when Tapari
bought the pig from Kundana, while Apaka had looked after it and
cared for it.

When Tapari was congratulated on his success that afternoon he
informed us with a sad face that after he had reimbursed his friends
the rest of the money was taken by his wife. She then explained that
if he kept the money he would be spending it when next he visited the
trade store, but if she took care of it there would be cash when the
family needed money.

One very important aspect which should be mentioned in addition
to the fact that the market itself was an innovation, was the language.
Nearly every person who was served or who requested to buy something spoke Neo-Melanesian. It was as if the transfer of the market
set-up to the village location brought with it a complete atmosphere.
As Akuna did not traditionally have such a pattern of interaction, so
the new system of communication came with the new economic institution.

The introduction of coffee by the Agricultural Station at Aiyura has
had a profound effect on the life, labor, and values of the Akuna com-

munity. When first offered, the people were not in the least interested in this new tree that took years to mature and required much tending. This was also true for the people of Akuna, except for a middle-aged man, Munowi-(at present about fifty years old and the father-in-law of Tapari just mentioned), who as a young man decided to plant some of these new plants. Today the villagers look back at their attitude and laughingly tell how they teased him and made fun of his new interest. They watched him laboriously clean the plants, and finally pick, shell, and dry the beans and then leave Akuna with a sack three-quarters filled, walking across the steep mountain carrying it. Napiwa relates that when Munowi got home that night, everybody was sitting around waiting and were absolutely amazed when he drew from his string bag eight crisp pound notes (approximately $17.50). "The next morning before the break of day nearly every man in Akuna was on his way to Aiyura and when Masta Orp (Mr. Orp Schindler, Head of the Agricultural Experimental Station) came to the office, we all asked for coffee."

Today coffee gardens are a part of every village settlement and new gardens are cut out of the bush. As far as could be ascertained, perhaps because it is of fairly recent introduction, this has not affected in any way the rights on land ownership or the length of time a garden would remain fallow. It has affected the social organization, as man and woman, and usually their unmarried daughters, work together, but the boy still is not expected to join the economic activities of the family to which he, during these years, forms an appendage. Young boys will frequently help, and upon marriage the man is both legal and economic head of his household, but between these two statuses is the post-puberty period of freedom to be referred to later in this study.[28] The men are mainly responsible for keeping these gardens clean, but for the first time men and women cooperate in the proj ect of planting, cleaning, tending, and harvesting a crop. It has also given the old people, who formerly were economically unproductive, a valuable niche to fill in that they still spend their day sitting in the sun around the house, but remove the coffee beans from the bamboo containers and keep turning them on a blanket or 'pandanus' leaf blanket and then quickly take them inside when a sudden tropical rainstorm threatens. The harvesting of coffee has become a means to an end, the need for money. While young men can go to the coast or work at one of the European settlements where they can use their knowledge of Neo-Melanesian, the older men who are not allowed to go on indentured labor have a means of maintaining their position and of securing

28. For an elaboration of this point please see our discussion of "those in transition" in the following chapter.

200

cash. This is needed in many contexts since the arrival of mission agencies, tax collectors, and trade stores. It was mentioned that the Aiyura villages, which have done exceedingly well under the influence of the agricultural station, expected a crop of coffee in 1962 worth nearly $14,000.

Chapter 7

SOCIALIZATION

Every community aims at the replacement of its members. This involves not only the biological fact of infants being born but even more importantly, the process whereby such infants are made into members of the community. It would logically include as well the socialization of any other individual whether in fact born in the community or outside of it. Whiting and Child point out that "in all societies the helpless infant, getting his food by nursing at his mother's breast and, having digested it, freely evacuating the waste products, exploring his genitals, biting and kicking at will, must be changed into a responsible adult obeying the rules of his society (1962: 63)."

But since we all have different community settings in mind when we read about this process of enculturation, we also think about a diversity of methods and aims in this process.

In spite of the importance of this subject for both anthropologists and psychologists, relatively little has been done along these lines. For Western cultures a number of studies have appeared which point at differences which are dependent on the social category to which the parent belongs (e.g. Davis and Havighurst 1946; Klatskin 1952; Havighurst and Davis 1955; White 1955). The information suggested by these studies led to a study of the Patterns of Child Rearing (Sears, Maccoby and Levin 1957) which analyzed data from three societies. While all three of these are within the United States of America, the Homesteaders of New Mexico were as different from the Rimrock community with its strong Mormon tradition as these were from the metropolitan dwellers in New England. While the sample covers three different communities the results are not presented nor the conclusions offered on a cross-cultural basis.

The most extensive cross-cultural study along these lines was conducted by Whiting and Child (op. cit.). Their study follows a psychologically oriented analysis and the terminology is strongly behavioristic. While often dealing with topics not discussed by the general ethnographer describing the culture of a community he studied, it nevertheless presents an admirable synthesis of the data regarding methods of child training and personality.

A number of studies, which aim more specifically at our general approach have been made. In her first Samoan study of adolescent girls Mead (1928) showed that biological changes at puberty did not necessarily affect the personality. LaBarre (1945) again saw the relationship between toilet training and personality development as a precondition for his study of Japanese personality orientation. In this chapter I would like to show how the method of child training — as it is practiced among the Gadsup in Akuna — contributes to producing a person well adapted to the flexibility of the social structure, with a strongly developed individuality and able to practice the freedom of choice offered in this part of the highlands.

CONCEPTION

The Gadsup recognize a direct relationship between the act of sexual intercourse and conception. While their belief of what takes place and the way in which the foetus develops does not agree with our own knowledge of this reproductive process, they nevertheless are conscious of the fact that intercourse is a prerequisite. Because of this belief, too, they see the father as contributing essentially to the child that will be born.

There does not seem to be any suggestion that certain times of a woman's cycle are more dangerous for conception to take place than others. While they recognize the relationship between lunar changes and the onset of menstruation, there is no attempt to see this in terms of specific time periods. Conception then may follow the cohabitation of a man and a woman at any time. At the time of impregnation, it is believed, a man's semen ('yipe?aka?i) enters the woman's womb and combines or mixes with her "fluid" — egg — ('yipimi?aka?i). Thus the foetus is formed to which the man contributes the blood ('ubo').

Most women are able to bear children it is said, but those who are unable have brought this upon themselves. This impotency or sterility can have various causes. Often times, it is said, a girl has occasion to witness or assist at the birth of a baby. Due to the unrefined and often unhygenic conditions in which this takes place and irregularities which may develop, she is so shocked that fear develops and instead of having to go through the ordeal herself she may react psychologically and physiologically by having repeated miscarriages. In other cases she may eat the plant medication 'a.yaye' which in overdose or prolonged ingestion is said to produce sterility. This plant medication is taken internally through the mouth to affect abortion. Informants state that the western Gadsup learned this medication from their neighbors in Pundibasa. 'A.yaye' is the bark of the 'yanapa?i', or 'wild taro' (Alocasia macrorrhiza) which they scrape off and eat. This is held secret from the men who value children and forbid their wives from taking it. Even today women practice this in secret, for fear of their husband but also of the medical officer of the administration. A number of women among the Gadsup claim that their inability to bear children is a result of this prophylactic.

Children are greatly valued among the Gadsup but they also feel that five or six are enough (see Table 5). If a person has too many children others will look at him and say, "doesn't he do any work, he seems to be playing all the time, look at his children!" A change, however, has taken place with the coming of the European administration. Infant mortality has been sharply reduced as a result of better medical and hygienic conditions and partly due to the penalty should a child die without good cause. This has also been influenced by the system of taxes which may be reduced, depending on the number of children a man has.

There is no traditional medication to aid conception or to counteract barrenness, but such a potion is reported from the Fore and the Gadsup are interested to learn more about it.

The Gadsup state that there is no preference for the sex of the child, but they all seem to want both sexes represented. An informant stated that if his wife had produced five sons he would be cross with her. The woman then seems to bear the responsibility for the sex of the child in spite of the fact that the male contributes the blood. In response to questions whether the woman or her husband could ritually determine the sex of the child, the answer was negative. The same response was given regarding twins, and while I never heard any beliefs about twins it was interesting that Akunans knew of only one pair of twins near Kuranka in Eastern Gadsup territory. The reason why certain women produce twins and others not is explained quite logically: "We cannot see inside the abdomen of a woman and don't know why one is different from another."

There is little thought regarding birth control, and methods of preventing conception. Little time is spent in theorizing about something which cannot be influenced or changed. This holds also for the pregnant woman. Pregnancy is caused by a man having physical union with a woman and can result at any time when the semen of a man enters the woman's body. Logically it cannot take place during menstruation as the woman is always in seclusion during this period and will therefore not have intercourse. The menstrual cycle is associated with lunation. When the moon reaches a particular position, a woman will think that her time is close now. She never attempts, however, to calculate the specific time involved in this cycle. They do not know why menstruation takes place or what physiological changes cause the flow to take place at regular intervals. Since anything that has been in close contact with a person, including feces and menstrual blood, may be used in sorcery against a person, it is considered dangerous to other persons and for these reasons a woman who is menstruating is isolated in the woman's house and all excretion is buried.

A woman may be working in her garden or gathering fire wood when the first symptoms of menstruation are recognized. She will hurry home, dropping her large carrying bag full of taro or sweet potato inside her house, fill a smaller bag with food and fire wood, and with her sleeping mat proceed to the woman's house. She remains inside during the night but may go about her business during the day but usually remains in or near this isolation hut. She may not enter her gardens, or anybody else's, and her younger sisters or daughters, if they are old enough, will gather and prepare food for her husband. Men fear and shun a woman during this period and will in no circumstances take food from her for fear of disease. Should she be unable to take care of herself, or become ill, she will be assisted by her mother who is responsible for burying all excretions.

While a woman is in periodic seclusion her husband is subject to various food taboos. She, however, may freely eat any product or dish. During this period the husband may not eat taro which has been grated on a grating stick and cooked in the earth oven. He is also forbidden to partake of 'moku', a green leafy vegetable; the red juice of the fruitbearing pandanus which is added to grated foods; and the meat of the tree kangaroo. All of these are considered soft foods, and he must eat only roasted taro and sweet potato.

The first signs of pregnancy which are recognized seem to be an abdominal protuberance, although informants also stated that menses no longer take place. When a woman suspects that she is pregnant it will be discussed with her husband while she also informs her mother of this change. If she has only recently been married, for instance only six months, her mother will reprimand the husband stating that they have been married only a few months and his wife is already pregnant. This same spacing is expected in subsequent pregnancies. No attempt is made to forecast the time of delivery but the woman will await the onset of labor pains.

Pregnant women are treated and act in the same way as other women, until the final stages (around the eighth month) when she relaxes more due to her lack of comfort. Under modern conditions, too, the administration excuses these women from the regular Friday road work. I have seen women, a day or two before delivery, spending the whole day in the hot sun while picking coffee or working, in a half sitting position, in her taro or sweet potato garden. Even at this stage she is not subject to food taboos.

Embryonic development is seen as a natural growth process in which the clot of blood, formed by conception, gradually develops arms and legs. In time a replica of a healthy human being appears. It seems natural that many of these ideas are the result of foetal observations following miscarriages at various stages of pregnancy. Pregnant women then do not require or receive special treatment and I could not discover information suggesting practices or rituals which were aimed at the insurance of healthy offspring. There is one preventitive, but this is not a taboo, namely that a husband is warned by his mother or his wife's mother that his wife is pregnant and he should discontinue sexual intercourse lest he harm the growing foetus. Should miscarriage take place it is usually blamed on the man for not abstaining from intercourse or alternately on the woman for having practiced adultery. Miscarriages take place at the same locality where normal births would have occurred and the foetus is buried there by the mother (real or classificatory) of either husband or wife.

A woman does not make any preparations prior to the arrival of a baby, but one frequently notices a pregnant woman making a new 'unami' (string bag) in which the infant will be carried. Her household activities and her role as supplier and preparer of foods remains unaltered. Her daughters may share a larger share of the fire wood collecting or even the preparation of meals; but she does not withdraw from these role expectancies. She remains mother and wife till the day when she withdraws to deliver her child.

BIRTH

Should she become conscious of intermittent pains one morning, she will not go to the garden but remain at her house. The day is spent performing regular activities of preparing material for a new grass skirt or preparing a meal. If she is in the garden when the pains are noticed, she will return home and await the onset of parturition. She will remain in her house until the last minute when she becomes conscious that the time of nativity is near and then leave the village, withdrawing to the secluded spot (usually a clump of bamboo) where birth takes place. Each village has such a specified place or places where women go for parturition and these are strictly taboo for men. In the case of Akuna one such a place is located south and the other north of the village. Upon leaving her house she will inform her husband or leave word for him that she has withdrawn to this spot. At these places of seclusion there are small lean-tos for protection in case of wind or rain.

The parturient mother is frequently alone at this stage, especially if this is a second or subsequent child; and depending on the rapidity of the birth process she bends over forward, gripping a tree trunk or piece of bamboo which allows her muscular control during the birth. At this time or shortly afterwards her mother or her husband's mother arrives to give her assistance in case she needs it. The older woman will also kindle a fire for warmth, especially if this takes place at night. There is no midwife or assistant during the act of parturition, but any women, whether they are related or not, may be present. The mother or husband's mother severs and ties the umbilical cord and also buries the exuviae. She then cleans the new born infant and may assist her daughter, but it is important that only she, in other words a kinswoman or affine, may act in this particular capacity. During the period we spent in Akuna a number of births occured, one of which was witnessed by my wife. On all of these occasions persons related to the woman either consanguineally or affinally were present, but in all cases only adult married women were permitted to assist. It is taboo for the mother who is giving birth to be assisted in any way by her younger sister, real or classificatory, as she may convey danger or disease to the woman's husband whom she may have to care for. This once again points at the fact that the process of and conditions surrounding a woman's seclusion, either during menstruation or child-birth, are held secret from and contain potential danger for the men. Should water be available the mother and infant may be washed but in many cases, as in those that took place during the period of research, they remain unwashed until the next day. During childbirth a woman is relieved of her grass skirt which she wears at all other times and under all conditions.

In the case of prolonged or fatal deliveries, the mother or mother-in-law of the woman will reprimand the husband, blaming him for having had sexual intercourse after his wife had conceived and thus injuring the foetus. In other words, the responsibility rests upon the man to take care of his wife during pregnancy, to abstain from inter-course, and thus to contribute directly to her normal delivery. A prematurely born child is referred to as 'ayampamana inta' (lit. 'bone (only) child'), or child without much flesh. When the child gains in weight and develops normally this referent is dropped. If the child is stillborn, the mother or woman assisting the parturient mother im-mediately departs to inform the husband that his wife did not deliver normally. The woman goes to the woman's house to recuperate. If the woman dies while giving birth, her mother or the woman assisting will return immediately to the village and inform the men, saying "so-and-so is dead, we will bury her now". The men remain in the village, clustered in small groups to discuss the deceased and may not approach the place of seclusion or see the corpse. The deceased woman's sisters (real or classificatory) prepare a grave and then carry the corpse and complete the interment. All the other women accompany the procession while wailing but are not permitted to assist or to touch the corpse. It becomes clear that upon death, the role of the deceased wife's sisters changes. No longer are they responsible for the bereaved husband. The ties which bound them in an affinal relationship have been severed. Since the sororate does not prevail, they no longer stand in a definite relationship toward the husband of their deceased sister.

A deformed infant is cared for by his mother. This question did not seem to be of great importance and it was significant that no deformities in children were observed. A number of adults who had deformities could readily explain that their misproportions were the result of accidents during childhood or early maturity. Upon being questioned what factors could produce deformities or influence the normal development of the foetus, informants exclaimed, "How do we know, we cannot see why one woman's inside is different from another's!" Regarding twins, informants were vague. There do not seem to be many twins, in fact they could only recall one set of twins living near Kurangka, and in earlier times twins did not live. Whether they were actually killed or simply allowed to die due to neglect, never became clear.

After the birth has taken place, the post-parturient mother dresses and with the infant in her arms, returns to the village. She takes up her recuperation in the women's house. In earlier days these places of seclusion were not situated in the village but in the bush close by. They were in fact subject to attack by hostile neighbors and accounts tell of many women who were burnt while in seclusion. It should also be stated here that these houses of seclusion were traditionally constructed by women as men should know nothing of or be associated with them. Under present conditions men are often called upon to assist in the construction of women's houses as these are on the outskirts or even form part of the village layout.

Shortly after being notified by the older women of the birth of his child, the father will leave his house and give an address to announce the name of his newborn child. There does not seem to be any ritual associated with this name giving. The parents had previously discussed and agreed upon a name and this is announced to the community. There are no ceremonies or rituals to align the newborn member either with the group or with the supernatural. The new member is thus announced to the community. While the whole process of parturition and seclusion is limited strictly to the female realm, the birth of a new member is of immense importance to the community as a whole. In all cases it is an addition to the membership of the community. This is of importance in itself. The first child of a couple is of still greater significance because it proves the man's virility, and, due to his fatherhood, he becomes a real man. It also acts as a bond between the man and his wife, stabilizing their marriage and binding their respective kingroups into a reciprocal relationship of assistance. Children then are a positive value to every Akunan family. They are valued in the familial structure but are also valued in a wider social context.

While in seclusion in the women's house during the period of post-parturition, the new mother fastens a small grass skirt to the waist of the infant. In the case of a boy this will be removed upon leaving the women's house, but a girl retains her formal women's garb. Male infants are always completely naked and remain so till their seventh or eighth year, but females are always dressed in the grass skirt which changes on various occasions during the period of maturation.

On the day that his wife and her infant child return to his residence, the father gathers his bow and arrows, his sleeping mat, and other personal belongings and moves to the house of a relative. He also enjoys his meals with these relatives while his wife's sister or her mother supplies her with food. This period of separation lasts for a further ten days to two weeks upon which he will collect food products and return to his residence offering them to his wife. Purification rights are absent, but sexual life may not be resumed before the infant is fairly strong, and independent of the mother's constant attention — for instance being able to crawl and sit by itself. This, it is stated, assures that the first child will receive proper nourishment and also protects the health of a second child for it is felt that the mother should not conceive again too soon.

NAMING

As far as could be ascertained the allocation of a name to a child was not accompanied by any ceremony. The father simply announces in the village square what his child's name will be.

While names do not necessarily have specific meaning many of them refer to situations or conditions at the time of the birth, or may in fact be the equivalent of our family names. In the following list a number of names are explained. All of these persons live in Akuna with the exception of the fourth who has married and left the community.

1. This is a young man whose name refers to the condition of his father when birth occured. Akunan males, unlike females who carry loads on their heads, place any weight they transport on their shoulders. The father had been carrying split fence posts to enclose his new garden and his shoulder ('apumi') was aching when he was informed of his son's birth. He gave the name Apune?o.

2. In the origin myth which was discussed earlier in this study, we mentioned a particular species of tree kangaroo which lives in the gras on the valley floor. This small-size kangaroo, with the easily discernable markings of hair down his back, is called 'ironi'. On this occasion the father had prepared a trap for the tree kangaroo which was eating his 'yomi' and had killed it when he learned that his wife had given birth to a daughter. He called her Irona?o.

3. A very beautiful and human reaction can be found in the name of my first interpreter in Akuna. His father, Aya, was married to three women. The first had two daughters and the second had one daughter. Now the third wife, Ara?o, was expecting and she gave birth to a son. The father was happy and proud and so he derived a name from 'wa?i' — the tiered grass skirt worn by men in the traditional society. He called his first son Wa?iyo.

4. It is not quite certain what the relationship is in the following example, even though the derivation is linguistically clear. Napiwa was named for a brother of his father's who had been killed in warfare with Kundana. Their name had been derived from the condition of relative license during the male initiation ceremony when the boys chase and beat their age mates of the opposite sex, with thin branches called 'yapi'.

5. Napiwa and his wife have no children and to assist them in the chores of fetching water and working around the house, they adopted Apipe, daughter of Ako and his wife Urake. Since the latter have a number of other children, Apipe, which refers to "strong legs", lives with Napiwa and Irona?o. The exact relationship between the meaning and the reason for allocating this particular name did not become clear.

It is unavoidable that with culture contact between Akuna and the Europeans who are present, the reservoir of names will be increased. A number of cases were recorded while we were in Akuna.

6. The day after a son had been born to Yajima, the kiap visited the Arona Valley for tax collection. Since Yajima and Ti?e had not yet decided on a name, the kiap's name was ferretted out and the little fellow now is known as Toni.

7. In an earlier section it was pointed out that some of the males have been away from the Arona Valley for various reasons. Two men from Akuna received police training. One of these is Torawa. When he was asked why the little girl his wife Tatome had presented him was called Katarina, he explained, "When I was away to Chimbu for police training I saw those large airplanes which land on the water. They are called Katarina." Obviously his linguistic habits were causing the mispronunciation of the Catalina flying boats which were especially familiar after the Second World War.

8. One last case will suffice. A short while after our arrival in Akuna Atino gave birth to a son. That evening, as I was finishing my meal, Wa?iyo came in. After some discussion, he asked where I was from. Not attaching any special significance to the question, it was suggested that he draw up a chair while I got a pencil and paper and for the next hour we had an "introduction to the geography of the United States". It was explained how America had various states, just like New Guinea and Papua; how various towns and cities occured just like Akuna and Amamunta and similar villages. When we had just about exhausted the topic and I was sure that this young man's intellectual appetite had just been whetted, I enquired into the reason for his original question. He looked up and said: "Oh, Yappe and Atino are looking for a name for their son!"

The significance of names, and the need for sons and daughters is also important in terms of the practice of teknonymy. By this the suffix — 'napo' is attached to the name of a man's eldest son, while — 'ano' is attached to the name of the woman's eldest daughter. The meaning being 'father of —' and 'mother of —' which permits younger persons to refer to and address their seniors without using their names.

In a number of cases it was pointed out that a person had more than one name. The other forms seemed to be nicknames but I did not go after this in any detail.

INFANCY

Infants, especially during the first six or eight months of their life, dominate a woman's life and her activities. Up to the time when the child is strong and independent enough to crawl outside or be cared for by a sibling, he hardly ever leaves the mother. In most cases, the child is actually in physical contact with the mother, being cradled in her arms, or lying on her lap as she sits cross-legged by the family hearth. On these occasions the child is never more than a few inches from the mother's breast. The nipple ('nanoni') produces the milk ('nanano') which is essential to the child's growth and development and is always available. A feeding routine is unheard of. When the child cries the mother's first response is to present the breast, saying 'ma! ma!' (here! here!). Also during the night one is often roused by a crying baby and will hear the mother stir to present the breast, with the same words. A mother stimulates the flow of milk by patting the breast in an upward movement while touching the nipple with the palm of her hand. Should the mother not be holding the infant, he will be in a carrying bag by her side, close enough to be touched when he stirs. He has the opportunity to feed whenever he wants to and this generosity on the part of the mother continues for nearly two year at which time she may be again pregnant.

There is little or no difference in the treatment of children, either due to birth order or sex, except for the fact that an only child may be treated with greater indulgence and affection than a child who must share his parents' interest, time and love with siblings.

One of the clearest examples of this relationship was the tultul of Akuna. He, Yerai, and his wife, A?adido, had only one child. This young lad, Tokawa, was neat and well cared for and he received the constant attention of his parents. On many occasions I found the tultul and Tokawa in discussion at the family hearth or saw them walking together. While the latter's boyish sense of humor drew a ready smile from the indulgent father, he could not go too far. When Tokawa was instructed to fetch water for his mother and forgot to do it while playing with age mates his sense of humor did not prevent his father from very severely reprimanding him. From this follows the logical conclusion that a youngest child shares this parental love and indulgence for a much longer period than his older siblings, male or female. This is also true as regards breast feeding. For the first eighteen months or two years, the infant is allowed the breast at any time and spends most of his life close to his mother. Should a second child arrive, the first is now suddenly refused the mother's breast. This, it is suggested, is a shock to the infant who has had access at every time and every place to his mother's breast. Suddenly there is a competitor, another infant that has taken his place, not only at the breast but also in the attention of his mother. The first shock at independence training is rapidly allayed by the interest of kinsfolk and the increasing degree of integration with Akunan childhood culture.

Regarding breast feeding it is believed that if an infant has fed at his mother's breast for a day or two or three, he will not survive should the mother die. In such a case a wet nurse will be found, preferably among the kinswomen of the deceased mother. Should an infant refuse to feed at his mother's breast, another woman will take the infant and nurse it. Should the mother die when the infant is approximately a year old, it can be fed and raised by another woman and it will survive. Usually a woman does not breast feed another's baby except in those cases which have just been referred to or when the child is already from nine to twelve months old. On such occasions a woman may have to go to the garden, leaving the infant for whatever reason, in the care of another. Should the infant cry, this temporary substitute mother may breast feed the infant. There are, however, certain legal complications which may develop from these situations. If an infant refuses his mother's breast, the woman may request

212

someone else to feed him. Should this continue for a long time, a few months, the biological mother may not reclaim the child as the second 'mother' will say: "No, look the infant has grown strong on my milk; it is my child now." When a mother does not have milk to offer to her baby she either requests another woman to assume responsibility for lactation, or she chews sugar cane, passing the sugary juice from her mouth into the suckling lips of her baby.

Infanticide is no longer practiced nor does it seem to have been very common. Informants explain that under traditional conditions, i.e. prior to the advent of European Administration, Akunans and the Gadsup in general occasionally practiced infanticide. A mother or father, having agreed upon this action in time of warfare when the woman's movements might be hampered by the additional weight, would kill the child by beating or pushing in the fontanelle. They would hurriedly bury the little corpse 'and run away'. This seems to have been a utilitarian and escape mechanism rather than an institutionalized form of population control or sex oriented preference. While it is no longer practiced and spoken of in abhorrence, it was culturally sanctioned within the hostile and unpredictable environment which characterized the New Guinea highlands.

Children born out of wedlock must be seen and defined in the Gadsup cultural setting. No child in Akuna, it seems, was ever born to an unwedded mother. As soon as the usual signs of pregnancy were noticed by or in a young girl, her elder sisters would ask her whether she has had sexual intercourse with a man. She admits while naming her lover. Her elder sisters immediately fasten a married woman's skirt around her waist and send her off to him. He seldom refused to admit his guilt or to marry her. Should he, however, claim to be innocent of the act, her male kinsmen would confront him with her allegations and summon expert witnesses — those young boys who see and hear everything. Confronted with this evidence he invariably took her in as his wife, concluding the usual marriage arrangements. Under traditional conditions as well as modern times, women may become pregnant by men other than their legal husbands. Should the man suspect his wife of extramarital relations, he will convene a court case forcing her to admit. She would then either be married by her lover or the latter would be heavily penalized in the form of fines, but hostilities frequently flared up and especially so if the lover belonged to a different community. Modern conditions have presented greater opportunity for this variation of extramarital intercourse. Every year a number of young men, usually and preferably unmarried, are taken from the Arona Valley on indentured labor. They go to various places on the coast or even as far as New Britain. Should any of them be married, he leaves behind a woman in her early twenties,

213

the desire and temptation of every able-bodied man in the community. He is no longer present to keep her under surveillance and it frequentl happens that she cohabits and occasionally conceives during his absence. In all cases a woman's sisters attack and beat her with their string bags, because she has been unfaithful or because she has left her husband.[1]

There is, of course, no form of diapers or clothing. The result being that an infant frequently soils the mother or the pandanus leaf mat on which he is lying. The mother will simply brush away the urine or scrape off the faeces with little attention to the company. This reaction and treatment continues for the first eighteen months to two years, but by this time the child is expected to start its independence training and a mother may react with annoyance when a young child soils her.

After about six months, the mother will gradually return to her expected role as wife and provider of her family's food supply. She will usually leave her house and the village for a few hours each day, taking the infant with her if the weather permits. Gradually she spends longer periods in the gardens, planting and cleaning those products which form the staple of their diet. On these outings she will be accom panied by a younger sister or daughter who may substitute for her at short intervals, holding the infant or cradling him in her arms. While she goes about her daily activities, her first responsibility is always to her infant close by. On these occasions the infant is always placed in the string carrying bag referred to, while lying on a mattress of sewn pandanus leaves. This carrying bag is hung around the mother's head extending either fore or aft, depending on the direction of the sun. While she assumes her responsibilities the bag will be attached to the garden fence[2] or the woven bamboo side of the house.

Parents, both the mother and to an increasing degree, as the child matures, also the father, play with the infant, tickling him while making gutteral sounds in simulation of those the infant makes. At about this time, the father may spend a few hours with the infant on his lap while the mother is away in the garden, visiting her relatives in some neighboring village, or attending a ceremony in her natal village. His treatment of the infant resembles closely that of the mother. Such motives as personal contact and reaction to the slightest sound from the infant characterizes his reactions. He constantly touches, strokes,

1. This point was taken up in a section dealing with rights and wrongs in Akuna in an earlier chapter.
2. During hostile raids such infants were frequently captured and taken away. In some cases an individual from a hostile village would creep up and remove the infant, taking it home to be raised by his wife or a fellow villager. Also mentioned by Flierl.

and pats the child. He may be relieved by a female sibling or daughter, but on the other hand, he may spend a whole morning caring for the infant.

The independence training of children starts relatively early among the Akunans. From about the seventh or eighth month an infant is increasingly left to play by himself or in the care of an older sibling. While a mother is preparing food, scraping 'yienni' for a grass skirt, or digging in her garden, the infant might play close to her, scratching in the dirt or busying himself with some object lying close by. While this stage is not marked by any active withdrawal by the mother, it is contrariwise a greater leniency on her part to allow the infant to withdraw. She no longer fusses over his movements and selection of objects he handles. She no longer watches his every action but on the other hand, is readily available when he cries. Immediately she will be there, holding him in her arms, speaking to him in a soft voice, or helping him to complete the action he is attempting. While the infant is allowed to become independent, he is not actively stimulated in this developmental stage. While the infant is allowed to move on his own, the mother does not withdraw. The result is a gradual developmental phase during which the infant becomes progressively more independent, but yet is always able to fall back and know that the mother, or her substitute, will be there, ready to satisfy his need of love, association, or assistance.

When the infant reaches the age when he can crawl and will thus be separated from the mother for short periods, the Gadsup state that the time has come for the ritual bathing of the infant. This is called 'nomkadankara' (lit. to bathe a child in water) as opposed to one bathing oneself, 'nomkadara'. Prior to this brief ritual the father will announce that he wishes to wash his child and will send a message summoning the child's mother's brother. The latter proceeds to the stream where the ritual is to take place and cuts a long bamboo which he decorates with colored leaves and 'kekeyan' (a white flower with a pleasant odor). At the opening of the bamboo he inserts a bunch of 'kamope' (small pink flower). When the parents arrive with the infant, the mother's brother will hold the infant close to the decorated bamboo while bathing the child and washing his whole body including the head. No prayer is offered and informants stated that this was not aimed at soliciting supernatural aid for the health of the infant. It was, however, essential as a preventative of disease, particularly skin disease. When the infant has been rubbed down with soft leaves the family group will return to the village where the father kills a pig for the feast. This meat is eaten by visitors who were invited while the mother's brother receives the head and legs of the animal. The bamboo remains in the house for a short period after the ceremony.

215

The infant spends more time away from the mother now. When she goes to the gardens an older sibling, male or female, may care for the infant, carrying it on the hip or astride the shoulders with a leg hanging down the back and the other down the chest. The infant holds onto the sibling's hair. This is also a common way for the father to carry the infant, but mothers will invariably place the child on their hip or back, where he is within reach of the breast. While the infant crawls around, it will return periodically to the mother for lactation. On many occasions, it was observed how a mother continued her domestic activities while the infant would stretch her breast under her arm and suckle while squatting next to her. Adults watch with interest and little interference the development of the growing infant. On various occasions I was struck by the freedom of a twelve month old infant playing with a sharp knife. While the adults might be sitting around the earth oven or discussing some topic in the village square, an infant with a long sharp-pointed knife would be crawling after a fruit seed or empty milk can from the trash heap, flicking it with the knife. The adults would watch with interest and enjoyment as the infant handled the knife and every time he flicked the milk can, they would respond with "Ai! Ai!", coaxing him on. In this manner children are allowed to learn by experience but yet under the protective watch of an older sibling or adult. Developmental phase, then, is marked by leniency of superiors and freedom in choice and action, allowing for the unfolding of individuality.

At this stage, too, the infant will first try to stand and move in erect posture. He starts on his own, rising and sitting down with a thump, then rising again. The parent may be sitting close by while watching with interest and enjoyment these first attempts by his child. Occasionally the parent may put out a finger to steady the faltering child, then moving back while smiling, and enticing the child to take his first tottering steps. Should the child fall, they leave it to him to get up again and start the procedure all over. When they were asked why they did not assist the child to get to his feet again, an informant stated, "I'm not always going to be there to help him; he must learn for himself."

CHILDHOOD

From the second year of a child's life, he is gradually weaned. This may be in a severe form as when the mother has another child, or more gradually in the sense of making the child independent. Under normal conditions oral training was not severe. As the child spends more time on his own, returning occasionally to feed, the parent or somebody who is present may attract the child's attention with some object, or by playing with the child. It should also be stated that children from a year upwards are gradually introduced to solids in the form of soft roasted sweet potato or the various grated tubers baked in the earth oven. A child who wants to nurse may be distracted by being given something to eat or offered a drink of water from the bamboo container from which water is used for cooking and drinking. The child on his own accord makes fewer and fewer demands upon the mother, but she still accedes at certain times as when the child cries. or when he wants to go to sleep and cuddles close to her. There is then little shock to the child at weaning except in cases where a sibling is born. On these occasions the elder of the two is denied the breast, and while the same process of substitutes and distraction is followed, the mother does not allow the child to nurse. The first few times this happens the child will cry fiercely but an adult or the father will take hold of the child, focusing his interest on something in a different direction.

Children eat exactly what, when, and where they like. If the mother is preparing food and they are hungry they have it pre-cooked, and if at the actual mealtime they are not hungry, they just keep running or continue with the game they are involved in. Parents never reprimand children for not eating at mealtimes, or attempt to discipline them into eating only at those particular times. They may be playing near the parents of one of their age group when she has prepared food and they will each receive a share. While this suggests that young children are not watched, it is only partially true. Children, both boys and girls up to the age of five or six, are free to play where they like or accompany their friends to fetch water or collect grasshoppers, provided the parents know where they are. They need not be in attendance at mealtimes, but they should inform the parents where they are spending the time. While Gadsup parents allow their children considerable freedom, there is always a limit and they always expect to know where the child is.

As regards sphincter training, we find a similar gradual process during which the parents at first take little notice, later suggest control on the part of the child, at still a later stage become visibly annoyed if they are soiled and finally might be punished. In the case of an infant, as was remarked above, there is no expectation of sphincter control nor does it cause visible embarrassment if a parent is soiled. It should also be kept in mind that while an infant is in actual physical contact with his mother, the latter learns to react to abdominal noises and abdominal and sphincter muscular movement. When a mother becomes suspicious of excretory activities on the part of an infant, she will quickly remove him from her lap, holding him to one

217

side. By the time an infant has reached the age of around eighteen months a number of developments have taken place. Usually independence training has started. While weaning might not have progressed much at this stage, the infant has already been introduced to solids. This means that the excreta are no longer fluid as when the infant was on a milk diet, making anal training an easier task. It should also be kept in mind that children who go completely naked, or are only covered by a pubic grass skirt as girls are, have less of a problem in relieving themselves and less discomfort when "accidents" occur. From about his second year a child is gradually instructed to go outside to relieve himself, or to inform the parent when an urgency develops. Soiling still takes place but to a lesser degreé.

In addition to the embarrassment caused by such an act, there are also certain social structural implications attached to it. In a later section such an example will be discussed.[3]

The initial indulgence of the infant's free excretory behavior gradually changes as parents start to expect responsibility on the part of the child. By the second year the child should start to practice self control. He should start to inform the mother when urgencies develop. Should a child of this age soil the mother she will in addition to being embarrassed, become annoyed, clicking her tongue and raising her voice while informing the child that he should have told her or should have gone outside. This also holds true when the child soils his sleeping mat. By age two and a half or three years most children in Akuna have completed both their oral and sphincter training. While the child may still occasionally urinate or defacate in the house, this is now looked upon in a different light and the child will be reprimanded or even punished. It should, however, be stated that physical punishment is not common among Akunans. Instead of beating a child, an elder might be more likely to speak to the child, informing him that he is now growing up and that a mature Gadsup does not act in that way.

As has been discussed elsewhere Akuna lacks any strict system of age-grading or definite stages in the life cycle. The Gadsup see infanc as extending from birth until the age of four or possibly five years of age. These infants are 'aninta' (boys) and 'akinta' (girls), and are treated as irresponsible and incapable of caring for themselves. They are cared for, cleaned and fed, and are under constant surveillance of a parent or an older sibling. While they are infants, they are, however, recognized as progressing through a stage of learning. They are seen as merely inexperienced adults. By allowing an infant to play with a knife, to sit near the hearth, or to tease a wounded dog, it is believed that by "bumping their heads" in various ways they will learn

3. See page 335.

and in time grow up to be experienced adults. There exists, then, considerable freedom for the infant, and later the child to learn by experience and exposure.

While sexual training in the true sense of the word does not start at this stage, there are a variety of instances in which these infants are inculcated with respect to the socialization of sexual behavior in a general sense. Infants are frequently seen fingering their own genitals or those of other infants in their company, and this is especially true for males who go completely naked. During the first year or eighteen months there is little reaction to this, but soon older siblings or parents will distract the child's attention by offering him something to play with. In the case of older children a parent may reprimand the child saying that "it will fall off" if you pull or play with your penis. It is important to keep in mind that while modesty in girls is strongly inculcated, the same is not true for boys and that the infant gradually grows accustomed to the fact that older boys who play close by expose similar physical features without much attention being paid to them. Should an older child of four or five finger his genitals the parent will again warn that touching of the genitals may cause them to become sick, but in private this will continue.

A boy of five may play with his penis and this may continue for years, but there does not seem to be an understanding of masturbation as discussed in American sociology books. The boy, it is said, enjoys feeling the stiffness and sensitivity in his genitals in such sexual play, and will continue this "to see the penis jump" under his slightest touch. A girl does the same thing, stroking her breast and genitals and again enjoys the sensitivity and warmth engendered by this action.

Children from about six years of age are told by the parents about the genitals and how they should be treated, about the fact that a girl never is seen without her grass skirt or allows herself to be exposed, and the fact that playing with the genitals is not good. At about ten or eleven years of age children of both sexes are informed, by both parents, that heterosexual play and intercourse which might follow, can cause trouble. Such play might result in pregnancy and "if she is the wife of another man you will cause great trouble", a father will say. This seems to have some influence on the sex play between children and certainly is a factor in limiting intercourse between young persons. Once young people have reached the age of about seventeen, possibly in anticipation of marriage and as a result of the slackening of parental discipline, there is a marked increase in heterosexual play and sexual intercourse.

Masturbation, as mentioned above, does not seem to reach the climax it does in many other societies. We do find a form of homosexual play. This is referred to as 'iyeranenu', simply meaning 'play',

219

as children play with marbles or any other objects. In this homosexual play among boys, one will assume the active and another the passive role, while the first places his penis in the anus of the second. This seems to be relatively common among boys as they become increasingly segregated from their female age-mates, and informants explain that it usually continues until the sixteenth or seventeenth year of life.

Among girls there is something every similar, in which two girls associate intimately with one another carressing and petting the breasts and genitals of the other. In this homosexual play they assume the position of intercourse with one lying on the other. In neither case, it seems, is orgasm reached. While a certain amount of exhibitionism seems to be present, and this is especially true in boys who wish to attract the attention of a sex partner of the opposite sex, it does not seem to be standard behavior.

While adults are conscious of these behavior forms and sexual deviations, they can do no more about them than to express their strong disapproval. These activities never take place in the house but in the kunai or bush through which children roam all day. The parent is therefore unable to act in a preventitive way but can warn the child that these irregularities, for they are seen as such, are dangerous. It should be mentioned here that they are not transferred into adulthood and that all adults are expected to and in fact do get married. We can therefore decide that homosexuality and sexual play are participated in during developmental stages in the maturation of the child. They form part of the Akunan childhood culture.

In their discussion of Child Training and Personality, Whiting and Child point out that "age at beginning of socialization is not a very meaningful concept for aggressive behavior. It was not generally possible, either for aggression as a whole or for separate aspects of aggression, to recognize any approximate age as the time when demands were imposed upon children to start changing from an initial freedom to a greater degree of control" (1962:99). This statement also holds true for the Akunan, but by concentrating on various aspects of aggressive behavior, it might be possible to single out particular aspects which receive preference in training. We will then discuss the aspects of temper tantrums, physical aggression, verbal aggression, property damage, and disobedience. While the first of these is somewhat limited in the age at which it occurs, the same does not hold true for the other aspects.

In general informants stated, and this agrees with observed behavior, that young children have to be taught that there is a time for aggression and that everyday living with kinsmen and fellow villagers was not that time. Should a person attack, strike, or offend you, you should be in a position to retaliate by striking harder, but be careful not to act without good cause or to offer reason for the person's attack. This is the somewhat philosophical approach taken by adults and it is the ideal which is conveyed to the young. While a certain amount of this is learned in the socialization process, it is inculcated more by day to day living with age mates and older children.

Feelings of frustration and aggression in young children usually find expression in temper tantrums. Should a child want something or feel that a mother should carry him, and not be satisfied, he will stand in one place, thumping the ground with his feet and yelling at the top of his voice, while dancing in a circle. He may also fall down and while lying on his back kick and beat the ground, crying loudly. If the child is very young, or has reason for this expression of his aggression, a parent or other adult might approach and comfort the child while speaking to him and allowing him to calm down. This submissive reaction on the part of an adult depends entirely on the situation and the age of the child. Because of the fact that most children between about the second and fourth year react in this way, the behavior is accepted. When, however, an older child about four to six reacts in this way, and a number of cases were observed, the parent will tell him calmly to quit acting up or simply hit him on the buttocks with the hand or a stick. Three cases come to mind in which physical action was taken against a child.

1. One morning while I was talking with a number of males, Ko?ana left his house, bow and arrows in hand, on his way to the bush. His little daughter, Ma?o, of about four years old wanted to accompany him, but he told her to remain with her mother Inuntino. The child insisted and when she was refused once again she started to howl at the top of her voice while stamping her feet. The father picked up a piece of plank which was close at hand and struck the little girl twice across the buttocks. He was very angry as he turned around but observed me watching him from behind the group of men with whom I was talking. After a few seconds, during which time the child had calmed down and was only crying, he turned to her and picking her up into his arms cuddled her against his chest.

2. On another occasion Ande was on her way to the gardens. She called for her son, Purosa, a lad of about three or four, as she was leaving her house. He remained where he was playing with some seeds in the sand. Again she told him to accompany her, and as he did not respond she turned around, pulling him up by one arm and laying into him with the other. While Purosa loudly bewailed his fate, his mother carried him off to the gardens.

3. This case occured during the Orande ceremony which was celebrated in one of the Omaura hamlets in the center of the Arona Valley.

The ceremony with sacred flutes and bull roarer takes place at the men's house. The men do not leave this place but are served their meals here and, then, as everybody withdraws the ceremony continues. On this occasion a young fellow of about seven years old had wandered off to one side and was partly obscured behind some corn stalks in the garden as the women and children left. The boy's father now noticed him and in a few long strides was beside him and grabbing him by the arm, helped him up the incline to the path and sent him after his mother with a very hard whack on the backside.

The expression of aggression by young children is permissible. When a little child that is learning to walk gets frustrated and stamps and screams, the parent will attempt to remove the cause of frustration or to comfort the child. These children, then, are in the initial stages of independence training and they are assisted. During the later stages of this period, the child is given to understand that adults must be listened to, that the individual is still subservient to the group, that accepted patterns of behavior must be heeded by day to day living with age-mates and older children in the community.

In the early stages of aggression training, punishment or frustration is often reacted to by physical aggression on the part of the child. Once again it is difficult to define exactly what is meant by this term. If a young child feels ill-treated, he might strike or bite a parent and the usual reaction is that the parent exclaims "Ai!" and ignores the matter. When the child should know better as is the case with children about the fourth or fifth year of life, the parent might strike back, not too hard, but in such a way that the child learns that an act of aggression leads to a similar act in return. In time the parent might strike harder in return while explaining that physical aggression, when not called for, is unnecessary.

While a parent is always responsible for the child's conduct and the principal factor in the socialization process, the child receives much of his training regarding physical aggression from siblings and age mates. While children are seen as inexperienced adults, young children are seen as inexperienced older children. In the rough and tumble of daily play activities, it is brought home to the child that he should contain himself. This does not mean that the strongest or the bully lords it over his playmates. While an older sibling might slap a youngster who quarrels or acts in an unacceptable way, it was observed how this same older sibling would warn an older boy who attempted the same thing. In other words, the older and more responsible children are factors in the socialization process but the siblings are mutually responsible and while an older brother might discipline a youngster, this same right does not pertain to other older boys who are not in this position of kinship.

Verbal aggression takes the form of teasing and in more serious cases changes to insults while pointing at previous unacceptable actions by the person being attacked. A child may raise his voice or scream at a parent, and depending on the situations, this is usually ignored or passed off lightly. Due to the fact that children from the sixth year onwards increasingly form a separate group, though this applies especially to boys, it is suggested that verbal aggression plays a minor role in the reaction of the children of Akuna. It is also suggested, and observation confirms this, that the freedom and pliability in personal relations allow for good natured needling to extend into adulthood. On various occasions a child would be seen teasing or heckling an adult, tugging at his arm, stealing his 'kote' from his string bag or tossing small stones or sticks at him. The adult would react good naturedly as though he does not expect anything else. He would return the heckling or return the sticks thrown at him, even running after the children. They loved it and would scatter in all directions only to return once more. When the adult has had enough or is busy he will turn away, and ignoring the children, let it be known that they should discontinue their game. In most cases they lost interest and would start a different pastime or would be sent on an errand to divert their attention. This discussion applies mainly to children up to the age of four or five after which they are usually engaged in some game, and are always in the company of age mates.

Regarding property damage, it is more difficult to write conclusively or even to make meaningful generalizations. In a culture where there are no glasses to break, books to tear, or flower beds to uproot, the amount of damage a child can cause is limited. Those objects which are liable to damage are usually in personal use, such as the man's string bag with all his belongings, or the woman's knife and household utensils. The child is soon taught that objects not belonging to him should not be touched. He cannot reach the decoratively carved arrows or bamboo container with feather head decorations of the father, or the mother's necklaces as these are attached to the roof of the house. He might, however, enter the garden to acquire fruit or vegetables. This again is permissible, even of adults, provided there is good reason for it and provided he does not damage the fence or unnecessarily uproot plants. The important point to be discussed again, is that a child of seven or eight is expected to be responsible for his actions and is punished for transgressions.

First impressions among the Gadsup suggest that there is very little discipline and very little punishment to enforce etiquette, attitudes of respect toward others, and similar virtues. This in fact is true and only on a limited number of occasions can I recall having seen a child being punished. Gadsup youngsters do not adhere to the biblical edict of 'honor thy superiors' in the way that we interpret this virtue. Reference has been made above to the way in which they heckle and tease adults, but this is always in a particular sanctioned way. On a number of occasions a child was seen disobeying his parents and allowed to get away with it. When a parent, however, instructs a child to do something, it is expected that he will listen. Should the instructions go unheeded a second or third time, the parent will raise his voice and occasionally punish the child physically. In cultures where scolding and 'shaming' are important these may be of greater embarrassment than physical action.

Above I mentioned the case of Tokawa, son and only child of the tultul in Akuna. The fact that he is an only child means that he must assist his mother and when he was told twice to fetch water and still 'forgot' his father reprimanded him very severely. Coming from the tultul who is a just but strict man this was harsh punishment.

But once again it depends on the particular adult involved. Old Duikama, a grandfather of about 65, instructed his grandson to take care of his baby sister while their mother was in the gardens. The youngster, however, was involved in some game with his age mates and did not respond. When Duikama had repeated his order a second time with no avail he went after the boy of about six. The latter ran off leaving the grandfather, stick in hand, to chase him. This was great fun for the other children who ran after him laughing and jumping. Adults looked on in amusement. After approximately ten minutes Duikama returned to his house grumbling but smiled at the adults who were present. It is important to mention that this old man is looked upon as a 'humbug grandfather' (Neo-Melanesian denoting weak, bad or not serious) but it also seems as though this is the appropriate behavior for children. A few minutes later the youngster returned to pick up his sister and went off to care for her.

When children are older, about ten or eleven years of age, a parent will shout at them for not doing something that was requested. She will tell her neighbors that her child is impossible as he never listens to her and will then talk to the child when he returns, pointing out that if he did not change his habits he would grow up to be like so-and-so who could not take care of his affairs or who was poor, materially and especially socially.

Children, even in this age category, start to spend much of their time with age mates. Boys live apart more than do girls. They sleep with the parents, have something to eat at the family hearth, but spend most of their time playing in or near the village. At this age they are not as mobile as their older friends and do not roam the bush or separate themselves too much from the adults. A number of little boys and girls will get a hold of some long pit-pit leaves, fold them double and pierced two at a time with a stick, thus producing a "propeller" with which they spend hours, running from one end of the village to the other. As the village of Akuna is located on raised ground there are a number of places where the high rainfall has caused erosion to take place. A number of boys will pick up the thick lower part of the leaf of a banana palm and take it to this smooth decline. Mounting it they will slide down the hill, only to return and repeat the act, while others await their turn. The resemblance with our children's tobogganing is clear.

During this period children do not see much of their parents, but this is even more true for the next age category when boys between seven and twelve hardly see their parents all day.

Formal education, which came in with the missionary activities and, while we were in the field, the European Administration, is still very limited. A number of Gadsup villages have Lutheran or Seventh Day Adventist missions and as an appendix to these a small school. Classes are not convened regularly nor is discipline strictly imposed. At this stage of the education children do not learn much except to count and to read Katé (the Lutheran Mission lingua franca into which hymnals and the Bible are translated). This influence can be detected among the children. A couple of yards from the village fence in Akuna there is a large tree where the young children frequently gather to play. On a number of occasions they were seen to congregate here, singing Katé songs or marching in procession while shouting "Wan, Too, Tree, Paw, Pipe, Sikis ..." One morning a group of small boys were gathered in the tree, perching on the branches as the leader (a little fellow of six) shouted their names at the top of his voice. "Tutupa! Punke!" and so down the line and each in turn would shout "Yes Sa! Yes Sa!"

On a number of occasions a group of small boys were observed playing on a clump of pit-pit of which the tops had been bent down to form a platform. The strongest boy would be on top while the others were attempting to displace him. This was fun and loudly enjoyed by the participants and I could not help thinking of American children playing King-o'-the-mound.

Shortly after the Lutheran Mission in Akuna had its flamboyant ceremony which culminated in the baptism of new converts, the boys in this village decided to re-enact the whole procession. It took a full morning for them to prepare their headdresses, make their handdrums (often empty milk cans), and complete the decoration of their bodies. Now approximately twenty small boys, with the same Punke from our "tree school" in the lead, between the ages of six and ten formed a procession, marching through the village in single file or in two's. Each small boy was carrying a dry branch which had been decorated with pieces of colored paper, feathers, cloth, flowers, or pictures from old magazines. Marching through the village they were singing in imitation of the way the adults had during the mission ceremony and the more traditional 'sing-sing' the previous December. While this was a very pleasant interlude, it also was an illuminating example of Gadsup childhood culture, an uninhibited imitation of their seniors, colored by the initiative and absence of restraint of the child.

While this procession would have produced an unbearable noise in a Western cultural setting, and this is true for many other cases of childhood activity, the adults were merely amused. Noisiness does not become a problem in a Gadsup village of one storey structures and wide open spaces. A favorite game among children of this age is for fifteen or twenty of them to chase a fowl through the village, flapping their arms as the terrified fowl does, and shouting at the top of their voices. Parents hardly notice this din and if the children get too close to where they are, they will simply be told to go somewhere else, or get a playful whack across the buttocks.

Children generally do not take the afternoon nap common among the adults, and especially the males. Sleeping at night they take the same posture as that of their parents, usually on the side. While children go to sleep a bit earlier than the adults, we lack any organized pattern that it is "time for children to be in bed". Community life here must be seen in its proper perspective. A child is expected to be in bed when night has set in, but this does not imply sending the child to his own room with a large bed, away from the family who are gathered around the T.V. while enjoying beer and popcorn. A child sits next to his mother or father in the cozy and homey atmosphere by the family hearth. Discussion takes place in low voices as the coals occasionally flare up, painting shadowy designs on the woven walls or thatched roof. The parents, sitting cross-legged on their pandanus leaf sleeping mats, chew their betel nut or dreamily puff at a home-rolled cigarette. Without knowing it the child becomes drowsy, lies down and loses consciousness. He is not forced away from the center of family activities because the hands of a watch on mother's arm happen to point at numbers he doesn't understand. Frequently while I was visiting with a family, allowing discussion to take its own course, the children in the company would grow drowsy and one by one stretch themselves on a sleeping mat, and drop off to sleep.

Akunans say that a person in the age range from about six to about eleven is a child. They distinguish between 'ammi' (boys) and 'anati' (girls). Children of this age group spend progressively less time in the company of the parents, but girls continue to be somewhat closer to the mother. They are expected to help her in all kinds of domestic duties, assist her in the gardens, and care for a younger sibling while

226

the mother is absent or busy. The relationship between mother and daughter, more so than between father and son, is one of comrades. This is generally true but not categorically so. It was striking that Atino and her daughter Wo?iya, a girl of about ten years old, often spent a whole day in the gardens together. Squatting a short distance from each other they would talk all day, and yet when they returned home to the village, they would still be talking as they prepared food or sat by the hearth. Parallel to this we have already discussed the relationship between the tultul, Yerai, and his son Tokawa. The latter, however, might not be typical for his age group since he has no female sibling to assist his mother.

As an age category, children in this age group form themselves into groups among whom the eldest and strongest is usually the leader. This does not imply that he lords it over the rest of his group. Perhaps because of a clearer demarcation of age groups among boys than girls and the fact that boys become segregated from the family to a greater degree, sisters get along with each other better than brothers do. Partially too this can be ascribed to the fact that a girl's training all along is role-oriented, she is trained as a mother even when she is still in her pre-pubertal years. She looks after younger siblings, she substitutes for an absent or ill mother, she assists her mother in food collecting and preparation or the preparation of material for clothing and forms an integral part of the household. This might be a major reason why girls make the transition into adulthood more rapidly and fluently than is the case with boys.

Boys in this age category can best be described by the designation of 'monkeys'. My field notes variously refer to them as 'coyotes', as the 'pack', or as 'running wild'. The boy between seven and twelve is hardly ever in the village during the day and is hardly ever alone. If it weren't for the connotation our culture had ascribed to the term, these groups would be called 'gangs'. They move in groups of two to six, all approximately the same age, visiting the gardens, roaming the bush, playing in the kunai, catching rats or grasshoppers, or, under influence of the trade store, playing marbles or cards. They are always present, and yet absent when they are needed. They see everything and know of everything that is taking place, and they disappear as rapidly and soundlessly as they appeared. The boy in Akuna is free of restrictions and regulations and fills every day with as much energy-consuming activity as is physically possible. They are allowed to run with "the pack" and fend for themselves during the day. They visit the gardens to pick a few cucumbers when the sun is hot, or roam the bush to collect wild fruits. They carry their little bows and arrows with as much pride as their fathers and create a world in complete contrast with their age mates of the opposite sex. When the shadows lengthen and night descends on the village, however, they must be in sight of their parents. Now permission must be obtained to play outside or to eat at the hearth of kinsmen or friends. It is to this group that we refer in discussing childhood in Akuna. Girls, too, are allowed freedom and independence is encouraged, but they seem to remain closer to the female realm and to spend more time with the older women. Children of this age category own their own property and have the right and privilege to do with these as they like. Most boys have their bows and arrows, a gift from their fathers or made in imitation of the latter, and many of them own their own knives. But as was stated, this is easier for females than in the realm of male activity. It seems likely, in the light of the aforesaid, that this conscious parental inculcation of cultural norms and expectations, activities, and techniques, is greatest during childhood and into the early years of puberty. Following this phase there are a number of years until marriage when both sexes are freed of responsibility.

These statements should not suggest that only boys enjoy their childhood or that girls are continuously involved in domestic responsibilities and activities. The girls have ample free time for games, group discussions, and informal activities, but much of this takes place near the house or in the garden while the mother is tending to her horticultural responsibilities. It does, however, point at the fact that from the fifth or sixth year the segregation of the sexes becomes more pronounced, leading to separate cultures for boys and girls.

When a child is about six years old, and it would also depend on the availability of a food surplus, the mother's brother performs the nose piercing ceremony. In agreement with the child's father a time is set when a pig is presented by the mother's brother and a feast prepared. This pork must not be cooked on the open fire or in an earth oven, but in bamboo containers where it is steamed. In some cases the mother's brother will be assisted by his brother who holds the child, but frequently he would sit by the hearth talking while encouraging the child to lay his head on the adult's lap. Meantime, he heats the thumb and forefinger of his own hand while holding a sharp needle of pig femur in the other. The expectation is that the child will fall asleep, and allowing the child's head to roll back slightly he feels the septum with his warm thumb and forefinger. In both positions, however, a quick move pierces the septum. He now places a length of thin pit-pit, which has been greased with the fat of the ceremonial pig, through the hole. This plug will remain in position for a couple of months, being replaced occasionally by a clean one to prevent infection. The ceremony is known as 'Atiankara' ('ati?i' — nose) being derived from the piercing action 'anka' which is the same used to denote a sharp stick which is driven into the ground. Traditionally, it

1. A young sweet potato garden
2. Grating taro and yams for the earth oven
3. Young banana leaf contains grated taro, green spinach, the red juice of fruit bearing pandanus and pieces of pork

1. Adding salt by
dipping water
through ash of the
pandanus proper
2. Banana leaf cov
containing the gra
tuber, is tied ty th
leaves before plac
in earth oven
3. The earth oven
containing a meal
for three or four
families, steams
under an earthen
cover as 'cooks'
look on

seems, this occasion was frequently used to also pierce the ear 'Ankamankemi' ('akami' — ear). When informants were asked why this was no longer practiced, old Daweka pointed to his torn ear lobes and explained that when an arrow was shot during fighting, it frequently got hooked on the ear of the man handling the bow. This hardly seems like an answer, but very few of the young people's ears are pierced.

With the operation completed the feast begins, and all the villagers gather to share the food and meat prepared by the female kinsmen of the 'patient'. The mother's brother, it should be kept in mind, is very often from another village, and he will now make a brief oration about the child that is maturing, asking the villagers to care for the child and treat him well. The child will now be presented with gifts which mark the change in status. Previously a girl would receive string bags, grass skirts, and beads while a boy got his first bow and arrow or something along these lines. Currently they are more likely to be presented with lap-lap, soap, a spoon or glass beads from the trade store while a boy may also get his first knife or some marbles.

Very often a small operation is combined with this traditional ceremony, namely the piercing of a small hole in the front of the nose. Here, especially females, wear a short sharp stick or the wingbone of a flying fox.

CHILDREN'S GAMES

There are a variety of games, some more formally structured than others, with which Gadsup children spend their leisure hours. While some are limited to only one sex, others are shared by both boys and girls, separately or less commonly mixed groups.

Walking through Akuna after a rainy spell, one often notices a group of children plugging at lengths of string to the end of which is tied a small brown object. Immediately one thinks of an American boy deftly handling a yo-yo. Upon closer inspection this small brown object proves to be a 'munani' beetle. These are caught in the long pit-pit in the valley, upon which a thin strand of bark or fiber is tied to one of the beetle's legs. Those which are still strong would keep flying around the child in a circle, while insects that have been in captivity for a longer period would hang limply on the string and only respond when the string is plugged a number of times. Both boys and girls join in this pastime as they vie with one another in keeping their insects performing. Each time the beetle opened its wings and started buzzing around its captor, the child's face would light up and a smile of triumph would cover his face. These are the same insects which are

caught in larger numbers and eaten by including them in the food
being baked in the earth oven.

On these outings, and in fact on most other occasions when boys
leave the village, they carry their bows in imitation of their fathers.
The bows ('anan'kumi') are exact replicas of those used by adult
males (and called 'itan'da') except that the bow itself is made of split
bamboo rather than hardwood. They are about 15 or 18 inches in
length as compared with the adult bows which vary between five and
six feet in length. The arrows are in most cases much simpler and
less durable. Usually they simply cut shoots of kunai by cutting it on
the stem and about an inch into the grass blade. By sharpening the
stem side they have a point but also a natural 'feather' formed by the
grass. These arrows are about twelve inches in length and are called
'owaro'. It is of interest that this 'feather-effect' is employed in these
children's arrows but never in arrows used by adults. The only suggestion of this in adult weapons are the feathers attached to the spear,
'wan'kami', but they apparently do not serve the same purpose. Children also use lengths of pit-pit to which a harder, and heavier, frontal
piece is attached. This 'yapito' is used to produce its effect by sheer
force of shock and weight rather than by piercing. Occasionally children will play with small 'takumi', or bird arrows, which are exact
replicas of those made and employed by their fathers. Here the shaft
is once again made of pit-pit while an inset and three-pronged point of
hardwood is attached. While the first two arrow types are used to
shoot at rats, beetles and other objects, the bird arrow is used only
when they are in the bush and notice one of the many species of brightly
colored birds. On these outings they frequently carry knives and spend
long hours playing or imitating the adults. I never heard of them shooting at each other or playing with these weapons in a fashion that might
endanger one of the group.

Girls of about the same age group will join a number of their
friends and go to the bush where they start their own gardens. Here,
or in their mother's gardens they will prepare tubers and other food
products simulating the family earth oven. One of the most pleasant
intervals occured once while I was visiting the clearing of a new garden in the bush. On the way back to Akuna we stepped out of the dense
bush to see a number of small girls, about eight or nine years of age,
scraping sweet potatoes and making an earth oven. Unnoticed we
watched as they placed the tubers among the hot rocks and closed the
earth oven. When the food was cooked they opened the oven and sat
around it eating their food while talking. When we stepped forward
their first embarrassment was quickly replaced by the typical Gadsup
hospitality of offering us a share of their meal. While this is children's play it is conducted to the last detail in imitation of their seniors.

230

Girls also spend their free hours hunting and catching 'kanki?i' (crickets), 'munani' (beetles) and other insects in the kunai. These are played with or cooked in their earth ovens. Often, too, they would gather wild pandanus nuts to eat. A number of girls would go up a hill while each selected a large rock. The size of these depend on the age and strength of the girl. These would then be pushed down the hill in a competition to see whose would roll the greatest distance.

Keeping in mind that the Gadsup did not make their own pottery but traded these up from the Markham Valley via Wampul and Pundibasa, the following childhood activity is of interest. In this pastime called 'yimani wareno', clay would be gathered and the girls would make dolls, pigs, rats etc. in imitation of the animals they see from day to day. On these occasions they also make imitations of the imported pottery but none of this is ever used.

Both girls and boys participated in an activity called 'Watendeno' (lit. in a string bag to catch). On sunny days a number of children, usually separate according to sex, would go to the river while taking a few string bags with them. Independently each would now attempt to chase fish into the string bag stretched between the legs and feet and kept open by a short stick. The body leans forward while the hands are in the water attempting to drive the fish into the string bag. These, when caught, are once again taken home to be placed in the earth oven.

Boys also fish by tying a piece of kunai or liana to a worm and dropping the latter into a pool in the river. They will sit perfectly still, perched on a rock or on the riverside, and as the fish starts to nibble at the worm, the boy will give a powerful tug on the line attempting to pull the fish from the water. This is called 'inideno' after the earthworm ('ini') which is used as bait. Most of these games are in some way connected with the challenge of making a living, but do not stimulate competition among community members. Boys spend hours in the bush hunting for birds, organizing rat-drives in the kunai, or pit-pit and in this they are often joined by adults since these pastimes border on the economic realm. In imitation of their seniors a number of boys will set traps or snares to catch birds or rats. It has become clear that there are very few pastimes of competitive nature. When the sun is hot or they have walked far they will take a swim, paddling around while yelling and laughing. Where the water is deep and wide enough they will see how far they can swim under water.

With increased culture contacts boys sometimes spend the whole day playing marbles, but when they have none they will join with the seeds of round wild fruit. As the boys grow older they start to play for a prize, and this is especially true in the case of card games played by the post-pubertal boys. On these occasions the stakes will

231

be anything from marbles, old bandages (from the medical orderly), lap-lap, knives, tobacco, newspaper (for smoking), and other articles.

While a number of children associate together for a particular game or activity, there is no formal or organized group formation, and no conception of one's own group in contrast or in opposition to other groups or to non-group members within the village community. While this association is based on age and sex categories, it naturally is the local group and especially kinsmen who associate more frequently than others. The socialization process produces familiarity and trust among those who live together, and in this way the local group gradually gains importance for the individual.

Children on the whole are not told to be courageous, to be cooperative, to be competitive, or to have high aspirations. This is learned from life; from the daily association with age group members where these factors are of prime importance. Should some older child bully a person, it is the responsibility of the latter's older kinsmen to take care of him, for by this action the person can be assured of the same assistance when he is in need. Should a person be slow in learning from life and through experience an elder person or a parent may speak to him, recalling from his own experience or citing some case in which it is illustrated that if a person does not act according to certain standards of expectation, such-and-such is sure to happen. This is particularly true in the highly valued field of mutual assistance. All of this training takes place as informal instruction. One of the only fields in which there is definite inculcation of the required behavior is in the modesty training of girls. Above it has already been mentioned that a girl is always and under all conditions dressed in her grass skirt. This is no more than a frontal covering tied around the waist. When she is about ten or eleven years old her kinsmen or older female siblings show her how to add to this garb by the use of a bunch of green leaves which are attached to the string girdle as it passes around her back. A girl who is growing up, informants explained, never goes without it. On various occasions I asked them about this practice of adding to the frontal skirt or 'wa?doya' worn by 'anati' or girls. Why, if they wanted the central buttocks and anal regions covered, did they not replace the 'wa?doya' completely instead of simply adding this bunch of leaves called 'moyami' to it? Would it not have been simpler to have a skirt which completely encircles the waist rather than having to replace the 'moyami' every day? No explanation could be obtained for this strange custom except for pointing out that this was the way it is done in Akuna.

There are certain attitudes and behavioral patterns which contrast male and female, especially in the ways they act, walk, carry things and similar activity patterns. We lack, however, any clearly defined

232

manliness in contrast to femininity. While most of these attitudes and values are learned by example rather than formal inculcation, this is the age during which such differentiation takes place along sexual lines. Men wear their string bags hanging from their necks and down their backs. Women hang these from their heads, the band crossing the forehead just below the hairline. Should a boy in innocence hang his string bag from his head (as I did shortly after my arrival to test their reactions) he is told that only women do things in that way. If it is repeated there is much laughter and shame. Living with one's parents and their associates, depending on the child's sex, places him in the correct cultural context for learning by imitation. This does not imply that all learning takes place by example but much of it is based on following the example set by cultural surrogates. Boys find themselves in women's company more than do girls in men's company offering a greater number of cases for transgression of these norms and an adult to admonish that "this is not the way a man acts".

Formal education in the form of missionary schools are limited within the Gadsup area and the little school in Akuna has a restricted influence on children. Education is not vocational training to any extent, but it does serve a dual purpose, that of teaching children to read and thus to become better converts. It, however, also equips people for the first stages of participation in Western ways of advancement and government which they are forced to deal with in any case. It teaches them to count, read, speak Pidgin and even write. It forms the foundation for them to handle money and widens their scope should they desire to continue their education at Raipinka and possibly be trained as teachers or policemen. The primary aim of this education is not an intellectual academic one, but rather one of fitting Akunans to face the changing world in which they live. While the various village schools run by the Lutheran mission are very elementary, selected students are sent to the Lutheran headquarters in the Eastern Highlands at Raipinka. A number of students, usually girls, are also sent to Raipinka but instead of attending school they work for the missionary personnel, thus being exposed to the same cultural influences in an informal way. After completing this schooling, a few students are sent to Kerowhagi in the western highlands to be trained as teachers. The Seventh Day Adventist mission has a large hospital and training school at Omaura in the Arona Valley, but their influence affects only certain villages due to the custom in this area of villages as a whole turning to one or the other missionary denomination.

As the period of fieldwork was drawing to a close a larger school under the sponsorship and auspices of the Australian Administration was being built at Onampa. This is fairly close to the medical Aid Post and would be manned by a professionally trained teacher, either

White or a Papuan, from Port Moresby. People were looking forward to this, not because they value education as such but because it places in their midst a representative of the 'kiap' (administrative official) and, they hoped, would in due course lead to a trade store being established. This latter factor is of great importance to the Akunans who have to cross two mountains and ten miles on foot to reach a large trade store to sell their coffee crops and to get any decent selection of merchandise.

Although it was discussed above, we should point out that Akuna and the other villages in the eastern highlands participated in their first election in 1964 (see Leininger 1964; Watson 1964a). The role of the school and the responsibility of these children will increase in coming years.

PUBERTY[4]

The onset of puberty in both girls and boys is not anticipated as they reach a certain age. In girls it is marked by certain physiological changes occurring. Informants explain that puberty sets in for a girl with the first menstruation, and for a boy when his father decides that he is to be initiated. The factors which prompt this latter decision seem to be the size and maturity of the boy as well as the initiation of his age mates. While we may speak of female rites of passage at puberty, it would be more correct for males to speak of these rites of passage at or near puberty.

For girls the first menstruation is known as 'akintamwaremi' (lit. the girl is sleeping, denoting her isolation and rest in the women's house). When a young girl finds that menstruation has started, she informs her mother of her condition. Now a male relative, usually the elder brother or mother's brother sets out to catch a small tree kangaroo in the bush which is near the village. The young girl, in isolation, will cook this animal in a bamboo over the fire. When it is well cooked, the bamboo container with the meat is given to her mother. The latter alone eats it, while her husband is preparing a feast for the girl's paternal and maternal cognates. At the conclusion of this feast the mother places the bamboo container in the thatch outside the door of her house. Informants suggested that this symbolized the attainment of maturity by the girl, thus signifying to prospective suit-

4. In a forthcoming study (du Toit, M.S.) this initiation ritual is analyzed in terms of the three stages of transition discussed by Arnold van Gennep in his classical study 'Les Rites de Passage' (1910).

ors that there now is an adult woman in the house. This same procedure, however, is followed for girls who are already betrothed and
gains significance in the light of a woman's right and freedom to marry
whomever she decided on. The practice of betrothing young girls — or
even infant betrothal — and attaching them by the practice of brother-
sister exchange was not final and the young girl could make her own
decision regarding a marriage partner even though this often implied
leaving her natal village. When a girl who is already betrothed reaches
menstruation, the same food taboos pertaining to adult males, mentioned earlier in this chapter, would also apply to her 'husband' as
though they were married, even though he frequently was no more than
a boy. The morning after the girl's menstruation starts, it is taboo
for the husband or the betrothed to smoke or chew areca nut until he
has 'oyunkare' (lit. 'cooked a stone'). This stone is about five or six
inches in diameter, and on the first morning after menstruation begins
he places it in the hearth of the men's house, or currently in his own
home. While the stone is being heated he collects some of the wild
ginger called 'wimi'. He removes the hot stone with a bamboo tweezer
from the fire and places it immediately outside the door of the house
while he faces the menstrual hut where his wife or fiancé is. Having
chewed the 'wimi', which burns the mouth and purifies it, he spits
this juice onto the hot stone and sees the steam rise. No words are
uttered, but it is believed that the 'wimi'-smell and the cooling of this
heated object will prevent anything evil or dangerous from happening
to the woman.

Just as it is taboo for the man to chew the areca nut before he has
performed this ceremony which pacifies evil forces, so, too, a man
who has had sexual intercourse may not chew areca before he has purified himself. This latter ceremony merely implies washing which
cools down. The justification for this practice is found in one of the
Gadsup tales about their culture hero:

A long time ago there was a man whose name was Mani?i. He and
his wife lived in a place called Iyunanapa?i. While his wife was in her
garden one day, he and his wife's elder sister went walking in the
bush. When they were alone she sat down on the grass and invited him
to have intercourse with her, which he did.

When they had completed the act they walked through the bush again
and came to a betelnut palm. She looked up at the large 'adnumi' and
asked him to go up the palm and get her some in order that she might
sit down and chew some betelnut. Mani?i climbed up the palm but the
branch broke and he fell down with the sharp end of a branch protruding from his chest.

The woman now returned to the village and while crying informed all the other Mani?i that he was dead. They all went to the bush while weeping and returned carrying the corpse of Mani?i.

Following her first menstruation a girl is now known as 'pumara?ini' rather than 'anati', and she dons a grass skirt which is much longer than the first ones she had. It extends well below the knee and is known as 'oyanini' but still has the bunch of leaves in the rear.

A boy is ready for initiation at about eleven or twelve years of age and the decision for an initiation ceremony is usually taken by four or five fathers simultaneously. This decision depended partially on the availability of surplus food products and seem to have been a regular, if not annual, occurrence. By the fact of group initiation the ties between age group members are strengthened. It should be mentioned at this stage that due to the changes in status undergone by boys at this time they also undergo a change of name. Boys now are known as 'pumara'.

This initiation ceremony with the accompanying ritual takes place during the summer (rainy season) when the agricultural activities reach a low point. Adult males open their bamboo storage containers and take out their colorful feather headdresses, their pigs' teeth and decorated stone plugs which are passed through the septum, their elaborately carved and painted arrows, and red clay with which they paint their faces. During this period the men's house is surrounded by greater secrecy than usual, and the pubertal boys leave their mother's home, first to congregate in the young men's house, 'pumara yi ma?i', and finally to withdraw from the village into the bush. As the atmosphere becomes charged with excitement and expectation, the women and girls spend long hours in the gardens collecting a surplus supply of food and their menfolk assist in laying in enough firewood. On the day agreed upon the women and children depart for the river where they wash themselves and remain until they are recalled by the males. As the shadows are lengthening, the atmosphere tense, they enter their houses and start to prepare food while the children huddle close by.

From the night that surrounds the family hearth there is heard a far-off wailing like a child in need, countered by a rhythmic hum. The women will listen quietly and say "The grandparents are returning"[5] while children stare wide-eyed into the moving shadows created by the flickering flames in the hearth. This playing of the sacred

5. The wailing sound is caused by the sacred flutes, played in pairs, and called 'tinapu' which is in fact the same term used in the kinship terminology for grandfather. It is joined in all ceremonial performances by the bull-roarer, called 'ireno? namu', which is a small piece of bamboo swung through the air on a whiplike string producing the deep asthmatic breathing sound.

236

flutes and swinging of the bull-roarer marks the opening of the ritualistic initiation ceremony. During this night which precedes the actual ceremony, the 'pumara' and adult males congregate in the 'pumara yi ma?i' while the adults instruct the young boys in the ways of adult Gadsup men. A large fire is made on the ground in the center of the house and the initiates are instructed to sit in a circle close to the fire while the adult males congregate around them. The fire is stoked until the temperature in the house rises and it becomes nearly unbearable to remain close to the flames. From time to time when the heat becomes unendurable the mother's brother of an initiate may relieve him by taking his place next to the fire, thus giving the boy a chance to cool off. An initiate's father or male siblings will never relieve him for they feel that the initiate "should grow strong and become a real man".

During the night all post-pubertal female relatives of the initiates congregate near the men's house carrying large quantities of garden products. They also bring plenty of sugarcane which has to be carried in long pieces and may not be broken except by men. The women now circle the fire which has been baking their sons and attempt to put it out by drowning it in greens and garden produce. While doing this they call to their husbands: "Help us to put out this fire which has cooked our sons!" The men remain where they are and the fire is finally extinguished by the women. These proceedings serve to deepen the bond between a boy and his mother's brother and also with his mother and sisters. These are the people who have helped him and relieved him, who feel sympathetic, and who protect and assist him repeatedly.

When the women have put out the fire, they leave to prepare the meal and a pig, donated by the father of the initiate is presented to the latter's mother's brother. This is in recognition of the assistance and care he has bestowed on the boy. The initiates are now beaten with branches of the 'karu?i' tree[6] and their bodies rubbed. Informants stated the belief that this treatment would not only produce clear and healthy skins but would also make the initiates into strong men. The treatment here seems to be a dual one, producing hardy fellows who can take a beating and will grow into strong men, yet simultaneously offering protection on the way toward these ends. At about this time the men are informed that the women have prepared the foods and a feast follows. The food is not partaken of in a communal meal, but little piles of food are set out where the initiated males eat separately,[7] and off to one side the initiates. This action emphasizes their tran-

6. These leaves also have medicinal value. A sample was collected and submitted to a pharmaceutical company but no identification was completed.

7. It seems that frequently, especially if they come from another village, the brothers of an initiate's mother will eat separately from adult male villagers.

sitional status. They are no longer boys who eat with the children and women, and yet they are not adult males. Their associations and emotions are much closer to the former, yet their ideals are with the latter. They no longer sleep with the women and children in the family dwellings, and yet they are not admitted to the men's house. They occupy at this time the 'pumara yi ma?i' (the young men's house) where the activities just took place. This ambivalent position gradually decreases as the boy moves further from the uninitiated and gradually passes the different tests on his way toward adulthood. The rest of this night is spent in dancing and singing as adults don ceremonial garb and bear with pride their symbols of adulthood.

Shortly after dawn the initiates are informed they are to accompany the men into the bush to kill rats for a feast. They now stumble through the bush and then 'discover' a footpath which the initiated males had prepared and 'decorated' with drops of blood. An old man will point at the blood and say: "Look, somebody has wounded your mother's pigs. Let's go after them. Run!" The group, led by the initiates, starts running but suddenly a number of adult males appear out of the bush and with blunt arrows, wild bush fruits, and seeds they start hitting the bodies of the terrified initiates. The latter keep running down the path which leads to a pool in the river but without warning their path is obstructed by a dense brush fence. From behind their aching bodies are pelted and beaten, and while they are trying to fight their way through the obstruction the familiar but awesome sounds of the returning ancestors are heard before them. Then as they break through and rush ahead closer to the sounds they heard so often from the safety and protection of their mothers' houses, they enter a clearing in the bush beside a pool in the river. Off to the one side there are two men, in ceremonial dress, with pigs' teeth nose plugs, feather headdresses, and painted faces, playing alternately on the sacred flutes and producing the wailing sound like a small child crying. Behind them stands a man swinging a very long whip to the end of which is attached the bull roarer which adds the deep thumping sound and increases the haunting atmosphere.

Right before them stretched out on his back lies a mortally wounded man in full ceremonial dress, an arrow protruding from his chest and drops of blood marking the wound. While the other males are still beating them from behind they are blamed alternately for killing the man or being irresponsible and thus causing his death. An initiate is now instructed to step forward and remove the arrow from the bloody chest before him. While he is continuously beaten on his back and buttocks, he must tug at the arrow which has cunningly been passed through the ceremonial regalia in such a way that it passes between the chest and the arm and is attached to the ground.

In the meantime a number of males have collected eight or ten pit-pit leaves which are folded and tied to form a number of sturdy piercers about five inches long. These piercers have been placed in a bamboo container filled with a strong solution of native ginger (called 'wimi') and native salt (the same ash used in cooking). Without any warning the initiate is grasped from behind by an adult man and the head forcibly pulled back forcing the boy's face to look upwards. The officiator takes hold of two pieces of pit-pit — gripping them in his fist the way one holds a dagger — and pressing his fists together he simultaneously jams these two piercers up and down the nostrils of the initiate till the blod flows freely. This is called 'upaiyami' or nose bleeding. The blood flows freely. This is called 'upaiyami' or of the initiate, and then he is allowed to rinse himself in the cool stream. An old man, usually older than the officiator at the 'upaiyami' now steps forward holding a short cleft stick. The initiate is told to put out his tongue and the cleft stick is used like forceps to grasp the tongue and hold it extended. With a sharp bamboo knife a series of small cuts are made on the tip of the tongue and the cuts are then rubbed with wild ginger and salt. The same operation is performed on the glans penis[8] while the foreskin is forced back and the same treatment with ginger and salt is administered. All this time the initiate — whom it should be recalled is perhaps twelve years old and who has gone through great psychological strains in the past twelve hours — screams at the top of his voice while calling out the names of his paternal and maternal grandfathers. His voice, however, would be drowned out by the chorus of male voices as they rhythmically stomped their feet singing 'Ho-ho-ho!'"

At this moment, after the ceremony had been completed, the initiate's mother's brother appears, attracted — he says — by the shouts of the boy. He places his arms protectingly around the boy's shoulders while scolding the other adult males for ill-treating his sister's son. Still holding the boy he leads him off to the safety of the familiar village surroundings. Approaching the village the men meet the women who are returning from washing in the stream and a mock battle ensues in which the males are scolded for being so harsh on the initiates. The latter return, however, to their new home the 'pumara yi ma?i'. Shortly after entering they are once again joined by the adult males, but this time they bear gifts. As it gets dark a fire is kindled but this one is only warm enough to prepare food. After eating their roast food, the boys sit in the center while the father of

8. This series of little cuts, though painful, is not the same as sub-incision and causes no permanent deformation of the genital organs. (See: Notes and Queries on Anthropology, 6th ed. 1954: 228).

each initiate passes a 'porawani' — semi-circular nose plug of boar's tusk — through the nose of his son. The males all this time are singing, rising suddenly and sit down just as quickly. The mother's brother of the initiate now steps forward presenting a string bag to the boy and then proceeds to cover the boy's face with grey ash from the fire, while drawing white lines of decoration with 'kote' — the lime chewed with areca nuts. Now they all lie down in the young men's house and sleep for a few hours.

Early the following morning the initiate is further decorated with the arm band ('yanka') which goes around the upper arm, at the wrist, below the knee, and just above the ankle. Brightly covered leaves are inserted in these bands and while the men sing, the initiates are led out of the young men's house and presented to their mothers and sisters. The mother's brother and father will now instruct the initiate to shoot an arrow into the carcass of a pig which his father killed the previous day. He is given a bow and a 'puyani'[9] arrow, but since these bows are tremendously rigid and the boy not very strong, he can only just pierce the skin and subcutaneous layer of fat. The older men shake their heads disapprovingly explaining that he will never kill anything in that way. He tries again, but still it is not good enough and he is instructed in the correct way to hold and handle these weapons; that the bow is drawn down from above the head while the arrow hand simultaneously pulls back. After this practical instruction initiates and adult males return to the young men's house for a ceremony referred to as 'ara?anke?u'. The initiate's father and brothers go outside to cut the pig and they return with the meat which is cooked with taro which the other males had scraped in the meantime. When the meal has been prepared the mother's brother of the initiate holds the head of the boy while another man in the same kin relationship to him puts food in the mouth of the boy, but he is not permitted to swallow it. He must use the food, mixed with his saliva to extinguish the heat and flame on a log of wood taken from the fire. Since the pig had belonged to the father of the initiate, kinsmen are not permitted to eat, and the meat is eaten by the mother's brothers.

On that evening when it is dark when village life has quieted down and the women and children have retired to their houses, the adult males call on the initiates again. They are taken to an area behind the men's house which is strictly taboo to the uninitiated. Here there is a large hole in the ground which has been covered over by logs

9. In the approximately forty named varieties of arrows used in Akuna, two consist of a flat sharpened piece of bamboo. One with a point of about thirteen inches is called 'kayemi', while a similar one with a point of nine inches is known as 'puyani'.

leaving only a small opening. An old man will take each initiate by the hand, holding it to the ground in the dark and allowing him to feel the hole. Next he will bring the boy's face to the hole and tell him to smell. The old man now explains: "You have been a child and you always thought that we do not defacate because no one ever sees us. This is the place where we relieve ourselves. From now on you are not to defecate anywhere or near women and children. You are a man now." They are also accustomed to defecate over streams where the faeces will be neutralized thus preventing others from using these excreta for sorcery.[10]

During this two-day 'upaiyami' ceremony the initiate may not eat the meat of the 'wari' (eel), the rat, the tree kangaroo, or the bird. It is also taboo for him to eat the thick black sugar cane with the short joints called 'akuni'. This is adult sugar cane, but he may eat the thin red variety. It is also dangerous to take food from young or middle-aged persons, but the old men and women of the village hold no danger. Furthermore it is taboo to eat food from an earth oven presumably because it is soft food. All of these things are dangerous for the initiate and may cause disease or stunt his growth and development. For these same reasons the parents of an initiate must refrain from sexual intercourse lest they affect their son during this transitional stage.

It remains for us at this stage, after the 'upaiyami' has been seen as a whole, to explain briefly the river ceremony. Informants explained that the cuts on the tongue would prevent the boy from loose and irresponsible talk. He was now entering a new phase of his life, he had been introduced to the sacred flutes and the bull roarer, the ceremonial aspects of much of the male realm was being opened to him. He should not reveal these secrets to women or to children. Only the initiated, a restricted group who had passed the requirements for membership (duToit 1965:23), shared in this knowledge. The operation on the penis was aimed at producing a true man, a 'wainta', who could procreate and would be strong. The initiate is no longer an 'ammi' (boy) but he is not a man either. For about six or seven years he will be a 'pumara'.

THOSE IN TRANSITION

Earlier in this chapter I remarked a number of times on the fact that the period after the first menstruation for girls and the 'upaiyami'

10. Flierl remarks on these Abortanlagen which often were constructed over streams and which he observed during the 1920's (personal communication, 1962).

for boys, was marked by increasing divergence: that each category was becoming more role adapted in terms of the female and male cultural realms. While girls remain close to the house, the garden, the mother and junior siblings, and the duties associated with all of these, boys after their initiation gradually increase the social distance between themselves and fellow villagers.

Under traditional conditions a 'pumara' was not permitted to go on war parties. Under modern conditions a 'pumara', unless he is about marriageable age, is not permitted to go on indentured labor. A number of years prior to this study, a group of young boys, 'pumara' from Akuna, wanted to go to the coast. They were in Kainantu ready for departure when the village elders arrived and told them to return to the village. They were told that "boys, 'pumara', do not go to do a man's job" and another group of older men, about the age of marriage, were sent. In Akuna, and other Gadsup villages, these are the people today who sit around in sun-drenched nooks while playing cards. These are the fellows who walk about without doing anything constructive. They do not assist in gardening because that is woman's work. They do not clear new gardens for that is the work of a married man and their time will come. The same is true of house construction. They do not join the logging teams who operate near the villages in the bush. Occasionally they will do an odd job at Aiyura, or when the 'kiap' comes through on patrol. When they can, they sit around in the sun, they talk or walk, and later when they are about to graduate through the 'Orande' ceremony and be eligible for marriage, they may start meeting girls secretly.

By this I am not suggesting that 'pumara?ini' (girls) are always active and assisting the household. Here, too, it became clear that girls who were about to be or in fact were betrothed also occupied such a transitional state of non-activity. Very clearly in such a transitional status were the two daughters of Opura (they had different mothers) namely Mauyampe and Tododu. The latter was betrothed while we were in Akuna. They were hardly ever in the gardens, but walked about holding hands, laughing or remained in the shade of the village. Approaching this stage was Ti'ande who spent less and less time in regular activity. But the point is that males, first as 'ammi' and now as 'pumara' have never been very close to the role and expectations of the adult male. It is different with girls who have cared for siblings or visited the gardens with a mother since they were six or seven. While they withdraw temporarily from these roles, perhaps to enjoy a rest in anticipation of their future, it is not hard for them to resume the roles they played and to be wives at the same time. While girls remain in their natal villages, frequently in a girls' house called 'pumara?ini yi ma?i', males are more mobile and it is not uncommon

for a boy at this stage to make a residence change and move to his
mother's brother.

One of the clearest cases of this incompatibility of roles was
observed in the case of Iripe and Apono. He had been one of the
'pumara' a few years earlier who wanted to go to the coast but was
sent back to Akuna. They had been married shortly before our arrival,
in Akuna. He was the youngest child in his family or orientation, she
the second oldest child but the oldest daughter in her family. These
are factors which might enter into the situation. Apono was always in
the company of married women, she was busy scraping 'yienni', making a string bag, preparing a meal in the earth oven. She spent long
hours toiling in the garden which Iripe had inherited and somehow
maintained. In clear contrast to her Iripe was always in the company
of young men who were still 'pumara'. For hours every day I watched
them play cards; young boys from fourteen and fifteen years of age,
joined in nearly every case by three married men: Iripe, Wanara
(whose young wife had divorced him because he could not share the
weight of married life), and Ropa the son of the former luluai who
gradually withdrew and by the end of the year occupied a full position
in one of the logging teams while spending his free time assisting his
wife in their coffee garden.

While we are here dealing with age categories, it should be kept in
mind that while the 'pumara' is occupying a role of transition, the
'pumara?ini' is frequently married early in this period. While the male
status must be ceremonially terminated, the female goes directly into
marriage. The male is a 'pumara' between his 'upaiyami' and the ceremonial 'orande', after which he may marry. The female is a 'pumara?ini' between her first menstruation and her marriage — there is
no ceremonial transition which qualifies her for marriage.

The 'orande' ceremony once again takes place during that part of
the year when garden products are plentiful and the daily work load is
lightest. This is between November and January. As in the case of
the 'upaiyami' ceremony the decision regarding time is taken by the
father of a 'pumara' — usually attempting to find a number of his son's
age mates who can be initiated at the same time. These young men are
about nineteen years of age when this ceremony takes place.

The preparation for and duration of this ceremony takes approximately three weeks. During this period there is a general taboo on
sexual intercourse by members of the community for fear of contamination and death of the initiates. While sexual intercourse never
takes place in the village, the danger always exists that a person
might inadvertently disregard the purification requirements thus
producing impurity. Since people transgress taboos, in any case, the
danger can be "cooled down" by eating taro which has been grated on

a kanna stick and cooked in an earth oven.[11] While the 'orande' itself lasts a week, two weeks are involved in preparation. The women of the community will gradually start building up a surplus of tubers, greens, and firewood. This will allow them a greater amount of time in the village during the ceremony. For a week or ten days before the actual ceremony men spend an increasing amount of time in the bush — even sleeping there — while they set traps for and collect tree kangaroo, opossums and similar animals which are permitted during ceremonial times. The catch is carried to the men's house, and hung on posts inside the structure.

Each 'pumara' who is to make the transition to manhood receives from his mother, or alternatively from his eldest sister, a new 'unami'. His father will present him with a new 'wanaire', the adult male garb consisting of the tiered grass skirt in front with a clothlike sash hanging down the back and made of the 'wandim' bark. It is called 'maro'. Word is also sent to the young man's clan members who form part of other residential communities, and especially his mother's brothers, that he is to be presented in the 'orande'. In the clearing which is found in front of each men's house a number of tall posts, perhaps twenty feet high, are erected. There will be one for each 'pumara' being initiated into manhood. To these posts, called 'onankarano', all possible varieties of food are attached as well as a display of valued objects. The posts which we witnessed at Omaura had taro, yams, tape, sweet potato, banana, pineapple, sugar cane up to thirty feet in length, 'adnumi', tobacco, pandanus (both the nut and the fruit bearing varieties), and greens (more for decoration than use). In earlier days, it is said, a young man would also have stone adzes, bows and arrows, and similar prestige and utilitarian objects on his post, but on those we saw brightly colored towels and laplap from the trade store as well as steel axes had replaced the traditional objects. This post, is like a coming of age gift from parents to their son; it marks his transition into adulthood. When I inquired why the taro and tape still had leaves and roots, it was explained that these posts carried the food for the feast but in addition had much wealth for the boy to set him up as an independent man. Here he got tools and weapons, and now clothing; here he got garden products with roots and leaves intact as if the parents were suggesting that he should get out there and plant his garden. This was in fact directly stated at Omaura, for once a fellow has been initiated in this ceremony, he may marry and must assume manhood with its responsibilities. Also in preparation for this ceremony the adults all braid their hair into

11. It will be recalled that it was this very kind of soft food which was taboo during previous ceremonies.

1, 2. Torawa and his
mother Wakano
3. A young mother assists
her brother in the ritual
strengthening of her infant
son (see page 215)

Orande ceremony. Women and children form a half-circle near the
men's house and present roasted and cooked food. The males remain
in the center of the semi-circle, holding or fondling infants

numerous long tussels smeared with pig's grease, and don newly completed grass skirts.

It should be kept in mind that the young men have been living in the young men's house. Since their mothers still prepare their food, they would usually enjoy it in the village square, perhaps huddled to the one side or else retreat with it to the sleeping quarters. The last week before the ceremony, however, these fellows simply gather food and take it into the bush.[12] They hide it somewhere and will then gather in a certain spot during the day to eat together. During this final week too, the 'pumara' rise while it is still completely dark and leave to spend the morning in the bush. When the women have completed their domestic duties they leave the village with all the children and the 'pumara' now return to assist their fathers and elders in constructing long platforms, now-a-days referred to as 'tables', on which food is heaped. On these platforms there is a special place for each adult male and it is marked in brightly decorative croton leaves.

Amidst periodic gatherings in the men's house when traditional songs[13] are sung, the men kill pigs and prepare the meat. Work and singing is interspersed by the wailing of the sacred flutes and the asthmatic breathing of the bull roarer. On the afternoon of the first and second days of the ceremony, each man takes a large leaf which acts as plate and sets out enough food for each of his dependents. Each plate also receives a share of pork. Food for the 'pumara' is placed in a central spot which has been agreed upon. The men now 'wiwirem' (call) the women and children who collect their food and then return, as the shadows lengthen and night sets in, to their houses. As they fall asleep, they hear the men in the distance, and further off in the bush the sacred objects remind the uninitiated that the ancestors are near. On each of these evenings, when darkness has set in, the 'pumara' return from the bush by a different path and return to the 'pumara yi ma?i'.

On the morning of the third day all the 'pumara' again leave for the bush while it is still quite dark. On this occasion they are led by an old man and a younger adult who was initiated three or four years before this 'orande'. They are informed that on this day they must go out and cut bamboo to cook large quantities of grated food. When they arrive at the bamboo bush one of them is handed a rough piece of stone and told to start cutting. In the meantime the adult males have gone by another path and started clearing a path about three yards

12. The adult males fetch their morning meals from their wife's home and then return to the men's house where their evening meals will be served.
13. The types of songs which they sing on these ceremonial occasions are 'yana', 'ompa', 'uwaki', 'wininda' and 'poponkani'.

wide and perhaps ten yards long. The sides of this clearing are beaten back and decorated with brightly colored leaves. Also the men are decorated with leaves, which are hooked into the 'yanka' which encircles the biceps, legs and similar body parts. At one end of this clearing two men take up position with sacred flutes. On the ground, at the other end of the clearing, a man lies prostrate with an arrow protruding from his body and his hands folded on his chest firmly gripping the arrow. The "corpse" is partially covered by a pandanus leaf blanket. All the other adult males hide in the bush along the sides of this clearing.

Meanwhile those in charge of the 'pumara' have decided that their progress at cutting the bamboo is too slow. They are instructed to rather cut the bark off a tree since the bark may also be used as a container for the food. Again they are supplied only with a 'yonkuni'.[14] These men who are in charge have instructions to keep the young fellows busy until they hear the sacred flutes. When these are finally heard far off, the 'pumara' are led in that direction. As they enter the clearing the old man confronts them with an accusation that they have been playing around for days and wasted their time this morning thus permitting the 'pumara?ini' to kill that poor man ahead of them. "Now go and remove that arrow and correct at least some of your wrongs!" they are told. As they approach the corpse, men appear all round them and proceed to beat them with sticks and leaves. This informants felt was essential because these fellows now were going to have to behave like men and bear hardships in fighting and work. As each 'pumara' in turn attempts to remove the arrow, tugging and pulling at it, the adult males keep beating the young men. Then again, summoned by the noise, he says, the mother's brother appears. Again he shields his nephew from further blows. As the sun sets he leads his nephew into the village, and in the 'pumara yi ma?i' he gives him his own portion of food which is raw. There is no fire in the house. As the sacred flutes are heard approaching the village, each mother hastily places a burning log outside her door and returns to the hearth. The 'pumara' collects this burning firewood and return to the young men's house for the night.

The following or final day is a day of relative license, though not complete sexual license as has been reported for other communities. The 'pumara' have the chance now to retaliate their beating of the previous day on their age mates of the opposite sex. They spend much of the day running after them and beating them with branches. The girls who soon realize what is happening run away and hide with the

14. This adze head has been sharpened but is harder to handle and manipulate than the 'okunta?i', or hafted stone adze.

246

result that much time is spent in this pursuit. In the late afternoon the women, who have prepared a large meal, bring it to the men's house. This is the final day and the feast. As on previous evenings every woman and child must leave the houses and gather at the men's house. They do not approach the entrance but form a horseshoe shaped semi-circle before the men's house. Before them on the ground they place their taro, yams, and sweet potato, their dark scorched bamboo containers of grated mush, their corn and other foods.[15] With babies on their laps and children huddled around them, the women form an outer circle. The men now emerge from the men's house, each going to his family, and, squatting down on the inside of the semi-circle, talks with them and may even hold an infant on his lap for a while. The men do not eat at this time or in the company of women and children. Each adult male takes his food and a share for the 'pumara' into the men's house and places these in that part of the ceremonial structure where he usually sits. When all the men had received their food, they all turned to the 'table' — which had been constructed earlier — and returned with long pieces of the thick red sugar cane. This they broke into single nodes and presented the women and each of the children with a piece. The fact that this sugar cane was broken is important, for while men tend to these plants and grow them to as great a length as possible for ceremonial and prestige reasons, females are never allowed to handle or carry these. Women and children may only transport or enjoy cane broken into single nodes.

With their 'payment' in hand the uninitiated now depart leaving behind the adult males. The men now stand in front of the men's house, turned toward the bush, and call once with a ceremonial swinging of their arms to the 'pumara'. The latter come to the men's house to collect their meal, but they do not enter it, and having taken their meal, they return to the 'pumara yi ma?i'.

While the women are preparing for the night's feast and dancing, the young men once again receive instruction regarding the sacred objects they had seen. The boys are now decorated while other adults take down the posts removing the food for the feast and the other objects for the young man being honored. As the men have completed their work and entered the men's house to put on their ceremonial regalia of cassowary feathers, pig's teeth, dog canines etc. the women appear in festive mood singing ('i?ireno') and dancing in the square in front of the men's house. Now the 'pumara' who have been honored

---

15. Changes in dietary habits and new prestige items are also affecting these gatherings. At one of these meals I was served a warm cup of tea by a lady whose husband was in the men's house but associated with S.D.A. mission personnel.

appear in adult male dress and are presented to their fellow villagers. The rest of the night is spent in festivities. The initiates, however, are still 'pumara' and remain in the young men's house until marriage. The point, however, is that the 'orande' qualifies the male for participating in war parties since he is now a man, and for marriage, and while courtship may be initiated by either sex — and most likely has been going on for a long time in the form of trysts — the final stages may now be entered into. Once married he is a man and will be admitted to the 'wadnama?i', and he will be expected to construct a house, 'inama?i', for his wife.

During the whole time which is taken up by the 'orande' men may not sit down or eat in the women's houses. They go there to collect their morning meal but return directly to the men's house. Nevertheless, dangers may be involved even when a man just enters the house of his wife and she is expected then to neutralize this danger by collecting certain plants. These consist of 'no?ana' (the 'ana' — leaves — of the 'no?i' which is used to make string bags) and 'pandini' (this is a small variety of grass with narrow leaves). The woman will rub these with pig's grease and place them on the floor at the entrance to her house. When her husband enters the house he wipes his feet on these leaves thus protecting himself and the initiates.

This period of ceremonial activity makes the 'pumara' especially susceptible to dangers since they are in a state of transition. Those who are being honored may not eat 'uwoya' (the soft tops of pit-pit grown in the gardens), 'moku' (small dark green leaves which are cooked with pork as a spinach), tree kangaroo, rat, 'wari' (eel which is caught in traps), 'kauya' (a small fish) and 'kaua' (a frog). Prior to the actual ceremony, it is taboo for them to eat soft food, tubers, grated on a kanna stick and cooked in the earth oven, but during the final week it is permitted.

Chapter 8

ADULTHOOD THROUGH MARRIAGE

The nuclear family is formed by a legal and recognized step during
which either a person or goods are transferred from the bride-
receiving to the bride-giving group, in exchange for a female member in the latter group. The vague use of 'group' in this context
rather than speaking of 'brothers', 'kinsmen', or some other defin-
itive term, is a purposeful choice. The persons who contribute to a
betrothal gift, a marriage gift, or to any other prestation, and among
whom counterprestations are in turn distributed, do not agree with
any unilineal or bilateral tracing of relatedness, but in all cases
include persons who identify with the individual concerned due to kinship, to affinal ties, or simply because of common residence.[1]

Age is no bar to marriage, only to the full expression of conjugal
rights. In earlier times pre-pubertal boys and girls were frequently
betrothed or 'married' to older adults. This was a product of the
practice of brother-sister exchange. On the average today, males
enter marriage for the first time around their nineteenth or twentieth
year while females are married two or three years younger. All marriages are ideally with persons to whom no relationship can be traced
bilaterally. It should also be stated that while there are no official
trial marriages all marriages are on a trial basis. While marriage
is preceded by a formal bethrothal these two should not be separated
by too long an interval. The justification for this warning was again
explained to me by quoting a culture hero tale.

There was a young man who lived a long time ago and he liked a
girl. She liked him, too, and told her mother that she would marry
him. Now this young man had already transferred his betrothal price
for the girl, but had not married her yet.

One day the two were in the bush collecting edible mushrooms.
They had already filled their 'unami' with mushrooms when he suddenly
turned on the girl and killed her. He took her corpse and wrapped it

1. For a discussion of this see our chapter on the village economy.

in leaves together with the mushrooms which they had collected. This he carried back to the village and gave it to the girl's mother. That night they all sat around the fire enjoying their evening meal when the 'wami' of the girl came out of the bundle of leaves and pinched the arm of her mother. While the mother did not see or realize it, the boy noticed the 'wami' of his betrothed and hurriedly left the house. He quickly called his mother and father and they fled from that village, but he forgot to call his sister, who was living in the 'pumara?ini yi ma?i'.

Later that evening the deceased girl's mother decided to look at the mushrooms which her future son-in-law had brought. She unwrapped the leaves and found the 'anda' of her daughter inside.

Struck with grief, she called her brothers and sisters who went to the young women's house where the unmarried girls lived and killed the girl who remained.

THE BASIS FOR MARRIAGE

Marriage to the Akunan is more than a union for the sake of producing children or uniting economic partners. It is based on certain desirable aspects in the other person, and either the man or the woman may take the initiative in courting. Some factors considered in this respect are the health and appearance of the woman, her abilities as a gardener, her disposition, and sense of humor. A man should show some promise as a mate and supplier and the qualities which make for leadership.

Usually, some previous contact has occurred between the partners by the time the betrothal takes place. In contrast to other highland language groups for which information is available, courtship is often initiated by the girl rather than by the boy. If they are both young, a great deal of premarital sex play and occasionally intercourse takes place in the bush, and each will evaluate the other as a possible marriage partner. While the girl very often initiates contact with the man of her choice, arranging trysts or other meetings, it is also the girl who in most cases presents herself to the man she wants to marry. This may be in the form of simply joining him or of arranging a marriage. In the tenth chapter we will discuss the status of women in greater detail but early observers repeatedly mention the fact that women may take the initiative in marriage arrangements. This very often leads to a disparity in age between the partners as when a girl of sixteen married the luluai's son, who was no more than twelve years old and the patrol officer suggests that she did most of the seducing (Blyton, Report No. 32 of 1944-45). This phenomenon by which a young boy might be married to an older woman, or the other way around, was also a product of the brother-sister exchange which seemed to operate to a greater degree in the past and perhaps more among Akuna's neighbors than within the community. Skinner (Report No. K2 of 1946-47) sees this as the reason a "lad of ten or twelve may be married to a young woman of eighteen or nineteen — sometimes a young man to an elderly woman". The same observations are found in Linsley (Report No. 7 of 1948-49) and Jackman (Report No. K4 of 1947-48), the first of whom points out that these arrangements are frequently intra-village marriages. This is important for it means that the brother-sister exchange is in fact uniting and binding the local community sub-groups into a strongly integrated unit.

The significance of marriage

For the people of Akuna the act of marriage is of great importance and sets into operation a complex of interrelated factors, affecting the marriage partners and their future in the social groups they live in.

By transferring valuables for a woman, a man shows that he has reached maturity and that the woman is mature and able to enter into the responsibilities of producing children, gardening, and maintaining a household. Though he has met all these requirements, he is not a true man before his wife has produced children, for only then can he compete with others in the realm of men. The status of maleness, ascribed as it is, leaves open the subsequent steps which each man must take to achieve recognition. For this reason it is significant that among the nearly 4,000 names recorded in the genealogies, not a single bachelor or spinster occurred.[2]

By contracting a marriage, a man also aligns with members of his wife's kingroup and can expect their assistance when he needs it. Once again this is a step toward the achievement of status, for in the choice of a spouse, both persons consider a variety of factors in addition to physical attractions or economic capabilities.

Marriage is important in one other aspect, namely, the fact that it legalizes children. Though children may have been conceived out of wedlock, they are never born outside marriage, and informants could not recall any bastards in the group. As soon as a girl realizes that she is pregnant, she tells her mother. Her kinswomen then extract from her the name of the boy. Confronted with the news, he always consents to marriage, thus assuring the birth of the child in a legalized union. This aspect of marriage may then be referred to as the

2. Male or female of over twenty-five years of age who had not been married.

251

principle of legitimacy, namely the embodiment of the child in a recognized, formalized group, the family.

The rules governing marriage

There is no age or birth order requirement for marriage,[3] and while some persons may be betrothed before puberty, others do not make a decision before reaching maturity and selecting a mate. There are only two requirements that have to be met before marriage can take place; a girl must have reached puberty and a boy must have taken part in the 'orande' ceremony; by these ceremonies the persons concerned are declared ready for marriage and the sexual and economic responsibilities which this entails. They are prerequisites for marriage, though not for betrothal, and in earlier times it was common for pre-pubertal boys and girls respectively to be betrothed to and subsequently to marry older women and men after they had come of age and met the ceremonial requirements. This betrothal of children cannot be referred to as infant betrothal even though very young children were often times betrothed by their fathers or elder siblings. The primary reason for this practice, it seems, is to be found in the brother-sister exchange which was formerly much more prevalent than it is today. On such occasions of marriage a man might promise his younger pre-pubertal sister to the brother of his wife, thus assuring that the latter will have a wife.

It is forbidden to contract marriage or enter into sexual relations with persons with whom one can trace consanguineal ties through either parent. Any woman who through your father or mother is directly related to you, and this extends beyond the fourth cousin, is seen as a sister.[4] No cross-cousin marriage is allowed[5] though this is reported for the area southwest of Gadsup country (Berndt 1955: 32 and 187) and was also recorded among the Ataya Gadsup. This bilateral exogamy is ideal, and is adhered to, but it is also possible that the cultural flexibility allows for weak memories[6] and that they actually use genealogical shallowness, but not in exactly the same way

3. Among the Busama on the northeast coast, the principle of primogeniture is extended to marriage (Hogbin 1946: 135).

4. Among members of the Samoan ''aiga sa' marriage is also prohibited between all persons ''who acknowledge common descent'' (Ember 1959: 574).

5. In terms of the structural implications of brother-sister exchange about which we have spoken, it should be noted that Diagram 9 contains in effect a cross-cousin marriage though such an exchange did not precede the union.

6. Something that Barnes discussed under the term ''structural amnesia'' (1947: 52-3). In this respect, see also: Gulliver (1955: 113-4).

as the Kuma do (Reay op.cit.: 34-6). From a total of more than 500 marriages recorded in genealogies, only eleven were between persons whose relationship could be traced, and we will discuss these below.

1. Man marrying his FaBrWiBrDa: Under this category there are two cases, one in the present generation of married adults, and one in the previous generation, both deceased.

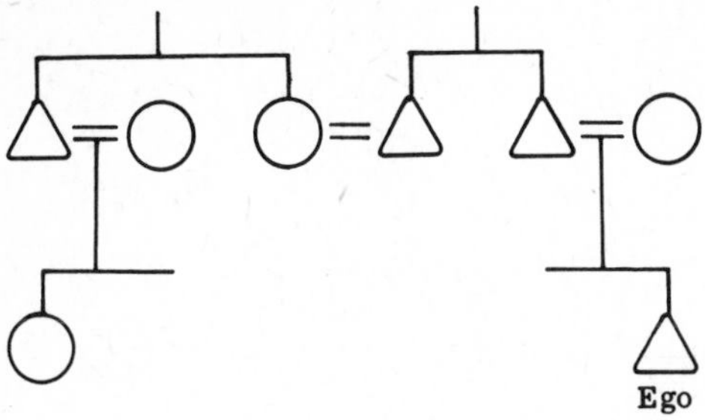

Diagram 9. Man marrying his FaBrWiBrDa

2. Man marrying his FaMoSiSoDa: This is a young couple of whom the woman's parents still live, and from whom children are born.

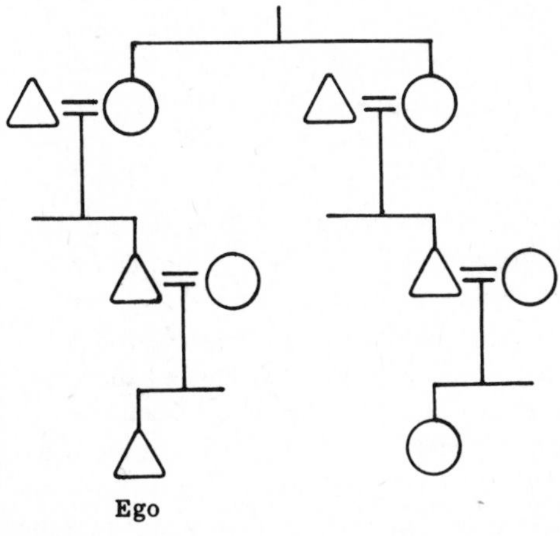

Diagram 10. Man marrying his FaMoSiSoDa

3. Man marrying Fa first WiFaFaSiSoDaSoDa: In this case Inde has two wives; both are in the same relationship to him, as their paternal grandmothers were sisters. Both of these co-families are young members from whom children have been born.

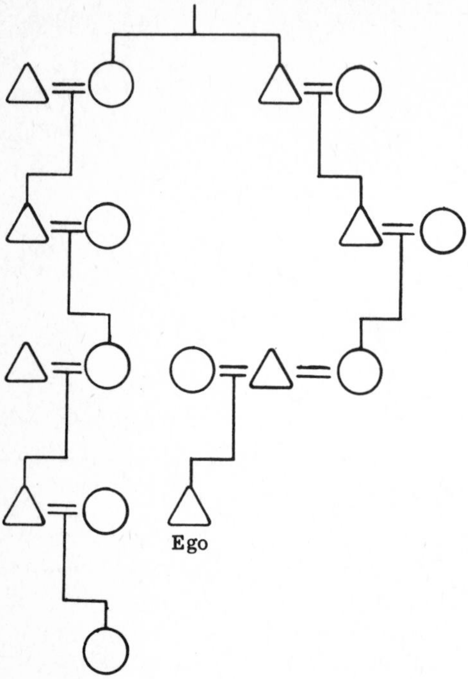

Diagram 11. Man marrying Fa first Wi FaFaSiSoDaSoDa

4. Man marrying FaFaSiSoDaSoDa: There are three cases in this category. Two of the women are sisters, one of whom is the co-wife with the third woman of one of the men.

In speaking of the riyer dwelling Mundugumor, Margaret Mead points out that these people "have a form of organization that they call a 'rope'. A rope is composed of a man, his daughters, his daughters' sons, his daughters' sons' daughters; or if the count is begun from a woman, of a woman, her sons, her sons' daughters, her sons' daughters' sons, and so on." (1959:127) Every person tries to use the ropes of a sister or a brother in a co-operative way by the establishment of mutual obligations between the descendants of an intermarrying pair of brothers and sisters. Van der Leeden suggests that this type of • "asymmetrical cycle through the generations" is more or less instit-

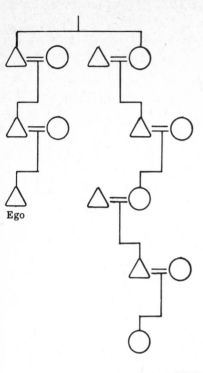

Ego

Diagram 12. Man marrying FaFaSiSoDaSoDa

utionalized in New Guinea and "is indicative of a certain lack of balance" (1960:142).

Now the alternation of generations is clearly observable in Diagrams 10-13. In the first three examples this 'rope' effect is particularly clear in the woman's line while the last diagram shows feature in the case of the male ego. It seems then, that the bilateral tendency on which I have remarked may also be present in this context by permitting wider association and affiliation with eg. the father and his family but also the father's mother and her family. Pouwer (1966) speaks of this as an infrastructure where there is little emphasis on filiation and descent (or genealogical depth) but maximum stress on horizontal ties.[7]

5. Man marrying FaMoFaMoBrSoDa: These two persons belong to the same age-grouping, though their positions on a genealogy seem to suggest a great age discrepancy.

7. This issue is discussed in greater detail on page 339.

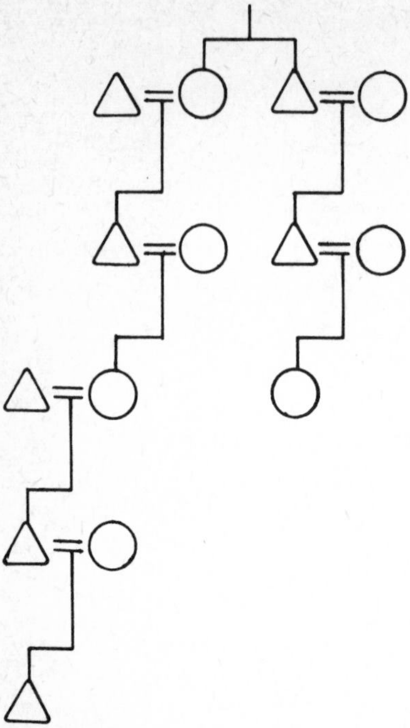

Ego

Diagram 13. Man marrying FaMoFaMoBrSoDa

6. Man marrying MoMoBrSoDa: This is a young man who was approached by the woman, and she simply moved into his house as a co-wife. The man's mother was upset, but his father told him not to worry about marrying his 'sister' as she came to him.

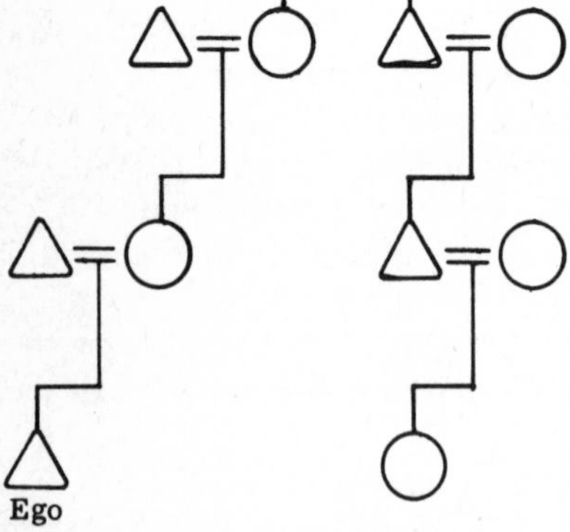

Ego

Diagram 14. Man marrying MoMoBrSoDa

7. Man marrying FaFaBrSoDa: Once again these are two young people, and they seem to be attracted to each other on a purely romantic basis. When the man's brothers told him not to marry his 'sister', his reply was that they liked each other and should they (his brothers) not be willing to contribute to the marriage payment, he would face it alone.[8]

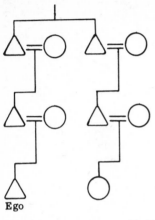

Ego

Diagram 15. Man marrying FaFaBrSoDa

In a recent publication Van der Leeden comments on the "complex forms of marriage resembling the Aranda System in Australia" (1960: 141) which are widespread in New Guinea, and which he recorded in Sarmi as well. Our categories four and five above, being a total of four of the eleven irregular marriages, resemble those which he discussed but could not be definitely classified. In personal communication, however, he makes a point which seems of great importance in terms of the small groups and the ties which they attempt to maintain. Referring to his own, and those listed above, he states that: "All these marriages indicate a tendency against too open marriage regulations. Somewhere in the descent lines, when kinship bounds are not 'close' any longer, one tries to continue marriage relations, which existed some generations ago between the linking ancestors. Perhaps this is the most basic element in many Papuan marriage systems, and perhaps it is better to refer to this tendency than to the 'Aranda' structural type, when starting a description of the situation in New Guinea."[9] It is then possible that the pattern which seems to emerge, even though

8. This reminds of the "Romeo and Juliet relationship" which, however, is institutionalized among the Tor (Oosterwal 1961:120-21).
9. Personal communication with Dr. A.C. van der Leeden 1962.

257

this is a very small number in the total of marriages recorded, suggests a previous lineal system of which these are the last occurring tendencies.

While it is impossible at this stage to discuss the possible connections of these four (and the other seven) cases, it should be noted that in all cases (except number three and number six) the relationship of the male marriage partner is traced through his father and in all cases through the girl's father.[10]

The forms of marriage

First marriages, and secondary marriages do not differ too greatly, are contracted among the Gadsup in one of four ways, namely:
1. Upon initiation of the boy or his interest group;
2. Upon initiation of the girl or her interest group; both of these being followed by the transference of marriage goods;
3. As a result of brother-sister exchange; or
4. As a result of wife capture.

The majority of marriages involve the transfer of goods between the bride-receiving and the bride-giving groups, the amount of these goods depending upon the group initiating the transactions. When it is the boy who makes the first overtures, payment is larger than when the bride's kin suggest the union and take the first steps in this direction. Should a girl favor a boy, she will make this known to him in some way, e.g. by a message via his sister, and then secretly meet him in trysts. If her family approves, her father and brothers propose the marriage. Indeed she may simply go to him and offer herself.

The latter practice is especially found in divorced persons or polygynous marriages, and the act of divorce where a woman leaves her husband and attaches herself to another man. The initiative, however, may also be taken by her male relatives without prior courting, as was the case during our field study when Opura appeared one morning at the house of Urinanda and proposed that his daughter, Todudu, be marked by Urinanda as a wife for Yantape, Urinanda's son, who was absent on the coast as an indentured laborer. The girl's father, Opura, related that he was getting worried because no young man had come to "mark" her as his bride. He was also worried because he and his wife were getting old, and found that whenever a ceremony or feast took place in some village, often a long way off, his daughter would go there and he feared the possibility that she might marry a man in some distant village. "Who will care for us when we are ill, who will work

10. Appreciation is expressed to Dr. Barbara S. Lane for pointing this out.

in the gardens while her mother is ill, and who will take care of the younger children if she lives in a distant village?" This statement points to the importance of the nuclear family, and the close ties its members maintain. It is the daughter of the household who aids her mother in the garden and in the care of her younger siblings, and it is she who substitutes in these realms when the mother is ill. Her children, and those of her siblings, will be responsible for their aged grandparents when these are no longer able to take care of themselves. The statement also suggests that while it is good to spread the ties of affinity, they should be kept close enough to be of service to the nuclear family members, an aspect neatly expressed by Schoorl, when he states that "the function of the marriage is to establish relations" (1957: 40).

Informants state that a man transfers payment for a wife to prove that he is not a rubbish man, and so that his wife will not accuse him of this if a domestic quarrel takes place. It is partly a purchase price, but insures only minimal ownership. By transferring marriage goods for a woman, the man is assured of her economic cooperation, her social importance as wife and kinsman of a different kingroup, her biological and procreative abilities to satisfy him sexually and to produce offspring which will prove his virility and assure him of children who will care for him in his old age. The rights which are transferred at marriage are difficult to define precisely because there is such great mobility in the society; and it is not always clear whether a marriage is dissolved because certain rights were misused, because obligations were not fulfilled, or simply because of the whims of the partners. This latter statement infers that predictability is zero due to the freedom of choice in behavioral contexts, but that is not the point made here. What we are saying is that within the permitted realm of behavior there exists such a latitude that it is difficult to specify the exact stimulus which prompts certain choices. One thing which is sure is that the transfer of marriage goods does not assure full rights over the person and her property. She retains her own private property, her right to visit and retain an interest in her natal village, and the right to react to ill treatment by the man or his kinsmen. She furthermore retains her land rights, and while he shares in the products, he does not acquire the right of ownership. It seems as if there are grounds for disagreement with the way in which Le Roux interprets bride price and its relation to land ownership. In dealing with the highland peoples of the former Dutch New Guinea, he constructed a fallacious paradigm regarding the ownership of land. He points out that males and females may own land (according to Van Eechoud), and that according to De Bruijn only the males own land, and he (Le Roux) then comes to the conclusion that in fact the data are not contradictory, "as

259

the woman is the property of the man, the landholdings are, therefore, his too" (1950:677). Though we enter again upon the old anthropological controversy regarding the extent to which the marriage gift actually 'purchases' a woman,[11] it is felt here that this statement goes to the extreme of not only 'buying' the woman, but also the privilege-rights and demand-rights (Hoebel 1954: 56-7) which she has due to her birth and membership in her group. In Akuna the marriage gift does not stabilize marriage, as it is a culturally accepted course during divorce for the second husband to give payment to the first husband. The interest group of both contribute to and share the payment. The marriage gift is not returned. The right of the first husband to claim his original payment from the second husband in fact seems to legitimize divorce. Once a woman has been married and her kinsmen have received a marriage gift for her, they move into the background and all future payments and claims are transacted between the successive husbands, and only then is the affinal bond re-established through visitations and prestations.

A girl may also decide that she would like to live some place else and desert her family of orientation. Such was the case with Wawentino, daughter of Ari and his first wife, Adudu. It seems that about four months prior to our arrival she had decided that she was old enough to marry and her parents had contacted a promising young man in Akuna who marked her. Wawentino, however, rebelled, saying that she did not like Akuna and that the people weren't friendly and she did not wish to settle there. She left for Amamunta and after a few days there went to Ikana where she met a young man, Tapapute, to whom her father married her. After a few weeks, she left Tapapute and attached herself to an older man named Kandi?o. About six months later she returned to Akuna, stating that she was tired of Ikana, nor did she like her husband who was getting old, and had "hair all over his body". Her parents and sisters, and in fact every other person in Akuna, tried to force her back to her husband (not because of the sanctity of the marriage, but because she had broken ties with Akuna and had refused to marry an Akuna boy), as they did not want her back. The first week back in Akuna Wawentino spent her nights in the women's house of seclusion. As tempers cooled and relations warmed, she returned to her parents' house. Some weeks later I no longer noticed her in Akuna and was told that she had decided to return to Kandi?o and visiting Ikana shortly before my departure from the field, she was one of the first to welcome us as visitors from her natal village.

About five years ago, Torawa of Akuna and Waropi of Amamunta left the Arona Valley for police training and duty at Madang, Mt. Hagen

11. See also Opler 1937: 210; Brandel 1958.

and Mende. Both were married at the time they left, but while they were away, Maka?o, wife of Torawa, because she was lonely, committed adultery, and then left Akuna to marry in Aiyura. When he returned to Akuna, Torawa found three women at his house; a young widow who later married Duka; a young woman who subsequently married Yayo; and the attractive and personable daughter of Munowi, whom he married. When Waropi returned, he too was a very desirable young man, and a young woman, Tiyorono, left her husband E?ananda of Wayopa, and married him.

It often happens that a boy is interested in a girl and will attempt to draw her attention by winking at her, by visiting her in her mother's garden, or by a message for her to meet him at some place outside the village. They walk alone in the bush and talk, court, and often engage in sexual play and intercourse. This is done alone, and even without the knowledge of their age mates. For them, even prior to marriage to spend time together in public and to sleep together as was the case in the Trobriand Islands (Malinowski 1929: 65-6), is unheard of. This is a period of trial and error during which a person experiments with different partners and makes up his mind about whom he will marry. Should a girl experiment too widely or too frequently, she often gains the name of being a Pamuk Meri (Neo-Melanesian for prostitute, also called in Gadsup 'Pamuku?ini', this being a Neo-Melanesian term with Gadsup feminine suffix), but traditionally other people refer to her as an 'umori?ono', literally: "One who steals habitually" in contradistinction to 'umonemi', "one who steals" (once or twice).[12] Boys, however fickle they may be, are never dubbed with a similar term. While sexual prowess and experiences are frequently boasted of among young men, it is not in such a way as to create the Playboy impression, nor were cases recorded where girls boasted of triumphs on the sexual level in contrast to the level where she had won a boy's interest or love.

In many cases a boy is not able to attract the attention of his girlfriend, and when his repeated winks and secret signs go unheeded, he will often resort to 'watuye', love magic. This is prepared by obtaining a leaf and a piece of bark of the 'Aweweyim' tree. These are rolled together and burned. The ashes are collected and for a period of time placed in the young men's house, immediately above the sleeping place

---

12. Regarding the Kamano to the southwest of the Arona Valley Berndt (1962:371) remarks: "In this sense theft and promiscuity (the same term, 'gumaja', being used for both) are regarded in much the same light and considered to require similar punishment". Also Pospisil, for the Kapauku, states that "both rape and adultery are termed 'oma magii', 'theft of sexual intercourse'" (1965:55).

of the boy. After a few days he will see the girl he likes and put some of the white ash in a leaf of tobacco which he will offer to her; he may also take a sharp piece of grass and inject some of the ash into a betel-nut which he will give to her. 'Watuye' may also be transferred, so tha he may give it to his sister and request her to dust it into the girl's hair, her grass skirt, or tobacco and betelnut which she would enjoy. 'Watuye' is said to be infallible and would cause the girl to notice his overtures or might actually cause her to initiate courtship as has already been discussed.

Girls do not have any kind of love potion like this, and in light of their position, they do not seem to need it. In the material which has already been discussed, and from much to follow, it is suggested that the woman in the community of Akuna has a high status. Her sphere of daily activities is of great importance to the maintenance of the social order, her rights and duties are in most cases equally impor-tant as those of men, the statuses which she may achieve, e. g. shaman are equal to or superior to those of males, and her roles as love partner and wife are as dynamic as those of men. She has the privilege of throwing herself on a man's doorstep, without overstep-ping the boundaries of permitted behavior, and for this reason a potion which is supposed to secretly affect the feelings of a man is not needed by these women.

The third form of marriage practiced in this community is that of brother-sister exchange which, it seems, is related to infant betrothal. This form of marriage is not very prevalent today, though, but often leads to union in which a great disparity of age exists between the partners.[13] While such marriages may have been more common in earlier years, it is also possible that age differences between the partners caused them to be noticed and recorded by officials.[14] Of the total number of marriages recorded in the genealogies, however, no more than fifteen are of this exchange type. In the context of the principle of reciprocity these marriages are significant, for, instead of a transfer of food and other valuables between two families, we have the exchange of one member of each group.[15] Akunans lack any suggestion of primogeniture in the exchange marriage, as is the case in Busama, though this seems to be an advantage and implies that

13. Patrol Report: D. Whitforde, No. K. 4 1948-9. Mention is made of a' tem-porary marriage between a young man and a woman of about forty, until a young woman becomes available.

14. Patrol Reports: Lt. D.R. Blyton, 1944-5, No. 32; R.I. Skinner, 1946-7, No. K. 2; G. Linsley, 1948-9 No. 7.

15. Levi-Strauss (1963:46) links this exchange of women to the universal presence of an incest taboo, which allows too for the role of the mother's brother.

elder brothers in Busama could manipulate their sisters. Among Akunans this exchange of partners seems to suggest the presence of certain elements characteristic of unilineal groupings in this social structure and should be seen as related to the eleven irregular marriages recorded above. We are dealing here with a social organization which lacks any group of primary importance outside the family, but yet we find the principle of sister exchange, a characteristic usually found with clans or lineages, for it is one form of preferential marriage set up and decided upon by others. We are, furthermore, dealing with a social organization in which a system of exchange of goods for a wife functions, a system then where the sister will be substituted for goods and valuables. The Akuna situation differs, however, from the Tor region where Oosterwal feels that exchange marriages are decreasing and being replaced by bride-price marriages because the group is too small and girls are not available (Oosterwal 1961:113-4). In both cases we are dealing with a change that is taking place in the social organization but which seems to be based on different reasons, because Akuna certainly and the Gadsup on the whole do not have a shortage of girls. There may be, however, a shortage from time to time within a specific district and with the practice of district ties and small group identification in mind, Oosterwal's explanation may apply among the Gadsup. It is felt here that this change is but one example of the trend away from structural unilineality. The purpose is a dual one, to do away with bride payment and to link the families together, but the latter aim has already been reached and the emphasis then is upon the first — an act of kindness toward a family with whom amiable relations have already been established. Exchange marriages are not contracted at great distances from the village, as the family wants the assurance that as they had given a daughter, they will in future receive one. As one had given food to one in need, so one was also sure of receiving food when needed.[16] Both this brother-sister exchange and the early temporary marriages which have been mentioned are seen here as adaptations to the hostile social environment. While hostilities continued between neighboring villages, a person might have been forced to form a temporary union with a person of a different age, or to give or promise his sister to another man in exchange for a wife.

In all three of these types, we are dealing with exchange transactions, and in two of them certain valuables, edible and durable, are transferred in exchange for certain rights and privileges over the woman involved. Note that we are speaking of certain rights and

16. This is the way the Moejoe in the former Western New Guinea express the exchange marriage (Schoorl op. cit.:41).

privileges, and that the girl is not bought by one group. The Gadsup express this transfer of rights very clearly when they state that "when a woman is married, her husband acquires all rights over her body (and reproductive abilities), but her family members retain her arms, hands, and head". The husband acquires a limited right over his wife, but she has the privilege of visiting and helping her own family of orientation whenever they are in need.

The last type of marriage mentioned is the case where a wife is taken by force and consequently no valuables change hands. This takes place during fighting, but no raid is organized with the express aim of getting wives, since sufficient girls are available. Even if hostilities do exist at a particular time with most of the neighboring villages, there still existed the possibility of brother-sister exchange mentioned above. In the case of younger villages where the available mates belonged to a person's own kingroup, this might have created problems, but it is suggested below that even at the point shortly after fission takes place there are sufficient non-kin who align with a leader to make each village self-sufficient even in this respect.

Informants stated that in earlier times during an attack on a village a man might see a girl or woman to whom he felt attracted and would inform all the others not to kill her as he wanted her. Some examples of this will be discussed briefly.

1. When Akuna fought their 'namuko' (traditional enemy), Tompena, Afita killed a man named Yuwarunke and then took his wife To?andi as his own. They lived in Akuna and she became part of a numerically and politically strong family and was the mother of Napiwa, one of the present leaders in Akuna.

2. Some years before the Australian Administration took over control of the area, Akuna attacked Onanika and scattered the men. Pumpua, then the 'yikoyumi' (political leader or strong man) who later became the supreme 'luluai' of the whole area, and a friend of his, Titipanawa, had their eyes on two attractive young women. Pumpua returned to Akuna with Koneno, who is the mother of Dapa?i (wife of Kau?a) and also Tiyo (wife of Waropi). Titipanawa returned with his wife Minokano, who, when he died, married Utawa of Wayopa.

3. During another attack on Tompena, the men captured a very attractive young girl, Minita, who was close to puberty. After she had spent some time in Akuna, she married Kika. In later years her brother, Tinuyi, left Tompena and joined her in Akuna. Both she and her brother are still living.

4. Mudendudu, as a girl of about fifteen years of age, who was also attractive and was a promising wife, was captured during a fight with Kundana (one of Akuna's archenemies). She was brought back to Akuna by Inunti. He later gave her to Nokame, the present 'luluai' of Amam-

264

unta. It is significant to note here that in such a case he does not require any bride wealth or counterprestation, but he does place the recipient in a state of obligation, and reciprocal actions are started in motion.

Leviratic marriages are permitted and do occur, but are not the rule, as the widow is usually either not interested in remarriage, especially when she is middle-aged, or prefers to make the decision regarding a new spouse. While the older brother of a deceased man has first claim upon the widow, a younger brother may also marry an elder brother's widow. Both of these may be secondary marriages, but depend in the last analysis on the decision of the widow and the person concerned, and not on any traditional rule.

Sororate marriage is absent, nor were any cases of sororal polygyny recorded. Informants explained in this respect that if a woman died without producing offspring, she need not automatically be replaced by a younger sister, real of classificatory. Occasionally, the deceased girl's mother feels 'ashamed' and she will replace her by a sister or female relative. This depends wholly on her and the man has no legal claim to a substitute wife. In the same breath one informant, Torawa, volunteered that another woman would cash in on the vacancy by placing herself at the disposal of the young widower. In many cases she will be an unmarried girl, but frequently a woman who leaves her husband.

Widows may remarry, but this depends largely on their age and whether there are male members of the kin to help them in the more taxing chores. Should a widow feel that she needs the help of a man to clear off new gardens, keep the fences intact, or help in some other way, she will remarry. This is especially true of younger widows who are either childless or have only given birth to one or two children. If her children are grown up or her brothers close by and willing to assist with unpleasant tasks, she may remain a widow.

A man who wants to marry a widow follows her to the bush or to her gardens, and when she looks around and notices him he calls out to her and requests a marriage. Should she consent, he will return to the village and that evening he will go into the village square, or some other central place, and announce to all that he had spoken to so-and-so and that she had agreed to be his wife and warning all the other men to stop following her around. If already married, he would have discussed the matter beforehand with his first wife, and asked her whether or not they might live together in one house. If the answer is in the negative, he must build a new house or arrange for her to move in with another family. If, however, she refuses to marry him, she makes the announcements, telling everyone in earshot that so-and-so followed her today and proposed but that she did not wish to become his wife, or did not wish to remarry. He would sit in shame listening to this while his wife, friends, and relatives sit around looking at him.

The marriage contract

When two people have decided to get married, the father of the prospective groom will approach his clan members and the members of the young man's interest group to prepare the bride price. This will include all the objects and products of value such as three or four pigs, garden products, objects of clothing, and among them a new grass skirt, bows and arrows, 'efara?yo' (the yellow head feathers and tail feathers of the cockatoo), stone adzes, and similar objects. The bride price is contributed to by persons associated with the groom and his parents, not only clan members.

The father or elder brother of the groom informs the bride's family when the prestation is collected. Whether the marriage involves a union between members of different villages or different parts of the same village, the bride's family remove the gifts to the house of the bride's father. The groom's interest group members do not follow. The bride, on the other hand, refrains from participating in the festivities which her wedding has prompted. Following this feast, and it is frequently two or three days later, a counter feast is initiated by food and gifts from the bride's interest group to the parents of the groom. Once again a preparatory feast follows where the bride's people are not represented.

The morning after this gift transfer the bride is taken to the river by her mother and sisters. This is the time when she is ceremonially purified for her marriage. They collect pit-pit leaves and force them up the nostrils of the bride.[17] The blood which flows is spread over her face and splashed over her body after which she is washed in the river, and rubbed down with soft leaves. The procession of women who surround the bride now return to her natal village and place her in her mother's house rather than the 'pumara?ini yi ma?i' where she might have been staying with her age mates. Purified and cleansed she is given a new grass skirt but not the one which formed part of the gifts offered her family by the groom's people. This skirt which she now wears extends all around her waist and is called 'anatiami'. For newlyweds the skirt is long, extending to below the knee, but as they grow older, the women's skirts shorten. At this time, too, the

17. This 'upaiyami' or nose bleeding is the same ceremony performed for the boy at puberty, and performed by adult males during ceremonies which preceded warfare. The flow of blood allows the physiological system to operate normally and the mind to think clearly.

women in her village prepare a small feast and present her with a new 'unammi', a new comb cut from bamboo, and a small pig which is given by her elder sister. The following morning all the bride's relatives, excluding the bride and her father, go to the village of her future husband informing the people to fetch their bride. During the early afternoon the two groups return to the bride's village squatting down in the village square.

As the villagers and visitors sit around in groups, chewing their betelnut and smoking, the bride's father rises. With all the fire and oratorical skill which characterize these people as public speakers, he describes the abilities and potentialities of his daughter, the fact that she is healthy, strong, and diligent. He asks the villagers of the bridegroom not to fight with her, but to treat her as one of their own group. "If I am ill, let her come and see me; don't be cross with her if she visits us then", he concludes, taking his seat. Now the mother of the groom goes forward and, taking the bride by the wrist, she says to the latter's mother: "Thank you for giving her to me!" She leads the bride away followed by her little pig attached to a string. As the gathering rises, the sister of the groom carries the bride's 'unammi' and the procession returns to the man's village. With them they carry all the food which was prepared by the bride-giving village and a feast takes place after their arrival in their own village. The bride, however, does not eat this food but eats her own meat and tubers which had been set aside, and carried separately.

For the next four or five months she lives with the groom's mother while the groom takes up residence in the 'wadnama?i'. He now collects food from his mother's garden or receives meals from her and other relatives, but he does not visit his mother's house. After this initial period of separation, he may return one day for a meal and so starts a new phase in his life. He now returns to his bride every day for his meals and starts the construction of a house for his wife. He is assisted by the male members of his interest group, while she is primarily responsible for collecting the 'kunai' and will be assisted by female members of her interest group. These will include the mother and sisters of her husband. He will also start a new garden which will be his wife's responsibility.

On a number of occasions informants remarked on the frequency with which adultery is committed between a man and the bride of his son. While I never recorded a specific case of this kind, one can see the conditions which might permit this kind of relationship. The bride, perhaps used to trysts with her lover is suddenly cut off as the groom is taken into the men's house and she into the house of her mother-in-law. He is not permitted to visit her while the father-in-law might return, and visit daily in the house.

It would seem as though the first few months of marriage contain the seeds for the growth of a healthy relationship. The 'pumara', as we have discussed, occupied a status of transition for the past number of years. After the 'orande', at which he was honored, and his subsequent marriage, these months allow him to settle down and to learn the role of the 'wainta'. They also seem to contain the seeds for divorce because these months permit the possibility of adultery and other relationships and attitudes which divide.

DISSOLUTION OF MARRIAGE

While the Gadsup mark the contraction of marriage with ceremonies and feasts during which bride wealth is transferred from the bride-receiving to the bride-giving group, the dissolution of marriage is not marked by similar ceremonies, feats, or transfer of goods between these two groups.

By adultery

In contrast to many highland people, the community of Akuna defines adultery as initiated by either male or female. Adultery is seen as sexual relations by either party with another person not bound to him/her by a recognized legal bond; in other words, the blame can be both on males and females. This section will deal briefly with this subject, and in the concluding paragraphs it will be tied in with divorce.

It will be recalled that in the early part of this chapter we discussed the Gadsup term applied to a prostitute, namely 'umori?ono', 'one who steals habitually'. This same term is applied to an adulterer, who 'habitually steals' the sexual rights and privileges of another. By this term an interesting suggestion is raised regarding the blame for adultery. The term is applicable to both female and male and implies the active partner, the one who suggested intercourse, the one who was responsible for the seduction. There is an ideal of marital fidelity, and once a girl is betrothed she is expected not to have another man, and the same holds true for the boy. Not very many people live up to this ideal for reasons which will become apparent. Traditionally marriage followed a period of relatively uninhibited sexual experimentation between the two partners, and then suddenly a complete taboo was placed on sexual relations between them, and in fact the boy did not see his bride for four to six months. He had taken up his abode in the men's house while she had moved in with the boy's mother or some other married woman. There was no restriction on the husbands of these married women visiting their homes, nor in fact of meeting the young bride. Informants explain that it was not uncommon for a newly married woman to find sexual satisfaction in her father-in-law or some other adult male. After about five months the young bridegroom visited his bride, ate from her cooking, and then proceeded to build her a house and to establish a garden. These rules no longer hold as strictly as they used to, but those regarding abstinence during pregnancy and the period of lactation have not altered. When it becomes known that a woman is pregnant, her mother will go to her son-in-law and tell him that from now on he must refrain from intercourse with his wife or he will seriously injure the foetus. Thus starts a period of abstention between a young mother and her husband that lasts for approximately two years, i.e. until the baby starts to move independently and is able to take solids. Children in this community are spaced at about two years apart. Based on these rules outlined above, a man must look for a part of his sexual satisfaction outside the legal union, and polygyny or adultery is a logical result.

The same rules regarding a pregnant woman and her husband obtain among the Jaqai, but it is strange to read the statement following the previous, namely that "by this it is explained why homosexuality occurs with both sexes" (Boelaars op. cit.:144). It is felt here that this is a case of fallacious reasoning, namely the joining of these two phenomena in a causal relationship. We are not denying the presence of homosexuality, but suggest that where heterosexual avenues are open for satisfaction of sexual needs, and these do not exist among the Jaqai, these needs will not result in homosexuality. This reiterates what has been said for the Gadsup and what was discussed above on the subject of polygynous unions among the Tor peoples. It is also important in the light of ethnographic data for the Gadsup, where homosexual play in the earlier age groups is said to disappear around the age of marriage.

The reader is referred to the discussion of rights and wrongs in our fourth chapter where sexual offenses and the relation between co-wives are discussed in the judicial context. A number of cases given there in full would clarify points made here.

By divorce

While the contraction of marriage involves a transfer of valuables, the dissolution of marriage lacks any kind of ceremony or the return of valuables. Akunans distinguish terminologically between 'weyana?i e?emi', in which a man sends his wife away, and 'weyawapumi e?emi', in which the woman leaves on her own accord. It would seem that the second was more prevalent traditionálly when the status of women gave them the right to leave, but under present conditions, with mission converts allowed only one wife, there are quite a number of wives who have just lately been divorced. Already we have referred to the case in which Tiyo divorced her husband to join Waropi, and the case of Arawe will follow.

The significant fact here is that divorce is recognized by all as a possibility and in fact appears so frequently that structural recognition is given to this fact. When an Akunan (A) divorces his wife (B) (or she leaves him), his affines do not return his bride wealth; he simply awaits the time when some other man (C) marries her and, at this time, the bride wealth the new husband (C) would have paid to the girl's parents is simply transferred to his predecessor (A). Among the Kuma a man can send off his wife and then claim back his marriage payment from her clan (Reay 1959:124). In this way Akunans face much fewer complexities than we usually find in the unilineal African patrilineages (the Zulu for instance) when cattle have to be brought together again and returned to the person who originally paid the 'lobola'.

It was mentioned above that by age twenty-five all Gadsup have been married at least once. There simply is no niche in the society for a bachelor or spinster, but at the same time it is felt that the marriage is very insecure for the first few years due to the state in which males enter into it, and due to the status and freedom of action of women. It should be remarked again that while a girl during adoles-

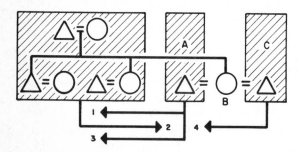

Diagram 16. Prestations and divorce

1. Betrothal payment between man's group and girl's group.
2. Transfer of a woman from girl's group to man's group.
3. Bride payment between groom's group and bride's group. Transfer of the woman to new husband.
4. Return of bride payment to first Hu.

Shaded areas represent the interest group of each person involved who contributes to and shares in prestations.

270

cence and prior to marriage helps her mother in gardening and domestic duties, thus being integrated with the women's realm and prepared for the role she will perform as wife and mother, this is not true for boys. They spend a number of years between their puberty initiation and the 'orande' festival, at which they 'come out', in a state of freedom with few rules and restrictions, during which time they are not expected to be productive or give assistance. Stepping suddenly into the realm of men and married life serves as a shock treatment during which many sex partners find that they are not ideal marriage partners.

In most cases where divorce takes place after children have been born, the man retains the children; but the birth of children gives much greater stability to marriage, and people do not usually separate once children are born. One of the few exceptions to this statement is Wio, who divorced Amo?a after they had had two children. She is now married to Arawe while the children live with their paternal grandparents. Amo?a, too, is married again. People who have been divorced live in quite good relations with each other, and often live in the same village, seeing each other from day to day. Napiwa is still worried as to why his first wife left him after the birth of Kanku?a, their daughter. She married Anokoro while the daughter was still a baby, and this was at least twenty years ago as Kanku?a has children of her own, but Napiwa is still trying to learn the reason.

Break-ups among younger people are often the result of such domestic quarrels. One morning Napiwa told his present wife that after she had done her share of the road work (this was rainy season and people were called out in the middle of the week), she should go to the garden and plant some taro which they had prepared for transplanting the previous day. At about noon he was preparing to leave for the bush and Apipe (an adopted daughter, as they have no children of their own) was cooking some sweet potato. Napiwa took one to eat as a snack on his way to the bush, but his wife Irona?o grabbed it. A quarrel developed, she laying into him with a fence post while he defended himself with an elongated sweet potato. While figures are not available to support this statement, it is important to remark that at least half the marriages in Akuna were not marked by domestic quarrels. When questioned a short while later Napiwa explained it calmly and observed that this was no reason to expel her or for her to leave. "Young people before the children are born leave each other, but after all, people argue in most families."

Marriage, then, may be dissolved upon the initiative of either partner, but with the birth of children and/or the increasing duration of the marriage, the likelihood of divorce decreases.

271

By death

The previous two topics involved voluntary (choice) means of dissolving the legal union between a man and his wife. While others may be instrumental, we are primarily concerned with the behavior of the spouses, as their decisions regulate the continuation or the termination of the marriage. In the discussion which is to follow we are concerned with involuntary (natural) termination of a union, namely by death, be this natural or irregular. Akunans, it might be pointed out, see all cases of death, except of the very old or the long ill, as the result of sorcery and therefore irregular. Such cases must be subjected to magical tests and the culprit discovered and punished, thus often leading to fighting.

Organizationally all marriages contain basically similar elements, and therefore the marriage in this case contains the first three steps which are regular for all contracted marriages, i.e. the transfer of valuables (or a woman) in exchange for a bride, thus placing the bride-receiving and the bride-giving groups at two poles in a continuum of reciprocal relations which are kept up while the marriage continues. When the marriage is terminated by death, the last step is forced on the original bride-giving group. Though this is by no means the only form or the only expression of the death payment, it is one way of squaring the reciprocity and of settling debts. This death payment should not be interpreted as a return of the bride wealth, for it is not, nor is it affected by the birth of children. In essence it is merely a retribution for the loss the husband suffered, and in a way of saying "we are sorry".

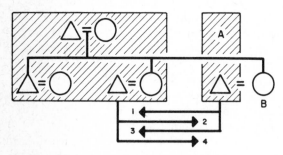

Diagram 17. Prestations and death

1. Betrothal payment between man's group and girl's group.
2. Transfer of a woman from girl's group to man's group.
3. Bride payment between groom's group and bride's group.
4. Return of bride payment to Hu.

Shaded areas represent the interest group of each person involved who contributes to and shares the prestations.

Akunans do not face death with confidence, because they enjoy living; nor do they face it with fear, for there is no afterlife, hell-fire, or any other consciously articulated philosophy. Their attitude can best be described as fatalistic, but this does not influence in any way their actions, for social sanctions substitute where supernatural sanctions are lacking. A person is not forced to lead a "good life" or be chaste in marriage, and all persons receive equal burials irrespective of the lives they led or the manner of their death. In contrast to the neighboring Agarabe (except for the Aiyura Valley Gadsup, who are said to have taken up the practice from their Agarabe neighbors) the Gadsup rarely practice suicide;[18] and widows are not expected to follow their husbands into death.[19]

Persons, both male and female, would often express their grief at the death of a spouse or child by self-mutilation. A woman would cut off the joint of her last finger for a favorite child, or even the third or second fingers' joints. As a rule, for a husband she cuts off the first joint of the index or second finger. A man might occasionally cut a finger joint for a favorite child, but never for a wife. As further signs of grief, women cut off small pieces of the ear lobe and allow the blood to flow and cover the breasts, as a sign of supreme grief.[20]

The corpse, after lying in state for a few days, is wrapped in a blanket, traditionally of bark cloth, and prepared for burial. During this whole period kinsmen visit and caress the corpse while wailing aloud. The group of mourners is constituted primarily of interest group members, namely kinsmen, villagers in general, and neighbors. These appear from time to time to cry and then depart to do their work. The close kin of the deceased will remain and will also be consoled by visitors. While a surviving spouse appears from time to time to express his grief, it is the deceased person's mother's brother who stays by the corpse and wails continuously. Should foul play be expected, divinatory action may be initiated at this stage by an invitation to all who are present to circle the corpse and lay a hand on it. It is believed that when the sorcerer touches the corpse a spasm will cause it to contract and to vomit, thus pointing at the culprit. If none of the

18. Flierl (1932:22) refers to a case at the Wanton river, on the Markham Fall and thus culturally marginal Gadsup, where in 1926 a man was so overcome by the loss of his wife and child who had been killed in an attack that he committed suicide by shooting "several arrows into his body" (how I do not quite understand) and then hanged himself by a vine tied to a tree.

19. Among the Kuma "suicide" was expected of women after quarrels with their husbands, or of siblings after the death of a brother (Reay op. cit.: 202-203).

20. This is a widespread practice in New Guinea. See especially Söderström 1938: 24-47; also Vicedom and Tischner 1943: 278-79; Le Roux op. cit.: 770 et passim; Aufenanger and Holtker 1940: 107.

villagers are suspect and the time has arrived for interment, the
mother's brother will leave the scene and with the assistance of his
kinsmen and other helpers, prepare the grave. The spouse or father
of the deceased will never assist in this task. While the death feast
and death payment take place in the village, the kinsmen quietly bury
the corpse and then return to the village and its activities. The spouse,
covered with white earth to express his grief, will now purify himself
by washing in a stream and then gradually return to normal life. A
period of about a month is spent in mourning, but this does not entail
isolation, any special dress, or food taboos to signify the condition.
If it was a young couple only recently married, the period may be
shorter while the surviving spouse returns to his/her natal village.
After a month or more young people remarry; but older persons often
remain single. During this period or after life has returned to normal,
the survivors may initiate any of the other means for detecting and
discovering the sorcerer. The prosecution is undertaken by the
brothers or close kinsmen of the deceased and involves a complex
series of tests which are beyond the scope of the present discussion.

Should the couple have been occupying a separate house, the survivor may choose to remain there or move in with some kinsman or
friend. This is especially true in the case of older persons who are
not living in their natal villages, and who prefer to remain where they
have so many social ties. If there are children, the surviving parent
will invariably move in with the child and his family while in turn
shouldering some of the economic burdens. In cases where the couple
had been occupying a household belonging to some other person or
sharing their house with another couple, the surviving member may
remain, or when strength begins to fail, either join a child's family
or accept an invitation to join another household, the members of
which may or may not be related. When a parent dies in a family
where there are small children, a kinsman or grandparent may adopt
the children or the surviving parent may retain them. The latter was
the choice of Unanata when his wife died two weeks after he married
her and he retained two children she had borne in a previous marriage.

Personal property is distributed among the heirs while land may
return to the family under certain conditions. There is no fear to use
the property of the deceased nor is there any belief that the spirit
remains in or near the dwelling.

Old age

As the brightness of the day must ultimately lead to dusk, and as a
hot fire must lose its heat, so too the adult man and woman filled with

energy and warmth gradually become 'ayokun' and 'ayokurini'. This
stage is reached when a person's bodily movements slow down and
when he can no longer keep up with the activities in the daily cycle of
events.[21] It is a stage which is reached gradually, but sooner or later
it must be faced by those who live long enough. It is often accompanied by grey or even white hair, by scanty beards in men, by the loss
of teeth, a weakening of the eyesight, and often a stooping of the body.
Those who are very old shuffle through the village leaning heavily on
a cane, or move around the house keeping the sun on the body.

The struggle for existence, and it is granted that this was not a
very harsh one when compared to that of many 'marginal peoples',
nevertheless left a mark on those who did survive. Infant mortality
was high and was accompanied by infanticide. Even at the time of this
study the Administration saw age thirteen as the point at which they
expected a child to live to maturity. We are now also dealing with a
condition where pacification has been largely achieved and where conditions of health and hygiene are improving. The life expectancy is
therefore rising every year, but this has only been true for perhaps
the past decade. Prior to this pax australia, feuds and warfare were
a way of life, and since only the death of the very old was normal,
every community was in constant state of planning relatiation or
awaiting it. Frequently women and children were the victims. Those
who survived, and this certainly applies to every adult male in Akuna,
who was about thirty-five or forty and older, have scars to show
where barbed arrows were removed from their bodies. These conditions made a person old before his years would qualify him as such.
He slowed down and visibly aged from strain and effort rather than
physiological deterioration. He died a natural death before his years
necessitated it.

At sixty-five a Gadsup is old and at seventy-five he is very old and
hardly moves from the hearth. Because senescence is gradual and
natural it is accepted with resignation. As a person is held back by
lack of strength and energy he begins to depend to an increasing degree
on the meals prepared by a daughter or a son's wife. By this stage in
his life a person has also produced offspring and cultivated the relationship with one grandchild in particular. This is the Gadsup 'old-
age insurance'. The grandparent took care of the needs of the grand-

21. This philosophical and yet practical approach to age reminds me of a
few lines read some years ago:
A man insn't old when his hair turns grey
Nor is he old when his teeth decay,
He's only heading for his last deep sleep
When his mind makes appointments —
    — his body can't keep.

child, and in fact spoiled him when the child was in need. Now the tables are turned.

For the women the menopause is seen as a definite sign that old age has arrived. There is a positive relationship between a woman's ability to bear children, her periodic isolation in the women's hut of seclusion, and menstruation. When the latter no longer occurs, seclusion is no longer required and logically she can no longer reproduce. They say: "My blood is hard in my body now, it does not flow, and children will not be forthcoming. I am old now." Since she can no longer bear children and is no longer subject to seclusion, the taboos do not apply to her. She may go into the gardens at any time. She may give food to a boy during his initiation. She may handle the food for adult males who are preparing for an attack. She is neutral. Both females and males in this stage of the life cycle are neutral and harmless.

Two groups of old people are recognized; those who are still members and take an interest in community and clan activities, and those who are too old and just sit. The first category still has some function such as caring for children, assisting in the preparation of food, returning with a piece of firewood from the bush where they went to relieve themselves, or occasionally bringing some food from the gardens. They do not actively participate since they do not have the strength to maintain a garden or to live alone in a house. The second group, traditionally at least, was an economic burden. They did not contribute in any way but had to be fed. One still sees these people who are no more than a bag of bones, turning with the sun to warm their bodies, but they do not receive much attention. Now at least with coffee seeds which need to be dried these old people just sit in the sun and when a thunder storm appears they can collect the seeds from the leaf blanket on which they are being dried, put them in a bamboo container and half drag themselves into the hut. They have some use now, but in Akuna I got the impression that clansmen and community members were simply waiting for them to die.

Chapter 9

THE FAMILY

The family in its basic form, the nuclear family, is the core group in
the life of an Akunan. It is within this unit that he first becomes conscious of his existence as a human being and within this narrow association he learns the forms of behavior which are appropriate for him.
At an independent age he may change his associations or his residence,
only to be integrated into another family which in turn is the core of
some kin group, the members of which are enmeshed in the ties and
obligations which bind them together or which link them individually
or as a group to a residential or interest group.

Linton distinguishes between two family types, namely that based
on a conjugal and that based on a consanguine relationship (1936:159-
65). The former emphasizes the role of spouses and only secondarily
recognizes kin relations, the latter places major emphasis on the
blood relatives linked incidentally by the union between spouses.
While the latter form is recognized to a certain extent, it performs
a minor role in Akunan social organization, mainly due to the freedom
which exists regarding residence after marriage and the fact that these
consanguineal ties grow weak or are completely lost in some cases.
Nonetheless, such consanguineal roles as that of the mother's brother
and to a lesser extent the parents' other siblings and ego's own brothers
and sisters continue to be influential.

For emphasis it is again pointed out that the nuclear, or conjugal,
family forms the core of the social organization and is a structural
feature of primary importance to each person individually and to the
community as a whole.

NUCLEAR FAMILIES

Every person, irrespective of his relationship to those closest to him
or those persons that are more distantly related, is a member of a
nuclear family defined here as consisting of a man and his wife and
their children, biological or sociological. The nuclear family is not

only the most fundamental group in Gadsup society, but also the most stable, especially after children have been born into a marriage. Even in cases of divorce — and the rate is fairly high — the children or child is immediately taken up by some other nuclear family, either within the extended family such as the grandparents, or outside of it through adoption.

In most cases the nuclear family lives in a dwelling constructed by the man. The separate men's house, where males lived, visiting their wives only for meals, is largely a thing of the past; and while a number of villagers in the Arona Valley still retain it in some form or another, it does not serve the purpose it used to and is falling into disuse throughout the whole area. This means that nowadays the man usually lives with his family. By living in the house he shares to a major degree the responsibilities of maintaining the family, of cooperating on a social and economic level, and of educating his children. It is common to find the modern Akunan man taking care of his children or assisting his wife in the preparation of a meal. By the very fact of living in the house and spending more time in it than was the case with his father, these male roles have changed and that of father now implies much more than it did traditionally. It furthermore allows for the individual choice to be practiced much more freely than was the case traditionally when all males were congregated in one men's house, as persons now can decide where they want to live and with whom they would like to associate.

Akunans have a clear-cut division of labor, with each sex being ascribed a particular status in the economic realm and performing this role as best he can. It was exactly the performance of this role which was of such great importance as regards the selection of a marriage partner and also in later years in a man's political aspirations, for by the surplus available he could exploit to the full the principle of reciprocity which is such a keystone in Akunan society. It is not suggested that the economic aspects brought man and woman together and formed the cornerstone of their union, but it is felt that the complementarity of the sexes transcends the sexual realms. As it is true that economic satisfaction may be found outside the marriage when a man assists his sister or widowed mother in the less pleasant tasks of garden upkeep, it is also true that sexual gratification may be found outside the marriage in some adulterous relation. Both of these are present and are recognized as distinct possibilities among the Akunans; they are thus alternate patterns of behavior and choices open to persons, but both state as preference that a marriage must first be entered into. Man and woman complement each other's needs and abilities and ideally are found in that relationship comprising the family of procreation, where each sex has its own roles.

Regarding the sexual division of labor, Lévi-Strauss has suggested that "the sexual division of labor is nothing else than a device to institute a reciprocal state of dependency between the sexes" (1956:276). This, then, would make a person of one sex unable to live without being in reciprocal union with a person of the other sex. For this reason he postulates that bachelors in so-called primitive tribes are exceedingly rare and ill-adjusted and the one "unclean, ill-fed, sad, and lonesome" man he met among the Brazilian Bororo was a bachelor (ibid.:269). In this regard it is interesting that Oosterwal describes a case of masculinization among the Tor of western New Guinea (West Irrian). Unlike the case among Akunans where female preponderance is balanced by a high number of male children, the "disturbed balance already starts at birth" among the Tor (1961:37). At the time of marriageable age we find an average of 47 per cent of the men who are bachelors. With such a large group of persons, who according to Lévi-Strauss' thesis should be the outcasts, we find in fact cultural adjustment. "A new society came into being with a strong ésprit de corps. Their common life no longer is broken off through marriage, but continues. The aggregate of bachelors has been transformed into a society with abstract functions in the community" (Oosterwal, 1959:833). This set-up reminds one of the African age-graded villages where the group of age-mates also develop this very strong group feeling and are responsible for certain duties. In contrast to the condition in the western part of the island which is also remarked on by Van der Leeden (1956: 35, 116), it is interesting to note that the Akunans do not even have a term for bachelor or spinster, and no person, male or female, over twenty-five years of age was recorded who had not been married at least once. Based on the suggestion from Lévi-Strauss given above, it would be interesting to hypothesize that the clearer and more strict the rules governing the sexual division of labor in a society, the lower the number of bachelors and spinsters in that society, and conversely the weaker these rules and the greater interchangeability that exists in this sphere, the greater the number of bachelors and spinsters. Does this hold true for divorce as well? The testing of these hypotheses will have to await a more detailed analysis of the data from Akuna and a comparative sampling of ethnographic data.

The nuclear family forms the smallest productive and consumptive unit in the community and, in fact, is self-sufficient. Each family, it is here felt, produces and collects enough to satisfy its basic needs, yet each family exchanges foods and other basic products on a reciprocal basis, not because they need the products, but because they need the reciprocity, the fact of always being linked to a number of other nuclear families in this respect. Women spend most of their time in agricultural duties, in the making and repairing of clothing, or in the daily chores of running a household, for even in New Guinea this takes time. A woman is assisted in this by her daughters, and frequently a number of women will spend a day together sitting in the sun and repairing rain gear or scraping and drying 'yienni' grass for skirts. Girls from about eight or nine years old may join the mother in making skirts, or weaving 'unami' from the strong white fiber of the 'no?i' (or also from 'repa?i').[1] They also may spend the day with her in the gardens, playing and cooking their meal in a small-scale earth oven which they prepare exactly like the adults.

Men will be busy on a garden fence, the thatching of a roof, or out hunting pigs in the bush, while boys simply run wild. During the field period the ethnographic notes often mentioned these boys as "running with the pack" or being as "wild as a pack of coyotes". It is therefore reassuring that an anthropologist who studied a neighboring group speaks of this "monkey band" (Watson 1960:136). Boyhood for the Akunan is a wonderfully carefree period during which a minimum of restrictions are imposed and boys are free to play as they like and go where they like. Between puberty and marriage (a period of about seven years for males) there are no inhibiting restrictions, a fact which makes the transition to married life harder for males than females.[2] This statement is made in full knowledge that the changes that have taken place in the Arona Valley have affected more harshly the male realm than the female and that the woman still lives and works much as her mother did, except that she is safer now. The changes that have been brought about, and especially the cessation of warfare, have affected the male sphere of action and given males more leisure time, and also more time to apply in a variety of ways to the earning of money. It was, however, only adult males, namely those who have been through the 'orande' ceremony (at about eighteen years of age), who were allowed to go on war parties. This change, then, would not affect the statement that boys between puberty and the 'orande', i.e. between twelve and eighteen years of age, do very little that is constructive, but loiter around the bush, and since the introduction of playing cards spend their day gambling for anything from marbles (with which the prepubertal boys play for hours on end) to wild fruit, and from an old piece of bandage from the Medical Aid Post, to tobacco.

The nuclear family is also ideally the group within which legitimate sexual relations take place between husband and wife. It has, however,

1. A discussion of these fiber cords, can be found in the sixth chapter of this study.
2. For a discussion of this period of transition the reader is referred to the section on socialization in the seventh chapter.

already been pointed out that premarital sexual relations do take place, and that extramarital sexual relations exist and may lead to the dissolution of a marriage. Of importance here is the fact that it is of course this legalized sexual intercourse which leads to the birth of offspring, the fulfillment of the marriage expectations and the blossoming of the nuclear family. In earlier times the Akunans, males and females, practiced infanticide, by pressing in the fontanelle, but this was not limited to either sex. The justification seems a legitimate one, namely that in the time of fighting a woman was often bogged down by a helpless infant when an attack was made on the village and under such circumstances they would kill the child so as to speed up their escape. It was culturally approved due to the hostile environment in which they existed, and due to the fact that women were deemed an object of capture.

Abortion was not unknown, though women never told the men when and by whom the bark of the 'yanapa?i' (Alocasia Macrorrhiza), also referred to as 'wild taro', was used. When this is scraped off and dried it is called 'a.yaye', and by eating it abortion could be effected. Informants claim — and give actual examples — that taken in large quantities or over time it will produce sterility. This action is the only escape for persons who do not wish to have children, as the culture does not allow the possibility of remaining single. Whether this is simply a folk explanation of sterility[3] or a practice which produces the results ascribed to it, is unknown, since this ethnographer as yet has not been able to have the composition of the 'yanapa?i' analyzed. This sterility which could be brought about, informants said, is the reason for Tea, wife of Domandi, not having any children. The same applies to Irona?o the second wife of Napiwa. Men were never told about this or of the persons who ate it, an old woman said.

The nuclear family is not only the core group within which a person finds his economic value, his sexual satisfaction, and the reproduction of his offspring, but it is also the social center within which he spends a large portion of his time, first in the family of orientation and in later years in his family of procreation. The Akuna man visits

3. As reason for the use of this abortifacient in large doses so as to produce permanent sterility, informants explained the following: Very often a young girl witnesses the birth of a baby and, shocked by what she observed, she develops a fear of childbirth and attempts to escape from it by causing sterility. This is a condition identified by abnormal psychologists as "hysterical sterility". It is not clear physiologically whether we are here dealing with a plant which poisons the system and causes the miscarriage (and possibly permanent after effects if it does produce sterility, as claimed), or whether we are dealing with this psychological condition which results in repeated miscarriages making it impossible for a woman to bear a child.

much more frequently with his family than one would expect, and very often upon entering a house at night, the husband and wife can be found around the hearth, smoking and peacefully discussing something that happened during the day. The husband is the head of the family, but he is not the stern fatherly figure toward wife and children. He jokes with his children or joins in horseplay with them and often spends a whole day in the company of his wife while clearing a new garden or constructing a new garden fence. He talks with her in public and eats his meals with the family (inside the house or in the sun when the weather is fair) unless some business is pressing.

While the mother is the person who spends the greatest amount of time with her children, the husband will very often stay home and take care of an infant or watch the play of a small child while his wife is absent in the gardens collecting food or paying a brief visit to her parents or siblings in a different village. It was certainly strange shortly after the arrival in the field to find Tapari one day in the deserted village of Akuna with his infant daughter. His matter-of-fact reply was this: his wife was visiting her parents in Wopepa, and that he was caring for the child. This is in fact a common situation, and it soon became apparent that Akunan fathers are very fond of their children, and that men are fond of children in general. They can often be seen playing with and caressing babies, joking with or throwing small stones or sticks at children of about five or six, or seriously discussing some happening with a child of about pubertal age.

A mother's duties to and company with a baby never stop. The baby is always within reach of the nipple and this is the first way a mother attempts to quiet a crying infant. Very often when a baby awakens in the quiet of the night one can hear the mother's voice as she thrusts the nipple into the baby's mouth. Children are weaned at about two years of age and from then on spend progressively less time in the company of the parents, but girls continue close to the mother, and are in fact closer to her than are the boys to the father. It is not surprising then to find that the relationship between mother and daughter, more so than between father and son, is one of comrades. While the mother's brother is tied in with the nuclear family to a great extent, and his role is a clearcut one regarding his sister's child, socialization still mainly takes place within the nuclear family. The result is that due to the conditions discussed, the daughters are more fully integrated with their expected roles than is the case with the sons. In this respect Linton has pointed out that 'conditioning' to social life begins so early that much of the groundwork of the personality is laid before such extrafamily agencies can be brought into play (1936 op. cit.155). Based on this proposition, it follows logically that because of this the female personality is more fully integrated to perform the roles which are expected of her than is that of the male. It would be hazardous to attempt factual or case material to support this claim after having spent only one year in the village, but along general lines the following seems to emerge from observations and case records.[4] Between birth and puberty, the child is primarily family-oriented and it is within the nuclear family that a child of either sex finds psychological, emotional, nutritive, and partly social expression. Even by this age the boy is further removed from the family and finds a greater amount of his needs satisfied outside the family than is the case with the girl. The ceremonies which mark the attainment of puberty act for girls as a gradual transition into the life they will live within their own family of procreation. To be sure, there may be a few years between these role changes, and if so, they are spent in relative freedom and unrestrained activities of play and courting, but even then the girls temporarily care for younger siblings or the children of siblings, assist in the preparation of food or the making of clothing, and in garden activities. While relatively free, they are not much removed from the roles they will perform. For boys who by puberty are already quite free of the family of orientation, the ceremony which denotes their transition to post-pubertal status is a harsh one filled with physical and psychological 'shock treatment'. In spite of, or possibly as a result of this, the post-pubertal boy develops even greater independence from his family and associates nearly exclusively with his age mates. Traditionally and in many cases today young boys now move to a young men's house. He very rarely enters into any constructive work the way his sister might. When marriage takes place and the girl enters into her new status, she is quite able to perform the roles as she has been doing these same things intermittently during the last number of years of her life. With the boy the situation differs markedly. From a period of carefree roaming and playing, with a few restrictions upon his behavior, he is suddenly jerked into a position of responsibility, both social and economic, and his adjustment to these new conditions is not always successful. Elsewhere in this study the case of Wanara is referred to, a young man who got married and divorced shortly before our arrival in the field. His wife had married and settled down elsewhere, while he was still running with the unmarried boys and playing cards all day. It is suggested here that he was simply not ready for the settled life and responsibilities which accompany married life. Another example refers to Iripe. He is in the same age-grouping as the boy mentioned above and also got married a few months prior to our arrival in Akuna. During the year there, this ethnographer

4. It is suggested that the reader refer back in this reading to the material presented in the seventh chapter.

studied a number of families closely, and one of them was that of Iripe and his pleasant wife, Apono. During the early phases of field work the notes frequently referred to Iripe as an example of maladjusted, irresponsible, and immature husbandhood while forming a contrast with his wife who was perfectly adjusted to her economic and social roles and performed them with proficiency.[5] While Iripe played cards or marbles, his wife worked in the garden. While he ran around the bush with the unmarried boys, Apono was making string bags or scraping 'yienni' with the other married women or caring for a sibling's newborn baby. Near the end of the field study Iripe formed a complete contrast with the immature boy we had learned to know so well months before. He had laid out a new garden for his wife and was starting one for himself where he could plant coffee. He was usually found in the company of adult males or at court cases rather than hiding around a corner gambling, and his wife was pregnant.

These are two examples from the same village. While the females adjusted well to their roles (even though one was willing to bear more in the process of waiting for her husband to settle down than the other), the males needed a time of transition during which they had to fall in with the behavior expected of adult Akunan males. One explanation is that the males simply have a wider range of choice regarding permitted behavior and greater latitude in action. If this is the case, then it also implies that males need greater adjustment to the roles they have to perform and need a longer time to adjust than is the case with females.

This might in fact be the key to that aspect of New Guinea cultures which led Held to speak of the lack of a "closed and logically analyzable system", and he then calls the Papuan with his "fragmented" social structure, a cultural improvisor (1951: 51-5). This lack of discipline which has been only briefly touched on here, during which a child follows his own whims, "runs wild" with his age mates and is free from strict rules governing his actions and behavior, is very important in the laying of the "groundwork of the personality" and also in the whole process of personality orientation and the choice of his own associates. This is but the beginning of a series of decisions which the person makes, all of which are based on personal and individualistic reasons and are made primarily for the advantage of the person. It is this matter which leads to that individualism we often remarked

5. This period which is needed for transition might in fact be related to Benedict's discussion of continuity and discontinuity in the conditioning supplied by different cultures (1938). The discontinuity in the conditioning of the Akunan male might also be a functional correlate of the high status of women in this community. This latter topic is discussed in more detail in the first section of our tenth chapter.

on while in the field, and which Van Baal sees as the Papuan's primary characteristic. It is not individualism in the Western sense of
initiative and a high degree of conscious reflection on life and condition, but rather a "deep-rooted undisciplined" attitude, which opposes
all kinds of rules imposed by others (1954: 441). These aspects of
decision-making and accompanying behavior, both by males and females, will be touched on again in the following chapter, but for the
moment we will return our interest to the family group and the attitudes which govern their behavior.

Two important consanguineal tendencies must be mentioned at this
stage. The only actual peacemaking ceremony that was recorded followed a quarrel between siblings. A brother and his sister had a disagreement, ending in a temporary rift between them and their families, but they sought out the first possible occasion when both were
in good spirits and presented a feast. On this occasion the brother,
Ya?e, prepared the feast and invited his sister Omu, her husband (he
had been away as he is a policeman and was home on leave), and the
latter's relatives to partake of the feast. On such an occasion the two
siblings ceremonially hold and caress each other in the fashion so
typical of these people when they meet after an absence. Caressing
in this instance consists of holding each other's hands and arms while
repeatedly stroking the other person's arms from the shoulder to the
fingers. While they in this way express their feelings for each other,
the brother says, "Let's not continue this quarrel, for when I'm sick
who will sit near me, and if I die, who will cry for me?" Following
this, the feast is partaken of by all, and the sister ceremonially hands
her brother a string bag. Normal life is resumed and each fulfills the
obligations required of him as regards the other — in this case Ya?e
made fences for Omu while her husband was away.

The parent's siblings show great interest in a baby and while visiting will always hold the infant on their lap or play and laugh with it.
It is, however, considered a great 'shame' and amends must be made
right away, should the infant nephew or niece urinate or defecate on
an uncle or aunt. Their feelings are hurt and they feel shamed at such
a disgrace happening to them. In all cases payment of a pig or some
other valued object must be transferred by the father of the child to
remove the 'shame'. This is especially true, and the payment required
increases proportionately, if this happens to be a mother's brother.
The curious fact is, however, that this does not apply to a father's
brother. Should a baby wet or defecate on him, everyone agrees that
it is too bad and that is where the matter rests. In spite of repeated
suggestions that the mother's sister may be treated in a similar way,
informants denied these allegations as well as the possibility that the
father's sister might be of greater importance than the mother's sister.

POLYGAMOUS FAMILIES

When speaking of polygamous marriages or families in the community of Akuna, we will always be dealing with polygynous unions but never unions composed by sororal polygyny. Polygamy is permitted by the society and is certainly not uncommon, but from genealogies it is clear that polygamous families formed only 12.1 per cent of all families, and that monogamy is the rule.[6] In cases where a person did have more than one wife the pattern was two or three and only very rarely more than this.

All that has been said for the nuclear family applies to the nuclear families composing polygamous families. Akunans see each of these as an independent family, where one man occupies a particular status in each of the families, but they are not graded as to priority or primacy, and each family functions independently.[7] It can generally be stated that among Akunans there exists a good relationship between the wives and only occasionally does it lead to sexual jealousy and quarrels. During the field study a number of fights were recorded between Tea and Tero, the wives of Domandi, but the circumstances were more complicated than simply the fact of being wives of one man. Tea was the first wife but was never able to produce offspring, and in time Domandi married a much younger wife, Tero, whom he settled in the house of a friend. Soon she became pregnant and by fulfilling the hopes of her husband got more attention than her co-wife. This increased the jealousy and led to various allegations of exploiting the husband and attracting all the attention to herself, finally backfiring for Tea when she was accused of adultery.[8]

Boelaars (ibid.:84) explains that among the Jaqai there is a direct correlation between man's power and prestige and the number of wives he has, for as his power and prestige increase he can offer safety and asylum to a woman and defend her in any quarrel which might flare from this. The number of wives, on the other hand, also increases a man's status, so that there is a direct relationship between prestige and polygamy though priority is given to power and prestige which lead to polygymy. The opposite seems to be the case among the Tor

6. Genealogies and informants' statements sketch such conditions in the traditional set up as well. It has, however, become more pronounced since mission churches require converts to be monogamists and to divorce the multiple wives.

7. Boelaars reports that the Jaqai of the former Dutch New Guinea distinguish terminologically between the first and subsequent wives, but that a wife can work herself into a favorable position with her husband (1959:123).

8. This case has been discussed in greater detail in an earlier section of this study.

peoples, where polygyny "mainly occurs with old men" and then women are usually widows who no longer bear children (Ooster-wal 1961:43), which suggests that women do not greatly add to a man's prestige since he is already an established person by the time further marriages are contracted. This same suggestion is made by Van der Leeden, who finds that among the Sarmi peoples polygynous unions are contracted by men "who are leaders in their groups" (1956: 138).

By and large our Akuna material supports the position taken by Boelaars, namely that wives are a positive contribution to a man's status, not by the simple fact of having more than one wife, but because they serve him in two ways. First of all they link him with a different interest group and village, and secondly, they contribute to the amount of food he has available to exchange and which enables him to set into operation the all-important principle of reciprocity. This seems to differ from the situation among the Kuma where polygynists are "admired", "accorded great prestige", and considered "impor-tant and wealthy". Reay places great emphasis on these aspects of polygyny, but as though she fears that her data do not support her position, she feels compelled to explain that "polygyny is more gen-eral than the small proportion of polygynists found at any one time suggests" and again that these unions are not very stable "despite the relatively low incidence of polygyny" (op. cit.: 84-5). Based on the Akunan and Jaqai suggestions it must be stated that among these peoples polygyny is a positive contribution to the economic position of a man and the woman's status in the family, and that, in fact, if we accept the proposition regarding the reciprocity of male and female roles and the division of labor as set forth by Lévi-Strauss, it is also logical to agree that a mathematical two-to-one relationship along the same lines simply doubles the man's position and wealth. "When the productive accomplishment of the two sexes is approxim-ately equal, and a small unit is as efficient as a large one, monogamy may be economically advantageous. When a woman's economic con-tribution is large, and a man can produce enough in his sphere to satisfy the needs of several women, polygyny fits the circumstances. In suggesting the basic importance of economic factors we do not, of course, disclaim the auxiliary influence of others, e.g. the prestige value of plural wives or the sexual outlet offered by polygyny when continence is demanded during pregnancy and lactation" (Murdock op. cit. 1949: 36-7).

Based on observations in the field and an interpretive analysis of genealogical data in terms of leadership and other factors, it seems possible to state that among the Akunans, polygamy is first and fore-most an economic factor and secondarily, but closely related, and integrated with this, of prestige value. This statement will become clear when we keep in mind that the woman is the gardener and the one who prepares food for festivities; she is the person who collects and prepares the 'yienni' which forms a part of every payment; and she is the person who bears children, thus raising the male to the status of man. The man's prestige is increased through the enhancement of his economic position. The fact of offering a sexual outlet is of little or no consequence in a society where the avenues exist and where extramarital intercourse is as frequent as it is among these villagers.[9]

The major distinction in these polygamous families is made among Akunans regarding residence, and in this respect we can distinguish three residence patterns: (1) where a polygamous family occupies a single house; (2) where the polygamous family occupies two or more houses in the same village; (3) where the polygamous family occupies two or more houses in different villages. The first case is said to have been true traditionally, but since the man no longer lives in the men's house it is becoming infrequent. In one room in Akuna lives Nona?a with her two children, and Autu (who simply attached herself to the husband) with her three children. These two women are wives of Eyo who was at the coast on indentured labor at the time of this study. When asked whether they got along well, the reply was, "yes, while Eyo is away", but it was explained that when he returned he would place each in a separate house because jealousy and fighting would start when he was present. He did in fact return a few months before the termination of the field period, but had not separated them nor had quarrels developed before we left.

The second type, in which the different nuclear families occupy different houses in the same village, is the most common and the type to which most of our discussion relates. It also seems to have been of greater importance traditionally when inter-community marriages might not have been as frequent as they are today. As a rule, each wife of a polygamous family occupies her own house where she and her children live, and from time to time they receive visits from their husband and father. The man spends time in turn with each family and takes his meals with them. Under traditional conditions a man lived in the men's house and would visit the women's houses to eat or visit and then return to his group, but at present he quite frequently lives with one of the wives, usually the first one. As a carry-over from the traditional situation, though the motivation might be a novel one, there are a number of men who have built small independent structures where they sleep. When asked why he did not

9. For this topic see the discussion under "adultery" in the previous chapter.

sleep with his family, Ropa replied that he could not stand the noises at night when the babies woke up and started crying.

In the third type of residential arrangement, with two or more families occupying houses in different villages, there are usually special circumstances which affect this decision. The examples which are to follow will illuminate two statements made above, namely that regarding the retaining of rights of land and that fact that plural marriages as far as possible are contracted in different villages to align a man with various villages and various kin groupings other than his own.

Men see marriage as a useful tool to make contacts and also to back them up on the economic level and make it possible for them to place others in their debt, which, based on reciprocity, assures assistance when needed. It is then not strange that a great number of polygynous marriages are between men and women of different villages but in this case they have gone even further and have placed their wives in the key position to care for their rights and justify their claims. The following are a number of cases which were recorded.

1. Yaya?o of Ikana has two wives. One lives with him in Ikana, but the other, his second wife, lives with her kin in Tompena where she justifies her claims on the land she inherited from her parents. In any dispute between the villages, Yaya?o remains neutral.

2. Waropi of Amamunta has two wives. He was married to one before he left for police training, but upon his return a young woman, Tiyo, attached herself to him and he married her but got her set up with her kin in Akuna. She now works the gardens she inherited from her father, while his first wife has her gardens in the Amamunta area.

3. Opura was one of the strong men in the old village of Akuna, but when Nori was appointed 'luluai', fission took place. At this time Opura was married to Monana who was originally from Akuna and had her own land which she had inherited. He left with his grown sons for Wayopa and developed a new village. When his father's brother's son, Aya, died, he married one of the deceased man's widows named Daitno and took her to Wayopa where she took control of newly cleared gardens in the valley.

4. O?oma, the 'luluai' of Wopepa, lives in Wopepa with his first wife and there works the gardens which he inherited from his father. Some time ago when Nanuwe of Onamuna died, O?oma married his widow Amayano, and left her in Onamuna to retain the family rights to the land there.

5. Kampuma, who followed Opura to Wayopa, was married to a woman, Didino, of Akuna at the time. He left her behind to work his land in the Akuna garden area, and then married two other women who live with him and work gardens that he cleared in the Wayopa area.

6. Wono originally is from Akuna where he married Umeno. When Amamunta started to prosper, he left her behind and went to Amamunta where he married a second wife with whom he lives, and thus has two gardens.

7. Yappa, a policeman stationed in the Western Highlands, has a wife in Wopepa and lives partially there with her and she retains and justifies his claims to his patrilateral land. His mother came from Akuna, and so he married Omu and she lives in Akuna and has her gardens there.

Note: It is not suggested here that all powerful men must have ties in more than one village, but it is significant to note that most people who have more than one wife have seen to it that they come from more than one village. This topic will be touched on again below when we deal with modes of alignment.

In polygamous families, the members of each family, i.e. a woman and her daughters, will have their own garden where they produce food. The man, performing his role as husband in each family, will erect the fences and help in the hard work that is to be done. A woman will also be able to call on the assistance of her brothers, thus calling into operation both conjugal and consanguineal ties and in turn strengthening them by a gift of food. The behavior between the children of one father, through co-wives, is the same as that between siblings, and they address each other as brothers and sisters. On the whole it can be said that children do not play together or become good friends because they are siblings or co-siblings and though ties of consanguinity do act as binding factors, relations of amity and comradeship grow out of unrestrained association with age-group members. This same fact is true as regards co-wives in that marital bonds which tie them to a third person, a husband, have little effect on their relationship, and this is especially true when they occupy different houses and till different gardens.

HOUSEHOLD FAMILIES

While the previous sections dealt with socio-jural groupings, namely unions between different members of opposite sexes, which we referred to as nuclear and polygamous families, this section concerns itself with those residential groupings that Fortes (1958: 1-14) calls the domestic group.[10] He sees these groups as constituting various

10. In a recent publication Carstens (1966: 70) has distinguished between four main kinds of families which he borrows from Wilson and Radcliffe-Brown. These elementary-, residual-, rejuvenated-, and extended families, however, refer to kinship groupings rather than residential groupings or household-families.

phases in the developmental cycle of the family and we can thus distinguish between the nuclear family occupying a residence, the extended family, and so forth, somewhat on a cyclical basis. Any person, however, who has dealt with residential patterns of urban Africans, or acculturated reservation Indians (and this is by no means the limit of the examples), will agree that Fortes' types do not adequately define the residential patterns of these groups. The material which will be offered below will show that among Akunans and neighboring Gadsup villages (who are not yet at the stage of urbanization or acculturation of the groups mentioned above) domestic groups cannot be typified in terms of conjugal, consanguineal, or kinship bonds only, but transcend all of these, and in some cases residence in a household, as is the case on the village level, is purely on a social or amicable basis. Before looking at Fortes' phases of domestic groups, we will have to set up a series of 'types' to deal with residential patterns in these Gadsup village communities.

In the six villages to be discussed here, we are dealing with a total of 1,126 individuals who reside in 219 residences — this includes a number of 'associations' such as men's houses, which will be segregated from the residential units in the study. In analyzing the household composition of this sample it was found that 'types' as such were not satisfactory, and each household was therefore compared with a similar household having a comparable composition, giving a total of some forty-five different variations of composition. These were grouped together on the basis of common elements into fourteen sub-types (A-N), and these in turn classified into seven (I-VII) major types. At this stage it will be necessary to delineate these types and sub-types and the justification which led to the groupings. While the types refer to families, we are interested in households as they are composed of these family types.

Type I   Nuclear Families
Sub-type: A. Nuclear family of a husband, a wife and their children
          B. Co-family of a husband, his wives, and the children of each

Type II   Compounded Nuclear Families
Sub-type: C. Two nonrelated nuclear families

Type III   Laterally Extended Families
Sub-type: D. A man, his brother, and their wives and children
          E. A man, his sister, and their spouses and children
          F. A woman, her sister, and their husbands and children

Type IV Generationally Extended Families
Sub-type: G. A man, his wife, and their children, and his/her father
and/or mother (and unmarried siblings of his or his wife

Type V "Broken" Families[11] (i.e. families of whom either the husband
or the wife is away; temporarily or permanently)
Sub-type: H. Wife (and children) — husband away on indentured labor
husband living in different house
husband living in different village
husband living in men's house
I. Wife (and children) — husband divorced
husband deceased
J. Husband (and children) — wife living in different house
wife living in different village
K. Husband (and children) — wife divorced
wife deceased

Type VI Association
Sub-type: L. Men's house
Young men's house
Young women's house

Type VII Childless Families
Sub-type: M. Husband and wife where children have not yet been
born to a young recently married couple
N. Husband and wife, where an aged couple is living alone
because they never had children; because all their
children are deceased; or because their children are
grown up and have left the house

Note: 1. The term 'broken' sounds very mechanical, but is used
here in the absence of any other term that can adequately
describe this condition where a household has 'been made'
incomplete due to social (choice) or natural (death) forces.
2. 'Associations' are included here for the sake of complete-
ness, but these will be disregarded in the domestic groups
as socio-jural and economic units.
3. When we speak of 'residence' in this study, it pertains to
either independent houses, or complex often 'apartment-
like' houses. 'Residence' is each separate room, be it in
an independent house structure or simply separated by a
divider from other rooms in a complex house.

11. Barth speaks in this same context of "incomplete families" among the
South Persian nomads (1961:12).

Table 14. Numerical composition of Gadsup household families
   (includes associations)

| | No. of persons per household | | | | | | | | | | | | | | | | | | | | No. of households |
	1	2	3	4	5	6	7	8	9	10	11	12	13	14	15	16	17	18	19	20	
Akuna	7	8	9	5	8	7	3	5	1	-	1	1	-	-	-	-	-	-	-	-	55
Amamunta	2	1	5	4	5	6	2	5	4	1	-	-	-	-	-	-	-	-	-	-	35
Manunampi	1	6	3	-	3	2	2	-	-	1	-	-	-	-	-	-	-	-	-	-	18
Tompena	2	1	2	8	8	1	6	2	-	-	1	2	-	2	-	-	-	-	-	1	36
Wayopa	1	4	5	3	1	3	2	-	-	1	-	-	-	-	-	-	-	-	-	-	19
Wopepa	2	8	9	3	10	5	8	2	4	1	2	2	-	-	-	-	-	-	-	-	56
Total	14	28	33	23	35	24	23	14	9	4	4	5	-	2	-	-	-	-	-	1	219
No. of Individuals	14	56	99	92	175	144	161	112	81	40	44	60	-	28	-	-	-	-	-	20	1126

Before we classify the households in the sample according to these family types, it will be necessary to discuss the numerical composition of the household-families. It must be remarked that while we speak of 'household-families', they are not always families and in a few cases are formal groupings rather than families, as in Type II above. For the sake of uniformity and because the term does apply in the majority of cases, it will be retained.

In Table 14 the total sample has been classified according to the number of persons occupying each residence (see note three above). Once again this statement must be reserved, for, while it holds true in terms of the total universe of physical residences and houses, it also includes a number of 'associations' where people of the same sex go to sleep. The men's house no longer acts as an integrating factor, nor is there much in terms of ritual or sacred and ceremonial objects to be preserved there, and while men sleep there, it is important in terms of the socio-economic or residential realm. At Omaura, which is not included in the sample, there were three men's houses where the men slept during the December festivities, during which they were having the traditional Orande ceremony. As soon as the ceremony was concluded, all males returned to their families and slept with them. Tompena has a men's house where twenty men sleep and a young men's houses for seven post-pubertal boys. Amamunta has two men's houses of six and five men respectively and a young men's house where eight boys sleep. Also in Manunampi there is a men's house holding seven and a young women's house which is occupied by four girls, all of whom were recently betrothed. By adjusting the totals given above in Table 14, to accommodate for these non-residences, it is significant that from the persons here mentioned in domestic households, a total of 743 (or approximately seventy per cent) dwell in households within the range of three to eight members. This figure is of particular interest in terms of the primacy of the nuclear family and the centripetal

293

function it performs in Akunan and also Gadsup society. The mode is five and the arithmetical mean is 5.05 persons per household-family.

Regarding the history of fission in this society, it is of great interest to compare the figures for both Manunampi (which recently split off from Amamunta) and Wayopa (an older village which hived off from Akuna). In these two, 82 per cent and 87 per cent of the people, respectively, fall within the range of from two to seven members per household-family, more than half living as nuclear families in a house. It will be noticed that we are here speaking of household-families with a two-to-seven range and not the three-to-eight category dealt with in the previous paragraph. This is due to a conscious attempt on the part of the writer to emphasize the fact that the smaller household-families, often composed of a nuclear family, are very prevalent and form an important phase in the evolution of these villages; a point to be elaborated below. At this point, though the discussion is to follow below, we might ask whether these relatively young villages have not yet reached the point in the cycle where they produce all the complexities in living arrangements? Are they on a village level, at the stage the families composing them seem to be, namely Fortes' first phase? This will be discussed below after establishing our household types and their importance in Gadsup society.

Turning our attention to Table 15, we are moving from the simple numerical composition of households, to the family types that occupy them. We have already pointed to the fact that the nuclear family is the basic social unit and the group of primary importance for the individual on a social, economic, and enculturative level. This is basically true for Akuna but also applies generally to other Gadsup communities. In this light, then, it is significant that the single nuclear family occupies a separate residence twice as frequently as the next most common family type, namely sub-type H. It will be recalled that this latter sub-type groups together all those women whose husbands are away on indentured labor. This latter group would always be a class of young women with one or two children only. In other

Table 15. Numerical composition of household families according to types

	Household type and sub-type														Composite households combining various types (to be discussed)	Total households
	I		II	III			IV	V				VI	VII			
	A	B	C	D	E	F	G	H	I	J	K	L	M	N		
Akuna:	19	1	-	1	1	1	1	9	2	3	1	-	-	4	12	55
Amamunta	7	3	1	1	-	-	4	10	-	-	1	3	-	-	5	35
Manunampi	2	-	1	-	-	-	1	6	1	-	2	2	-	2	1	18
Tompena	13	2	3	-	-	-	-	7	1	-	1	2	-	-	7	36
Wayopa	9	-	-	-	-	-	-	1	1	-	-	1	1	1	5	19
Wopepa	18	7	4	2	-	-	8	1	1	1	1	-	1	6	6	56
Total	68	13	9	4	1	1	14	34	6	4	6	8	2	13	36	219

Table 16. Composite household varieties

The right-hand column in Table 15 mentions a number of composite households[†] which combine a variety of these sub-types in such a way that their composition cannot be expressed in table form. Those households that are marked with an asterisk involve basically a nuclear family and will be discussed in the text. The composite households will be briefly mentioned:

Akuna:	*1.	Combination of sub-types A and H	3
	2.	An HG combined with I	1
	*3.	A and grandchildren but not the parents of these (they were divorced)	1
	4.	BH combination	3
	*5.	An A and H and I	1
	6.	An FK combination	1
	7.	An H combined with I	1
	*8.	A and BH combined	1
Amamunta:	*1.	A combined with H	1
	*2.	A combined with I	1
	3.	C combined with I	1
	*4.	H combined with H	1
	5.	DG combination domestic group	1
Manunampi:	1.	An old widow has her granddaughter living with her	1
Tompena:	1.	BG and A and H combination	1
	2.	BH combination domestic group	1
	*3.	An A and H combination	1
	*4.	An A and I combination	1
	*5.	An A and H and I combination	1
	6.	B and I combination	1
	*7.	An AK combination group	1
Wayopa:	*1.	An A and H	1
	*2.	A and H male kinsman and man betrothed to younger sister of head	1
	3.	An I and H	1
	4.	K and H and parents of the young woman	1
	5.	J and unrelated boy that was orphaned a year ago	1
Wopepa:	1.	M and H	2
	2.	B and H	1
	3.	DG and I	1
	*4.	An A and I	1
	5.	An FG composite domestic group	1

† This agglomeration which fits no pattern is suggestive of what Fortes calls a "miscellaneous assortment" of people (1949:64).

words, sub-type H (and sub-type J) deals with nuclear families which are temporarily disrupted. For the purpose of this study, these categories will be treated with sub-type A, the nuclear family, thus giving us a total of 106. In Table 16 all those that are marked with an asterisk involve basically a nuclear (or temporarily 'broken' family), and by adding the fourteen composite types to our nuclear family subtotal, we find that for the six villages there are 120, or 54.8 per cent of all the families which are primarily nuclear families.

While our first table[12] refers to the household types found in Gadsup villages, there is no neat correlation between that and the second table which enumerates the numerical composition of these households. The one pole is occupied by Type V or Type VII households, but the other may be filled by any one of a number of types. Examples appear in the survey data of households in Type I A where there are nine people living together. On the whole, however, the pole of the numerically greatest households is filled by families of Type II, III, or IV, while the other types, namely Type I A and I B, fall somewhere in the statistical middle.

In the introductory paragraph to this section reference was made to the types briefly mentioned by Fortes. He goes beyond the setting up of types to a consideration of the dynamics of residential patterns when he states that "residence patterns are the crystalization, at a given time, of the developmental process" (op. cit. 1958:3), and in terms of this process he proposes three main stages or phases which any domestic group goes through during its developmental cycle.

I. First of all there is the phase during which expansion takes place. Two people get married, produce offspring, and thus set up a family. It is concluded when the wife reaches menopause and can no longer add by natural means to the family. Though Fortes does not state it, he presumably makes provision for a sociological family here, where the children are adopted or removed to live with a mother's brother, while the father's sister's children take their places in this family. "In structural terms it corresponds to the period during which all the offspring are economically, affectively and jurally dependent on them" (ibid.: 4-5).

II. When these children grow up and start marrying out we pass into the next phase which continues until all the children have left or at least married — one may remain with his family to take care of the aged parents.

---

12. The reference here to first (Table 14) and second (Table 15) applies to our present discussion and is used because it is less cumbersome than table numbers.

III. The latter soon die, and a child takes their place, thus replacing and occupying the status of household head, protector of
family wealth, and so forth.

These phases, on a generational level, can also be distinguished in
the sub-types which have been set up to deal with the Gadsup sample,
as they are expressed below, the second being classificatory and the
first processional. This classification fits particularly well in the
case of Akuna where residential changes have been effected by contact
with missionaries and Australian administrators.

Table 17. Household types as phases in a continuum

Phase:	I	II	III
Sub-type:	M, A, B, C, H, J	D, E, F, G	N, M
		Sub-types I and K may of course appear at any point in this development.	

Every family naturally starts off as sub-type M in which we are
dealing with a married couple but no children. Comparable to this
early phase in the history of the family are those sub-types which
deal basically with the nuclear family though it occasionally is incomplete due to choice situations where the family members have for
some reason temporarily separated. When permanent separation has
taken place as in the case of death, the sub-types representing it
(namely I and K) might be found at any of the phases discussed above.
Our sub-type G is the only family type which definitely belongs in the
second phase, but three other sub-types have been included in this
phase for reasons which will become obvious. In all of the sub-types
under discussion, namely D, E, and F, we are dealing with cases
where more than one sibling is married, and by extension we can
then presume that we are dealing with persons whose parents are
middle-aged or old. This, it will be seen, agrees with the condition
sketched for the second phase. When we reach Fortes' third phase
we are dealing with the aged couples discussed in sub-type N and we
are also starting off on the next cycle, namely where "a child takes
their place" and the sub-type M initiates a new series of families.

The household-family among the Gadsup, and we have discussed
this as regards Akuna, is an economic unit. While each woman may
have her own garden, they will usually contribute to meals, and this
is especially true if they are related. They may either cook together
or alone, depending on the relationship between them. Occasionally
one finds that within the one household temporary or permanent rights
cause women to function independently, but this is not common. In a

quarrel one party may move in with some other relative or sleep in the women's house for a night or two. Most families occupying one house together form an ecnomic and social cooperative. The children will play together in the house but will not necessarily become friends as they grow older, spend an increasing time outside, and join into some age-grouping. Men who are household mates are very often related or good friends and either visit or work together. It should be reiterated that choice and individual freedom is very strongly expressed in this society, and that the reason why people consort together is often the simple explanation of friendship or a reciprocal act responding to a kindness shown to him or his father years before.

Co-residence may be based, then, on an invitation from a friend, and this is usually an age-group member with whom strong friendship has already been established so that co-residence is the outflow of rather than the reason for comradeship. The house is of course the property of the man, even though women do their share in its construction, and he will usually extend any invitations should some person seem in need of a residence. It frequently happens that the woman suggests the move or that the request comes from outside. When Anamaka was rebuilding his house he suggested to the tultul that the latter's house be made available to them. In another instance after a quarrel caused a temporary rift in their family, the wives of Anokara moved out, one accepting an invitation from Duka to settle in his recently completed house while the other requested accomodation elsewhere, and Anokara was left alone in his house.

The household group has other functions as well. It bears to some extent an educational load. The mother may often leave her child for a time with a household member as sitter. The members who live in one house are the primary contact group and the group with whom a young person interacts from childhood. It is within the nuclear and extending families that a person learns the appropriate terminology and behavior which are expected of him, and through it becomes familiar with the patterns expected of him as a member of the society. For the Gadsup, then, we can fully underwrite Fortes' contention that "the domestic domain is the system of social relations through which the reproductive nucleus is integrated with the environment and with the structure of the total society" (ibid.:9).

At no time is a person, living either with another person or in another person's house, expected to remunerate him for the use of his residence. Where joint residence is permanent, the persons may cooperate in the construction of a house, but a second family or part of it may move in with the owner after the construction has been completed. On the interaction level, the household constitutes a person's most immediate social universe. While days are spent by women in

their gardens or by males in the bush, during which little social interaction takes place, other days may involve domestic activities or pig
hunts which are cooperative ventures and activities in which interaction takes place with household and community members. Evenings
are usually spent with the household members, while discussions or
arguments take place. Because of the fact that co-residence is an outflow of amicable relations, quarrels usually do not involve household
members.

The fact that there is a correlation between filiation on the household level and fission on a village level has been suggested above and
is clearly inherent in the data presented here. This would hold mainly
for Phases I and II mentioned above. Fission usually takes place when
a strong man feels that he cannot reach his ideal in the local setting
due to a more powerful leader, and he moves away to set up a new
village, commonly drawing with him those relatives and friends who
support him. Those who move do so as family groups and settle in
independent households. The second phase (which correlates with
the stage where more than one family lives in one house and they
represent three or more generations) is reached when individuals join
this village, settling in the already existing houses, or when families
within this village have reached Phase II (or our family type IV). The
best example here is Wayopa, founded by Opura, when he took um-
brage because Nori was appointed luluai of Akuna. Both in household
and in village structure Wayopa is now approaching Phase II.

Chapter 10

COMMUNITY STRUCTURE

In every community there are basic premises which assure each member of the ways he may behave and the probable reaction of others. These premises are different for persons in the various stages of the life cycle; they are different for males and females, and they differ too in certain regards from situation to situation. Many of these expectations and reactions have been discussed or demonstrated with cases in the preceding pages. We need to look at some other features which characterize community life in Akuna.

THE STATUS OF WOMEN

Shortly before we left Akuna, I had the opportunity to read through a number of the reports which were made by patrol officers who visited the Arona Valley. At various places in the reports comments had been entered by the Assistant District Officer of the time. In one of these patrol reports[1] the A. D. O. Captain T. G. Aitchison had written: "From my observations, woman is not 'the down trodden slave of man'; if she were, administration would be much easier."

Due to physiological differences there are times when the maturing female withdraws into the women's house, but the rest of the time she lives in her house and is visited by her husband. Due to these factors, too, the division of labor exists which designates certain kinds of duties to each of the sexes. Thus the basic difference between the sexes is found in these statuses they occupy. The accompanying roles make one into the child bearer-gardener-cook, while the other hunts and wages war. I am not suggesting that the one and only distinguishing factor is a sexual one, but on this and associated psychological variations the roles are divided. If we look at the legal status, there is much less distinction, and females in fact may practice the right of ownership and inherit property. The right of the male to discipline

---

1. Patrol Report: Lt. R.J. Stevenson, No. K. 7 1945-46.

his wife — a right recognized by her male relatives — must be practiced with care, for she has the right to leave if she feels that she does not deserve it, while her brothers may react if she is ill-treated too harshly or too frequently. Generally, though, a woman can give as well as she can take and domestic hassles may go further than words. In an earlier section I mentioned the case where Oturo left her husband Arawe only to return many months later to lay waste the gardens she originally planted. One morning I met Napiwa as he was preparing some food. I asked him where Irona?o was and why he was preparing his own food. It turned out that they had had an argument the previous day after which some scuffle developed. She told him to do his own work and left for the women's seclusion hut. When I was somewhat shocked and asked what he would do, he responded very philosophically that she would return. When people were as old as they were they did not leave each other. On this and other occasions it was noticed that the woman's house served as more than simply a menstrual lodge. It was also a safe retreat for women to cool off.

In the everyday cycle of village life, too, the women participate fully in discussions and court cases. It is usually the older women who attend and comment at cases being discussed in a court session, but younger women do not necessarily play the role of the quiet, submissive cook. On various occasions while visiting a family at night, an announcement would be made in the village square and a woman would respond in a way best described by the colloquialism as 'back-chat'. This would draw forth other comments and in the stillness of the night little ripples of laughter could be heard spreading from house to house around the dark village. On these occasions of informal visits the women were as responsive and relaxed as the males, tendering information and posing questions of their own.

In Melanesia generally, women play little or no part in the realms of religion or magic. For this reason it is significant that Akuna has only a female 'u?wata'. Here is a ceremonial status with an important role which is occupied by a female and can only be performed by one. This does not imply any greater share in the realm of the supernatural. The men alone see and handle the sacred flutes and bull roarer. Only through a rite of passage does a boy reach the status where he may see them, but women's status never allows this. During the 'orande' before I was introduced to these ceremonial objects, the old men gave me a long lesson on the secrecy and value of these objects and when they permitted me to photograph them I had to promise that the photographs would not be shown to any of the women or children. But while the men have their ceremonial objects and men's house, the women's house which traditionally was even constructed by females held as much secrecy for the males.

In the social realm, however, women had their highest status. Here she was for all intents and purposes the man's equal. She was different and performed roles which he did not, but the statuses were not inferior. In traditional times, and so today, it is usually the girl who initiates courtship, who leaves her husband and attaches herself to another male who offers her services and pleasures. A number of these cases come to mind as well as the normality with which they are accepted.

1. When the son of the previous 'luluai' reached maturity, a young woman, Tau?a, informed him by way of his sisters that she wanted to marry him. The betrothal was arranged by her father, Nonuri, and the young couple were married shortly before we arrived.

2. From an earlier section it will be recalled how Ari's daughter Wawentino refused to follow parental guidance and presented herself to the older Kandi?o. After leaving him for a few weeks she returned.

3. An interesting case of the normality of these procedures is illuminated by Napiwa's first wife Upokandudu who left him and is now married to Anokara. Both these men are now happily married and settled but Napiwa still worries about the reason he was divorced by her. From his court case we discussed it will be recalled that Anokara and Napiwa are good friends and the former raises the pigs of the latter.

4. The case was discussed of the woman who left Arawe and later returned to wreck her gardens which his second wife was cultivating. This second wife, Wio, had simply left her husband Amo?a and moved in with Arawe when he was left by his first wife.

5. A similar case occured when Torawa went to the Western Highlands for police training. His wife Maka?o felt lonely and joined Pawiko. When Torawa returned he had no wife and was joined by a young widow who subsequently married Duka; a young woman who moved on to marry Yayo; and a young woman, Tatome, daughter of the influential Munowi, whom he married.

6. The same was true with Waropi of Amamunta who returned after police training. Tiyorono, wife of E?ananda, deserted him and joined Waropi. He liked her and she lives in Akuna working old family gardens while he lives with another wife in Amamunta.

With this high degree of female mobility and the status.they occupy, it is no wonder that they do not employ 'watuye', the love magic which males need. Furthermore, it becomes clear why Akuna has the cultural loophole for divorce in which bride wealth is not returned, but simply reimbursed by the second or subsequent husband.

The patrol reports mentioned above also contain some suggestions along these lines. Blyton[2] comments on a case where a young woman

2. Patrol Report: Lt. D.R. Blyton, No. 32 of 1944-45.

of about sixteen was married to the luluai's son, a lad of about twelve. She had done most of the seducing in arranging the union. Lindsley[3] is even clearer on the subject. He states "It is the woman and not the man who takes the active role in courtship. A young man must not indicate by any public act or hint that he is interested in a particular girl and to be caught in the act of making advances to a girl is to be covered in shame and confusion and to have ridicule heaped upon one ... it is the girl who indicates to the young man that she is interested in him, who induces him to meet her in some secluded spot and who finally asks him to marry her."

In the first part of this study I remarked on the importance of observations made by Leonhard Flierl during the early part of this century. In his first visit to the Akunans — on August 7, 1928 — the missionary made the following observations in his diary: "The forwardness of the women almost caused us great difficulty ... This forwardness made all the more impression, since the women are usually reserved. In many areas one is not allowed to see any women on one's first visit" (op. cit.: 50).[4]

While these field observations and recordings by early visitors do not deny the male status or the role of men which contrasts them with females, they do not suggest a clear picture of women's status. This picture furthermore is quite different from that sketched by Read (1954: 24-32) for the highlands. The suggestion then is not one of male and female status equality, but quite a clear demarcation of these statuses. For each status a particular sex is role-educated with full freedom to perform it as the individual desires.

Yet the woman cannot go alone to her gardens, nor would a man be willing to perform the task of gardening. The man will construct the house, but the kunai is supplied by his wife. This complimentariness is to be found in all spheres of life in Akuna, but not only a complimentariness between the sexes, also one which sets each members of this small community in a reciprocal relationship to everybody else. In some way each person compliments another.

THE PRINCIPLE OF RECIPROCITY

Every human being exists in an ego-centered universe, but for some the reality of living and social interaction becomes more alter-orien-

---

3. Patrol Report: G. Linsley, No. 7 of 1948-49.
4. I have used forwardness as a free translation. Flierl speaks of 'Zudringlichkeit' and 'Aufdringlichkeit' which literally mean obtrusiveness, intrusiveness or annoying.

ted than for others. Partially this is a result of the social system within which they live, resulting in the fact that social expectations and social behavior emphasize the dividual or the group.

It seems appropriate, then, that we may distinguish in this respect between societies where centripetal forces dominate and others where centrifugal forces are at work. The distinction can be applied to societies such as Gadsup and other New Guinea groups in contrast to strong lineally oriented societies such as Tallensi, Swazi, or Zulu in Africa. While the latter are primarily oriented towards the welfare of the lineage, the clan, or the tribe; the first mentioned emphasize the individual, not at the expense of the group, but rather for and through the individual to the existence of the family and the basic social group, usually limited to the village or 'ankumi'. The individual then receives the primary emphasis because of his culturally defined freedom of behavior and latitude in choice. This uninhibited action pattern does not lead to anarchy as might be imagined, but rather emphasizes and enhances the local or residential group rather than the kinship group.

CENTRIPETAL FORCES

It can be stated as a universal that every person potentially occupies a marginal position in respect to other members of the family, the group, or the community to which he belongs. Furthermore, the degree of integration which exists and the social equilibrium that is reached on the interaction level depend largely on the individual concerned. The duty of every member of a group is to perform those roles which are expected of him, for only by performing expected behavior can others be expected to perform their roles, and only when mutual responsibilities have been fulfilled will persons be satisfactorily integrated with the social group of which they are members.

We are faced here with two concepts, namely the need to belong, and the actions which are prerequisites for this integration. Every person in this respect must discover a degree of balance between individualism and social behavior and must maintain the ties of reciprocity which assure assistance in times of need.

In his valuable little book on The Cultural Background of Personality, Linton notes that every person has certain psychic needs which require satisfaction and which form the basis for an understanding of human behavior (1945: 7-10). The first is that every person needs the emotional response which results from interaction with other persons (Linton uses the term 'individual', but it seems more appropriate in this context to speak of 'person'). Because of the fact that a

304

person lives through time, conceptualizes this, and plans for the future, he needs the security that will be available on the long-term basis, namely on those occasions when it is required. The third psychic need, Linton notes, is that of novelty experiences, and like the previous two, behavior in social context suggests the limits within which this may be satisfied without disrupting responses to the first two needs. It will be noted that 'suggests' has been used in this context rather than 'prescribes', for it is felt here that the very fact that persons overstep the boundary of permitted or expected behavior leads to innovations[5] and therefore to culture change.

Mention was made above of the individualism which is so clearly present in Gadsup behavior. This, it is felt, is a response to the need for novelty and a reaction against too firm a structure which prescribes and enforces certain behavioral patterns. It is also allowed for in the cultural flexibility, or "looseness of structure" as Pouwer (op. cit. 1955; 1960(a), and 1960(b), Van der Leeden (op. cit. 1956: 1960), Brown (1962) and du Toit (1962; 1964(c)) have spoken of this characteristic of New Guinea cultures. Akunans, it was noted, need and seek integration, but they guard against getting bound to the extent that they cannot make individual choices. Residence on an ambilocal principle, marriage with the deceased husband's brother, or bilateral exogamy are valued and are leading features of the social organization, but individual choice is still retained, and a person may choose to follow his personal values rather than those values which are accepted by the group among whom he lives and with whom he interacts. It should be kept in mind at this point that individual values are not diametrically opposed to group values, and that individual values in most cases are incorporated in the values accepted by the group. In such cases, then, these values are chosen because they satisfy the individual rather than because they are accepted by the group. This cultural flexibility referred to is a functional correlate of the value for individual choice, as Embree pointed out for the Thai, the term "loosely integrated here signifying a culture in which considerable variation of individual behavior is sanctioned" (1950:182).

By and large, however, this individualism is subject to the behavior patterns which are accepted within the group, and while extraneous controls and social sanctions are not strong, the two basic needs lead to compliance with group expectations and group behavior. Every person needs to belong to, and is in fact integrated with, a nuclear family. Within this social context the biological needs are met and security is felt for the future. A person who does not have a sister to

---

5. For the processes in which innovations are made and become accepted, see Barnett 1953 passim.

make string bags for him, or a woman who lacks a brother to aid her in gardening if her husband leaves her, is 'poor'. So, too, is a child without a grandparent and vice versa, for within this social and emotio context a person is assured of long-term satisfaction of the need for emotional responses, as well as the satisfaction of biological needs.

In addition to the nuclear family, every Akunan belongs to a wider extended group, either a kin or a residential grouping, for again both of the primary psychic needs are satisfied by interaction with others and being assured of their assistance. While these may be relatively impersonal groups, we find that the cultural flexibility allows for a person to choose his own interest group and to associate with persons with whom he finds the greatest personal satisfaction. This interest group, as was pointed out above, is largely of one's own making, for by his behavior, his loyalty, and the use of his goods he may attract or make persons indebted to him, or, conversely, he may isolate himself to an increasing extent by not assisting others.

The maxim by which an Akunan lives, with the exception of hostile relations, is that he should do unto others as he would have them do unto him, for the principle of reciprocity is basic to every act. It is, as Lévi-Strauss has put it, "the most direct form in which the opposition self-other may be neutralized" (DeJosselin de Jong 1952: 7), and the status quo or advantage of the self be assured.

On the physical level a child very early in his life learns that one act calls for a similar act, and striking at an age-mate leads to a return blow. He also learns that the sharing of wild fruit or some choice tidbit brings in returns at some future time. This results in the friendship with an age-group member which may continue throughout life, and may lead to residential changes out of loyalty. The husband finds that the status of a woman allows her to treat him the way she is treated, or to leave him should she so desire.

Marriage is primarily the union between a man and a woman, but at all betrothals and marriages goods are transferred. These goods in most cases include objects of value. The nuclear family, it was pointed out, is economically self-sufficient, and yet we find that at marriage and during other prestations large quantities of sweet potato, taro, bananas, and sugar cane are transferred, products which each family has in its own garden. We feel that it serves a dual purpose: First and primarily it acts in the social realm to bind the members of two interest groups together (see also Firth 1936: 564 for an example of this in Tikopia), for by betrothal a complex and lasting system of prestations and counter-prestations is set into operation. We agree, therefore, with Nadel when he states that "the institution of marriage, over and above regulating sex relations and procreation, also serves that other end, the strengthening of social solidarity" (1953:269). Secondly, the transfer of goods is a redistribution of the economic surplus, or a redistribution of wealth. But wealth does not have social significance in a community such as Akuna, and a person's worth is measured largely in what he does with his products and how he acts toward others.

By his gifts and by his behavior he places other persons in obligation to him, thus anticipating reciprocal behavior. The continuity of the group and the degree of integration is proportional to the degree to which such expectations are met and satisfaction is derived from the system of reciprocal relationships. This may then be taken as an index of group cohesiveness — satisfaction leads to solidarity, but a lack of satisfaction disrupts the ties which bind people and will cause people to terminate these relations. The result, then, will be discord and even greater individualism and fragmentation of the social group. The fact of whether individualism is emphasized in a society, or the fact of whether we are dealing with a bilateral or a unilineal system, does not imply greater or lesser group cohesiveness, for this characteristic is the result of other forces which are prior to and which influence the kinship ties. The principle of reciprocity assures a person of assistance when he needs it, gives him the right of use over the tools and products of the other person, leads to a sharing of food and the continuous transfer of goods.

While people are eating they will offer food to every person present "for it is not good that when I am hungry you will not share food with me". Should a person enter a house with whatever request, he will always be offered food first. Should a person be making a garden fence and another person offers help, advice, or some tobacco for a breathing period, the fence builder will tell his wife that when the garden produces, a share of the produce should be given to the wife of the person who helped. Every gift and every act requires a return gift or act of equal value. While this is not expressed, it is fully understood by all, and an invitation to a feast or a gift with no suggestion of return stated, nevertheless implies such. "It has become evident, indeed, that it is not so much the nature or quality of the exchanged goods as the act of exchange itself which is considered to be of paramount significance" (Marcel Mauss as paraphrased in De Josselin De Jong op. cit.: 5).

This very principle of the social ties created through the transfer of goods and the giving of gifts forms the basis on which the political system rests. Through gifts and the social ties which result a leader gains in prestige and in the number of persons that support him in his aspirations, and through gifts other villages may be enticed into a temporary alliance. But acts of hostility in this same principle call for retaliation. A kinsman killed by another village because of suspected sorcery leads to a counterattack at some future date, and if by chance more than one person is killed, this has to be revenged. Thus an intermittent state of war develops between villages. This is easy to explain for Akunans state that certain people, a good warrior, an 'uwata', or a 'yikoyumi' are worth two, three, or even more regular persons. Should an enemy kill such a person the victims would have to retaliate two- or threefold to even matters.

In all of this, the individual is central. All actions are designed to enhance the person's prestige and to secure his future. "Social behavior is an exchange of goods, material goods but also non-material ones, such as the symbols of approval or prestige. Persons that give much to others try to get much from them and persons that get much from others are under pressure to give much to them. This process of influence tends to work out at equilibrium to a balance in the exchanges" (Homans 1958:606). Conversely, the person who does not honor his debts, who flouts his kinship obligations, and who does not offer assistance when others seem in need is a "man no gud", or a "rubbish man".

Informants were insistent on the fact that a person should be dependable, and such persons are admired. Admired, too, are persons who do not 'kiaman' (deceive), who are loyal to friends, villagers, and kinsmen, and who help where they can. If such a person walks in the rain and finds somebody mending a fence where the pigs had broken through, he will help him. If he comes upon a person doing some hard work in the hot sun, he will offer to help or at least offer some tobacco and talk to the worker in order that he may rest a while. Such persons are industrious and always have plenty of food which they share with others. In Akuna there are a number of persons who are valued in these respects, but there are also those others who are not admired and are despised. Such an individual is Tipuna. He is generally considered to be too much of an individualist, one who flouts his kinship obligations and who hardly ever works. He decided to get married. When he was collecting his bride wealth to transfer for Oya, he approached a number of his kinmen requesting their help. One of these was Tapari who, however, replied that he was not in a position to contribute now, since Tipuna had done nothing to ensure the ties between them. When this happened, Tipuna informed his father, an old man of about sixty-five years of age, and both of them entered their houses and started to cry, refusing to come out until Tapari and some of the other kinsmen promised to help. Finally Tapari appeared with two pounds in cash, an axe, a knife, a box of matches, two wrap-around laplaps, and a bar of soap. The men emerged and the marriage was contracted, but Tapari is still awaiting the return of his gifts.

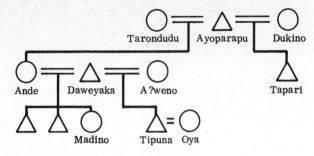

Diagram 18. Genealogy of Tipuna

What happened here is that Tapari responded for the sake of the
old man, rather than for Tipuna, and the impression one gets from
Tapari is that, for all he cares, Tipuna could still be crying. The
old man on his part was not so much concerned whether kinsmen
assisted Tipuna, for he too realized that the latter was considered
an outcast and 'man nuting' among his kinsmen and therefore also
had a limited interest group, but was concerned with himself and the
need he had for grandchildren. These are very important to an old
person, as we have seen, and he was interested in reaching his goal
of having grandchildren even though Tipuna was the vehicle. The latter then achieved his aims, not for personal reasons, but because of
the interests of his father and his villagers-kinsmen. This is one way
of aligning yourself with others and them with you, but it is not the
preferable way. Ideally alignment, loyalty, support and similar sentiments are not enforced but follow individual decision and personal
commitment.

MODES OF ALIGNMENT

"All forms of groups are based upon some principle of recruitment
whereby individuals are made members, that is, are made to assume
the implicit rights and obligations" (Nadel 1958:151), and in
doing this, their roles change or new roles are added. In an earlier
section of this study we dealt with membership by birth, thus ascribed
status. We will now look at achieved statuses, and the roles which
accompany these. Nadel distinguishes 'open' and 'closed' recruitment,
and in the latter, certain conditions and prerequisites are set up for
entering a group. Such limited or closed systems are common in
Melanesia, and played a relatively important role in traditional Gadsup society. The men's house, and the realm of the sacred flutes
(Tinapu) and bull roarer (Ireno?namu) to which boys were initiated

at puberty, were such 'closed' systems, where only persons of certain sex and age requirements were admitted. These have to a major extent fallen into disuse.

In contrast to such systems with closed recruitment are those based on 'open' recruitment, "so that individuals can join at will, for the purpose of participating in the group activities" (ibid.: 151). This study proposes to extend Nadel's terminology to allow for reciprocal action. By this we mean that membership is open to all who meet certain requirements, but that this membership entails reciprocal action on the part of all members. A member may participate in village activities as long as he shares in the burdens which this entails. A member of one's interest group is considered such as long as he returns interest, assistance, or prestations in which he was allowed to share. This open recruitment will specifically involve the choice of residence and association, as it is a major thesis of this study that the degree to which this freedom is expressed among Akunans places individual freedom, the wide latitude of choice in behavior, and the importance of the residential groupings on a par with, or above, the importance of kinship reckoning and cognatic group membership. We will then deal with achieved statuses, contrasted to those where membership is ascribed on the basis of birth, which depend on individual decisions, and also the behavior of others which influences such decisions and especially the role of the 'yikoyumi' in aligning others with him and his village. It will also be recalled that the interest group is based partly on such freedom of choice; these being the aspects which lead us to speak of Akunan community as being open and having great cultural flexibility.

By ties of blood — By birth, each Akunan belongs to a patri-oriented 'ankumi', concentrated in one village but by no means localized. Each individual, male or female, is free as an adult to choose a place of residence and thus to align with a smaller or larger part of the clan or 'ankumi' to which he belongs.

The simple fact of belonging and the possible choice this entails, has an important psychological effect in a land where sorcery is common and fighting frequent. Freedom exists for each person to align with his kinsmen, his affines, or simply with friends with whom he could best express his personality and satisfy his desires. Cases have been recorded where a person is living and associating with persons related, but so vaguely that they can best be described as falling in a "twilight zone" (Murdock 1962, personal communication). It is within this group that marriage takes place, and though disapproved, it is not forbidden. A person can always receive assistance from a member of his 'ankumi' and will not attack, with the intention of killing, a village where such persons reside. Furthermore, he has the right of use over property belonging to members of his 'ankumi' or may freely enter their gardens, provided he informs them of liberties taken.

Based on the alignment by birth, each person, then, is associated with a village and with a nonlocalized kingroup. His decisions and subsequent behavior are influenced by village members and by the loyalties and actions of kinsmen, and only when these reach a harmonious state do we have persons who are wholly aligned in the village of birth or in the 'ankumi'. This frequently is not the case. "Primitive communities do not conform to the simple pattern conceived by earlier writers, but are often segmented according to several coexisting principles. To determine the comparative influence of the several resulting loyalties is one of the most important problems of social organization" (Lowie 1935:141). In these sections it will be attempted to measure the loyalties and to suggest those that receive priority.

By ties of affinity — There are no preferential marriages among the Gadsup, and freedom of choice allows a person to contract a marriage with the spouse of his own choice. This freedom often leads to residential alignment, and it is significant to note that plural marriages are usually with spouses from different villages, a fact which served the leadership aspirations of potential leaders. In the past this was especially true and of the utmost significance. It will be recalled that the intimate affines are usually included in a person's interest group, thus sharing in his profits but also in his losses. Closely allied to this is the fact mentioned that Akuna would never wage outright war with a village in which they have cognatic or affinal ties, or if a part of the village did, those who were so related would remain neutral. This meant that by marrying into and thus aligning more than one village with him, the 'yikoyumi' was assured that these persons would materially assist him in times of festivities or prestations, but also that such villages would not attack his village for fear of killing their kinsmen and affines. It might even go so far that this loyalty would produce allies for an attack on some common enemy. Once the person is linked with such a village these ties exist for as long as he lives. When a betrothal takes place, a person enters into relations with a family and kingroup, which demand of him loyalties and assistance, and usually offer him the choice of residential change. Actually these forces are stronger and more formalized during betrothal and nearly border on avoidance, but as marriage is contracted persons settle into the expected behavioral patterns which allow for latitude in behavior and for a fair amount of interaction and cooperation. These ties then are enhanced and personalized at marriage. While this act sets into operation institutionalized forms of address and action between the person concerned and the spouse's parents and siblings, we lack the institutionalized avoidance, or expected joking so characteristic of many societies (Lowie 1953:84-6).

Once marriage has taken place a person is assured of the close relationship with his affines. This, however, requires reciprocity, and by their actions and behavior the affines may align the person with them and assure cooperation. A very close relationship exists between Munowi and his children's spouses with the result that they all cooperate in fence building and house construction and share in garden products from each other's gardens. This has also brought the different sons-in-law into a close relationship which borders on institutionalized friendship. On the contrary we can refer to Ari who has three married children. The eldest, a daughter, rarely brings her husband to her parents' house. While the second, a son, was away on indentured labor, his wife left his parents and returned home.[6] The third, a daughter, refused the man suggested as a spouse by her parents and moved to a different village; and after divorcing her first husband she remarried in the same village.

The degree to which affinal relatives join the interest group of a person influences and is an expression of the reciprocal relations between them while having a definite effect on the role they (say, as matrilateral kin) will play in the life of the offspring. It depends on their behavior as well, whether a leader can expect political alignment in time of defence or attack, and for this reason relations are maintained through prestations and counterpretations which continue for the duration of the union. These are made at various times, such as any of the rites of passage of a person's offspring, during village festivities, after any large payment as in court cases, and on all similar occasions. It should be kept in mind that reimbursements for losses or payments after killing a person's pig do not take the form of a few dollar bills, as they would among ourselves. In one court case over the accidental killing of a pig, the person was reimbursed as follows:[7] one large pig (slaughtered), one small pig (alive), one chicken, thirty-four elaborately carved arrows, three large bunches of bananas, twenty ears of corn, three long sugar cane stalks, thirty-

6. The case of Ma?e appears more fully in an earlier chapter as does a discussion of the third daughter, Wawentino, who left her spouse and returned after a few weeks in Akuna.

7. This material is also treated in context in our chapter on village economy, while the case from which the action flows was described with a number of other court cases.

five taro, five bunches of betel nuts, three fruit-bearing pandanus fruits, six rolls of tobacco, fifteen bunches of 'yienni', three lap-lap for the wrap-around men wear, two pounds and two shillings in cash (approximately $5.00), four enamel bowls, five spoons, a string carrying bag, newspaper for smoking, pandanus leaf blanket, a knife, matches, peanuts, soup, and a cowry shell. This rather extended list of items has been recorded here to show that these payments are not mainly in compensation for losses incurred. We may speak in this context of a manifest and a latent function, the first being the transfer of valuables from the guilty to the injured party while the second is the redistribution of valued objects. It was gathered by the interest group of the person who was responsible for killing the pig, and was then distributed among the interest group of the pig owner. Each one of them in turn will share his surplus when such payments are made to him, and in this way both the economic and the social balance is maintained while the legal requirements are satisfied by the guilty party.

By ties of friendship — Institutionalized friendship with its extensions into the realm of kinship and exogamous rules (Pouwer 1955:55) is lacking among Akunans, but strong associations based on amicable relations do develop. Table 19[8] includes Anapa?iyu and his wife, Timawareno (household 12), who have settled in Akuna because of such friendship with the luluai. We have also mentioned the case of Anamaka and Nanono with their sons who moved to Akuna because of amity ties with Napiwa and the tultul. As far as could be ascertained, they cleared new garden land, rather than getting it from some family. In the case of both Anapa?iyu and Anamaka, the decisions to settle in a village where they lacked cognatic or affinal ties served to enlarge the interest group of the leaders with whom they associated.

By and large this was the realm where the 'yikoyumi' operates most effectively. Because of his position in a numerically and geographically extended kingroup, he has a certain backing for his aspirations and forms of wealth for exchange and may call on his interest group (which would include affines) and villagers for assistance in payments and feasts. In marriage he has exercised his free choice to align himself with affines and other villages, and possibly faced a decision of village change (fission or schism) in order that he may better reach the ideal relationship of leader and followers living about him in his village. These decisions and the important process of aligning others with himself, either on social (temporary allies) or residential basis, were the bases for leadership.

8. See Addendum C.

By ties of residence — This section deals with that all-important aspect
of the Akunan community, namely its residential groupings. As Leach
remarked for Pul Eliya, we can state that "it is locality rather than
descent which forms the basis of corporate grouping" (1961:7), and
in all cases this locality of residence and association is based on the
choice of the individual concerned. Basic to any group formation, and
of primary influence in any decision which such a person is faced with,
are two prerequisites of a community. "One is that they have certain
common possessions or interests to share, defend, or promote, in
order to give their living together a purpose; and the other is there
exist between them bonds of sentiment, in order to give it continuity"
(Leach 1954:13).

It is proposed that we have already seen these factors at work,
namely the 'possesions' and 'interests' which affected village formation
and which we have been considering here as the basis for alignment,
and the 'bonds of sentiment' which form the basis for the interest
group. However, they are not completely separated, and one in fact
involves the other.

We will trace these factors here and follow their influence as they
affect decisions from childhood to marriage, also considering the
various choices which are available at particular times.

Premarital residence

Commonly children live in the same houses with their parents. At
this stage there is little choice, and rarely does a person have to
decide among various possibilities. Should the mother or both
parents die, the child will be adopted by either persons in the same
village or by a relative in some other village.

Once a child has reached puberty, however, boys and often girls
leave the mother's quarters to sleep with their age-mates and associates. Boys will sleep in a 'pumara yi ma?i' or young men's house,
from where they later are initiated into the men's house proper. Girls
sleep with their friends in the 'pumara?ini yi ma?i' or young women's
house, from which they move after marriage to their own households.
Should these associations not be present in the particular village, and
it seems that the latter was frequently absent perhaps due to the early
age for marriage, the children will remain with the mother. This is
the stage, too, at which a person often made the choice between two
or more villages. In nearly all cases that were recorded, the latter
decision and the resulting change of residence pertained to boys, while
girls remained with their mothers or in their natal villages.

The schism in Akuna and Opura's departure to settle the new village, Wayopa, has been discussed above. His sister, Wainano, had

314

a number of children by Pumpua, the youngest being Werora. Now a boy of about fourteen years old, he fluctuates between living in Akuna with his mother and living in Wayopa with his mother's brother, Opura. This is not strange in view of the role a mother's brother performs in the life of a person, but we lack any strongly patterned avunculocal residence, either before or after marriage.[9]

Trained as a teacher by the Lutheran Mission Society, Nonuri of Akuna left with his family for Mamerain among the eastern Gadsup to open a mission school. Fearing that his children would marry locally, while he was temporarily located in this village, he returned to Akuna and asked Yifoka (brother of Werora discussed above) to marry his daughter Tau?a. It will be recalled that girls frequently initiate courtship and this was initiated by her as she asked her father to betroth her to Yifoka. The son of Nonuri, Yayopuma, now a boy of about twelve, at present spends most of his time in Akuna living with his sister, Tau?a, and both of them are retaining the claims to land in Akuna which belongs to Nonuri and his wife. In other instances, Wira, whose father is deceased, joined the household of his father's sister and her husband, while the son of Apaka and Minitano moved from Akuna to live with the son of his paternal grandfather's sister's daughter in Wopepa. In another case, upon the divorce of Tapari and Orepano, their son joined the household of his mother's brother in Amamunta. These cases, then, do not fit any set pattern, and residence change lacks any firm system.

Ayatiwa, an Akunan man, married and went to live with Undaino, his wife, in Wayopa. Their sons, Wanara and Itu?wa, returned to Akuna where Wanara married (and was subsequently divorced by his wife). Both of them now live there, forming part of the mobile, irresponsible group of young men between puberty and marriage briefly discussed earlier in this study as 'those in transition'.

Diagram 19 outlines the instances in which a man, Arupenamu of Wopepa, married Tompenano of Akuna and they went to live in Amamunta. Upon the husband's death, Tompenano married Ampo?ima, luluai of Asarenka, and went to live with him while her children remained behind. She had a number of other children, of whom Rapinda is about thirteen years of age. He has left his village and spends most of his time at the Summer Institute of Linguistics base

---

9. The fact that we could not discern a pattern of avunculocal residence does not remove the importance of this form of residence change. It is particularly true in the case of post-pubertal males and a number of cases have been discussed (see pages 243 and 333 and this page). The residence change reflects very clearly the highly emotionally laden relationships which are associated with the avunculate.

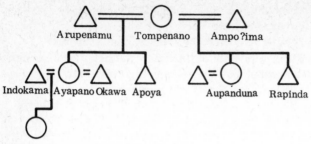

Diagram 19. Genealogy of Tompenano

of Ukarumpa working for Whites. His co-sister, Ayapano, married a
man, Indokama, of Onamuna and went to live with him. After a few
months she divorced him and returned to Amamunta, where she married Okawa. Some seven months later a daughter was born whom
Indokama claimed, stating that the mother had been pregnant by the
time she left Onamuna. Okawa disputed the claim, but under pressure
of the Amamunta residents, agreed to the removal of the girl; and she
is now living with Indokama and his second wife in Onamuna.

From the examples discussed above we can see that once a boy has
reached puberty he may select his residence. This decision is not
taken randomly, but definite motivations are involved. These are
usually tied in with land rights that have to be maintained, the association with 'ankumi' or an interest group, and lately, the possibility
of earning money.

Post-marital residence

Though the previous section dealt with shared interests and bonds of
sentiment, they are not of such primary importance among the younger
age group as they come to have among married persons. Upon contraction of marriage, two persons from different 'ankumi', different backgrounds, and very often from different villages come together, and on
the basis of these criteria the culture provides, must choose their place
of residence. We have been attempting to isolate those motivations which
lead to this decision, but must always remember that their reasons
might be completely different "from those used by the anthropologist in
classifying their culture" (Goodenough 1956: 29).

Anthropological terminology, though greatly refined in recent
years, is still only able to cope with major tendencies, and not with
every possible consideration which entered into the decision. If a
Gadsup from Amamunta marries a woman from Wopepa, and they
settle in Akuna, it is quite possible to decide that this is neolocal

316

residence. In tracing the genealogical background, however, it often comes to light that either or both have returned to land which was worked by grandparents or even more distantly related kinsmen. It is in this respect that the genealogical data of nearly 4,000 persons proved of greatest value. In the compilation of demographical data, this and other considerations had to be taken into consideration. One of these was the occurrence of secondary marriages. As an example, one case will be briefly mentioned. Napiwa, as a young man, had accompanied his father, Afita, to Kundana, a village in the Eastern Gadsup, and returned with a bride, Upokandudu. Shortly after the birth of their first child she divorced Napiwa, married A?ano, and upon his death married Anokara. All three of these men were from Akuna. The important fact here is that marriages between Eastern and Western Gadsup very rarely occur, and also that bride wealth for a secondary marriage does not involve the woman's parents, but is transferred to the first husband and/or his interest group. Since the woman's natal village has not changed, and even though her village of residence might have, should we list three marriages between Akuna and Kundana? The decision in this and similar cases was that one man (Napiwa) had the audacity to go outside his area and one woman (Upokandudu) was willing to sever the ties that bound her to a village and her 'ankumi', and that the secondary marriages were between Akuna men who married a woman in Akuna and transferred bride wealth within Akuna.

Residence rules are not strict, nor are persons required to choose between only two possibilities (ambilocal). At marriage persons may reside with the natal kin of either mate or they may settle neolocally.

The expressed ideal is one of virilocal residence, but this is not universally followed if the person does not have amicable relations with his kin or if other considerations enter the decision. Various cases have been referred to above where persons moved away, and in many cases, where they settled with affinal relatives. It is suggested that while relationship by blood exists per se and cannot be altered, affinal and non-kin relations must be achieved by conscious effort. This, then, would lead to a lack of emphasis on cognatic ties with possible selection from among these and an effort for persons to align themselves with others, and others with them, who are not so aligned through a natal core. This lack of emphasis, it has been shown, accompanies the great freedom a person has and the importance of such factors as land holdings, residential core, and interest group members, which enter into any decision. While lineal structures enforce alignment with the kingroup (though this usually corresponds with the residential community), a structure which is only nominally lineal such as Akuna places primary emphasis upon the residential community a person chooses (though this may include a greater or smaller part of the kingroup). It is suggested that this characteristic which is so markedly present among Akunans, and generally in New Guinea, is closely linked to the feature which has been vaguely termed "looseness of structure".

This freedom of choice in residence, freedom of alignment, and freedom as regards association is also at the foundation of the political system. While a Polynesian became chief because of his position in a lineage and clan and mana associated with his status, the family background of an Akunan does not make him a likely leader, except possibly through the number of pigs he inherits and the land available to him for gardens. These, however, are used as any other person uses them in the system of perpetual gift exchanges which draw others to the leader, strengthening his village and in turn spreading his power. This statement does not suggest that such increase in power and fame can lead to permanent political allies being made, for as with the Garia, "the internal structure does not emphasize group formation" (Lawrence op. cit. 10) and every village remains an autenomous unit often divided internally by potential leaders.

Every village has land for gardens surrounding it, and this land belongs for all practical purposes to individual families.[10] From this land a tract can be made available through one of the resident families for a new resident, a son-in-law or a political associate. Families could also decide jointly to make land available as was the case with O?e and Apopine following their parents' and sister's death in Ikana.[11]

Adoption is important as a means of establishing ties but also for the care of children who are in need. While the latter is very often the case in a society such as this where life expectancy is low and mortality of women at childbearing age is high, it is suggested that the first case was of greater significance. It is entered into by persons who have no children of their own, as well as by those whose children are grown up and have left the parental home. Above and beyond these reasons, adoption was also a means of alignment, for with the ties between individuals are also established ties between kingroups. In all of these cases land can be made available to the new group members. This form of alignment is common in the Eastern Highlands, and there is good reason to believe that Table 19[12] does not contain

10. Patrol Report: G. Linsley, No. 7, 1948-49 states: "Land is owned outright by individuals or families, as the case may be, and the right to land descends equally to male and female children."
11. These are the brothers of Daunamari — the case being discussed more fully above.
12. Vide Addendum C.

318

all adoption cases; however, children that are adopted are like real
children while kinship terminology is applied to them and the genealogical data rarely if ever refer to a child as 'adopted'. Adoption is
termed 'enaka?i mande windemi' (lit.: child of another take care of
it). The 'mande' means 'to take', as a person picks ('mandena') a
fruit which is very ripe and which drops into his hand when he
touches it. Should a person pick a fruit that is strongly attached,
forcing him to cut or tug at it, this picking is referred to as 'kide'.
Adopted children then simply 'drop' into the 'ankumi' of the adopting
parents and become one of them. An adopted child inherits from both
sets of parents and belongs to both kingroups, through his biological
and through his sociological parents, while exogamous rules apply to
both these kingroups. When an adopted child marries, the betrothal
and marriage payment is transferred to the sociological parents, who
give a share to the biological parents. Adoption, then, in addition to
being a form of insurance for persons who have no children to care
for them, is also a means of aligning two families, and by extension,
two kingroups.

This alignment through adoption, however, is based on the symbolic union which seems to accompany the sharing of food, or at
least to form part of it. This symbolism is expressed in many contexts and a few will be listed below:

1. It seems that during World War II, when air attacks were made
at Kainantu and Aiyura and the Japanese had occupied the Arona villages, a large communal house was built at Yatarorapa?o between
Tompena and Batanapura where women and children of both Gadsup
and Tairora villages took refuge together. While surveying the extent
of bilingualism in border villages, a Tompena informant explained in
Neo-Melanesian: "Ollageder i-kai-ka hap kai-kai bilong Tairora, na
em iharim lik-lik hap tok tasol" (They all ate a share of the Tairora
food, and now they only partly understand the (Tairora) language).

2. It will also be recalled that when a person farms out his pigs,
the person who feeds them is not allowed to eat of their flesh.

3. Once a child has been adopted "and has eaten our food", he/she
becomes one of the new kingroup and exogamic rules apply as they do
in the child's own kingroup.

4. Should a mother not be able to nurse her infant, and a wet nurse
is allowed to take over, the latter has a claim on the child after a few
weeks and will reply: "Look, the baby has grown strong on my milk,
it is mine now."

While the previous examples dealt on the group level with statements of general practice, there are two cases involving non-Gadsup
individuals who are considered group members because they have
lived with Gadsup families and eaten the food enjoyed by the other

319

family members. There is then clearly a symbolic tie between the eating of food and the resultant social relatedness.

1. In Tompena there is a young man of about twenty years of age called Nera?iyu who lives and works with the household of Kua?o, speaks Gadsup fluently, shares in prestations and counterprestations, and would never be distinguished as other than a kinsman. Many years ago, Kua?o related, he had been among the party of Tompena men who attacked Kamukera, a Tairora village. As they came close to the village he noticed a woman working in the gardens while her baby slept in a carrying bag hung upon a fence post.[13] As he and his wife had no children he took the baby home and they raised it. Today Nera?iyu is as much a Gadsup as any other and cannot even speak Tairora, though, like a number of Tompena people, he understands it when spoken. This is a phenomenon widely encountered in New Guinea, and something Wurm has referred to as 'passive bilingualism' (1961:14). When asked whether he does not feel an urge to return to the place of his birth, his reply is, "No, I have grown up here and have all my friends here; we speak one language and when I die I want to be put underground here."

2. As a young boy who had not yet reached puberty, Munoka ran away from his natal village of Yoyanakeno in the Agarabe language area. He arrived in Amamunta and was cared for by the villagers. After initiation he took up his place in the young men's house ('pumara yi ma?i') and when reaching maturity married Ayankino, a Gadsup girl from Akuna, and seems as much a part of the village organization in Amamunta as any other adult.

As mentioned, ideally post-marital residence is virilocal, but we are faced with a problem in dealing here with a society in which clans are associated with particular villages, but this never results in a localized kingroup and therefore village exogamy. Akunans then are not faced with this rule, and the result is that many persons marry within their natal villages (this practice is particularly true of Tompena it is said) and then either remain or move away. In such cases, are we going to call the marriage virilocal, uxorilocal, or what? When genealogical data did not clarify, it was simply tabulated as residing in the natal village of both husband and wife. By and large it is only upon marriage or at least during adult life that decisions are made regarding the place of residence, and only at this stage do motivations and sentiments become of prime importance, for only at this stage is

13. It is common practice for women to place a pandanus leaf mat in the string bag and lay the baby on it. They walk for miles with the baby in the bag, either fore or aft, depending on the direction of the sun, and when working or resting hang the string bag on a tree, fence post, or against the wall of a house.

a person faced with personal sacrifices or gains. "The fact", says Linton, "that a woman goes to live with her husband's people is less important, for practical purposes, than the degree of isolation from her own family which this entails" (1936:164). Or as Leach (1961: 11) has pointed out, what we need to know about a society — and this also holds true for the community of Akuna — is not whether it is labeled patrilineal or matrilineal or neither, but exactly what this label stands for. I have attempted in this study to use as few labels as possible, but to give a picture of life in Akuna.

While virilocal residence is expressed as the ideal in Akuna, demographic data show that we are dealing with a case of ambilocal (Murdock, ed., 1960:157) and often neolocal residence. Thus individuals freely depart in practice from the favored ideal, consulting their person advantage in the decision. This was also the basis on which persons could align themselves with a leader and move to his village in his support.

During the field period an interesting case occurred where the expressed ideal was reversed by village leaders. For some time a competition for power had been developing between Akuna and Amamunta. In spite of informants' reports that Amamunta originated from Akuna through a process of calving — and the Akuna people like to point to Amamunta as an offspring of theirs — the latter village was making rapid growth (partly due to the presence of a Summer Institute of Linguistics couple who had been living there for the past three years) and already counted nearly 200 residents compared to the 250 of Akuna. The luluai and village elders in Akuna repeatedly pointed to the fact that young people of both sexes from Amamunta found mates outside the village and then brought them to reside in Amamunta, thus strengthening that village. Three court cases developed from relatively insignificant matters, namely the killing of a pig, the cutting of a tree, and an attempted seduction. Hostilities were mounting when Apoya (see Diagram 19), a young man of Amamunta, had repeated sexual relations with Ma?e of Akuna.[14] One afternoon the Akuna luluai, Nori, and a number of the older men of the village called a meeting for all the Akuna residents. This was irregular, as most meetings are attended only by adult males, or adult males and females, but this one included all persons of post-pubertal status. After some initial discussion of the fact that Amamunta was "our offspring", the fact was pointed out that they were clearly attempting to increase the population and size of the village, while Akuna itself was decreasing, had very few young couples, and was in a state of decline. Then the new rule was laid down. From this time onwards, any young person of Akuna

14. The case is discussed in detail in our chapter on judiciary forces.

who got married had to settle in Akuna and strengthen the numbers and prestige of the village, and this would also apply to Apoya. He could marry Ma?e, but would have to live in Akuna,[15] which would clearly contradict the expressed ideal and result in uxorilocal residences

Residential change, however, need not follow immediately upon marriage. After some years of marriage, and with children already born, Apayu from Wopepa, Anamaka from Amamunta, Ana?o from Tompena, and Munowi from Wopepa moved to Akuna. After three children had been born, Yapananda and his wife Tana?o left Akuna during World War II and moved to Ikana. Their eldest daughter married the tultul of Ikana, and the third child, Daunamari, also married an Ikana man. Within a year of each other both parents died, and during the field study the death of Daunamari occurred. The Akuna people were very upset by these deaths in such a short period of time and asked the Ikana people whether they knew how to take care of 'visitors'. During the funeral ceremonies, Akuna men repeatedly urged the two remaining children, both post-pubertal boys, to return 'home', and though their parents had not retained their land claims it was made clear that land would be made available to them.[16]

One further case deserves mention. Shortly before the field work started, a man in Amamunta died. His younger brother, Dapaka, was at the coast on indentured labor at the time, and the luluai of Amamunta suggested to the young widow that she remarry one of the luluai's relatives. She refused, insisting that she would await the return of Dapaka and marry him. The luluai and his male relatives attempted to press her, but upon failing in this, and witnessing her marriage to the returned Dapaka, the luluai reported the case to the kiap in Kainantu, explaining that this man had married his 'sister'. Dapaka was promptly placed in jail, but when the matter was explained to the officials in charge by his relatives, he was released. The young man returned to Amamunta just long enough to rather his belongings and his bride and then moved to Akuna, where he is now living.

From the foregoing discussion it is clear that we lack a firm structure, and that individual freedom and latitude in choice are present as well as the possibility that existing rules might be reversed. It is this flexibility of the culture which allows personal interest and sentiments to play such an important motivational role in actions and decisions. It is also this flexibility which assures social order in the community, for where a rule or pattern might be too stringent, the pliability allows for compromise and choice.

15. Note the similarities in terms of structural change between this case and that of Awiitigaai, the Kapauku headman (Pospisil 1958:832).
16. This case has been treated in greater detail above.

FORMS OF ADDRESS AND REFERENCE

Each Akunan has one name by which he is addressed and which is used
in referring to him. Exceptionally one finds a person with two names,
but no significance is attached to this, and informants state that a
mother or father uses the second name as address, this being the
closest to a nickname the people have. A child who is very thin and
small is called 'ayampamana inta' (lit.: bone-only child), or the
equivalent of our 'scarecrow'. However, this does not affect the treatment the child receives, nor does the name remain once the child has
developed normally. There is no form of reference which corresponds
with our surname, a group name, or even a place name, and each
person, then, is known by his personal name only.[17]

Adults are addressed and referred to by a teknonym. When first
the author arrived in the field, he addressed adults by name, and
frequently observed youngsters to turn away their heads and snicker.
It soon became clear that they were simply not used to the idea of a
young person of my age using the names of the village seniors and
other adults. Needless to say, the ethnographer adjusted by using the
teknonym where he knew the name of the children, and the youngsters
became used to the idea that he occasionally did not know an adult's
family and had to address him by first name. Frequently they recognized that I occasionally needed help in this respect and would whisper
the correct name. A person two or three age-groupings above the
speaker is never addressed or referred to by the personal name, but
this teknonym deduced from the kinship term is used. Thus for a male
we find that the suffix '-napo' is added to the name of the eldest child
(preferably a son) and for a female the suffix '-ano' is added to the
name of the eldest child (preferably a daughter). Thus the mother of
Nori, the luluai of Akuna, is called Noriano, while his wife Timandudu
is called Dudinano (i.e. mother of Dudi, her eldest daughter). The
luluai is called Yajimanapo (Yajima being his eldest son). Informants
explained that "we young people can't speak to old people by using
their names". There is no preference for age and sex or leadership
in the use of reference terms, except that of 'yikoyumi', the traditional 'strong man'.

In the following section we propose to present the terms employed
in addressing and referring to kinsmen and affines with the appropriate behavior expected in each context. The terms are seen here as
an expression or formalization of the social positions occupied by the
persons involved, and similar terminology would suggest similar
treatment (Lowie 1948: 61). Following this presentation, an analysis
of the system will be offered.

17. See also the discussion of naming which appears above.

The terminological system

An outline of the Gadsup kinship system is presented graphically in Diag
20, while the letters with their corresponding terms are listed below in
Table 18. This does not include the majority of ego's affines, and these
well as the female's terms will be discussed below.

Kinship terms are applied over a very wide range, and their use is
discontinued only where memory and recognition of relationship ties
falter. Thus a person may know that children of persons that his
grandparents called cousin live in a different village, and these persons will be classified with the parents and their children as brother
and sister. However, these remote ties are of far less than formal
weight, and while they are recognized as belonging to ego's 'ankumi',
marriage may take place with them.[18] Once again we face the fact
that relatedness is recognized where it serves to promote the advantage of the person concerned.

Table 18. Akunan kinship terms

1. A anapu	6. F a?o	9. K anonta
2. B akaka	7. G apa?i	10. L akai
3. C apo	H awa?i	11. M aramumi
4. D ano	8. I anano?i	12. N anai
5. E amama	J umi	

These terms, when used by the ego, are prefixed by the first person
singular marker 'tenti-'. Frantz (1962: 50) distinguishes between 'ten'
and 'ti' namely noun stem referring to the whole and first person singular possessive. When the prefix, however, is attached to a stem
which begins with a vowel, certain morphohonemic changes occur
between the two vowels, namely /i/ before /a/ remains /i/. In this
way then a Gadsup speaker in Akuna will address his father as 'tentipo' and his mother as 'tentino', meaning my father and my mother.

In order that we may trace the primary and secondary meanings of
these terms and partially note their extensions, each will be discussed briefly.[19]

1. 'Ainapu' is used for both the father's father and the mother's
father as well as for their siblings of the same sex and is thus mainly
reciprocal with 'anai', except in so far as this term denotes sex.

2. 'Akaka' is used for both the father's mother and the mother's
mother as well as their siblings of the same sex and is also mainly

---

18. See the rules governing marriage in our discussion of how adulthood is
reached through marriage.

19. For a complete list of the Gadsup terminological system see Appendix A.

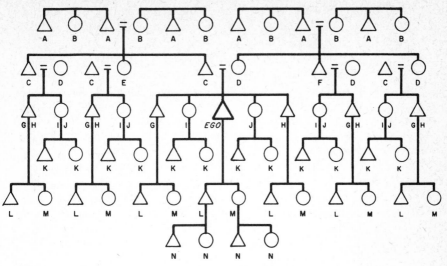

Male ego speaking — For Akunan terms see Table 18

Diagram 20. Akunan kinship system

reciprocal with 'anai', except for the distinction it makes in sex.

In the second ascending generation, Ego employs two terms only which distinguish between sex. On various occasions informants would refer to their ancestors and used the composite 'tinaputikaka'. This was also the generalized term Akunans used when they mentioned the third ascending generation.

3. 'Apo' is used for father primarily, and those males who are in the same generational level as father, except the mother's brother.

4. 'Ano' refers to mother and those females who are on her generational level, except the father's sister.

While genealogical data were compiled, informants would frequently speak of a father's brother and mother's sister as 'apo?i' and 'ano?i' respectively, thus distinguishing between the biological father and mother and classificatory parents and their siblings. When this was pointed out to them and the suggestion made that there was a real difference, it was repeatedly denied and they would again use 'apo' and 'ano'. Gadsup nouns use suffixes of -i, -ni, or -mi among others, and it is here suggested that we may be dealing with the distinction between 'apo', or 'tentipo', my father, thus a real biological father, and 'apo?i', a class of persons grouped together as fathers.

5. 'Amamu' refers to the father's sisters, both junior and senior to father.

6. 'A?o' refers to the mother's brothers, both junior and senior to mother.

No extensions were recorded in the application of these terms, and both these persons are of primary importance in the life of a child. They are used reciprocally with 'anonta'.

7. 'Apa?i' and 'awa?i' are used to address and refer to brothers; the first of these to a person who is senior to Ego and the second to a person who is Ego's junior. By extension these are also applied to all males on Ego's generational level as widely as relatedness is recognized.

8. 'Anano?i' and 'umi' refer to sisters; that is, to all females on Ego's generational level. The first is applied to persons who are Ego's senior, and the second term to females who are younger than the speaker.

9. 'Anonta' applies only to sister's son and sister's daughter, whether the sister is junior or senior to Ego. While it ignores the sex of the child it is reciprocal with 'a?o' and is therefore used only by males forming the basis of the nepotic-avuncular relationship, which extends throughout a person's lifetime.

The relationship is especially strong and enduring between the mother's brother and his sister's son, but is not balanced by an equal bond between the father's sister and her brother's daughter. The fact is that the avuncular-neoptic relationship does not include the father's sister; she refers to and addresses her brother's children as 'my son' and 'my daughter' respectively. We furthermore lack any special relationship between the father's sister and the mother's brother themselves, for though they may be bound by the relationship with the children and by the marriage of their respective brothers and sisters, nothing prohibits them from getting married, and under the brother-sister exchange marriages discussed above, it frequently happens.

10. 'Akai' is used to refer to all males of the first descending generation, except the sister's son. Its primary use is for son, but secondarily it is used for the sons of siblings, thus all the male children of 'apa?i', 'awa?i', 'anano?i', and 'umi'.

11. 'Aramumi' primarily refers to daughters and by extension includes all females on the first descending generation, except the sister's daughter.

12. 'Anai' is used for both males and females on the second descending generation, thus the offspring of 'anonta', 'akai' and 'aramumi' and is used reciprocally with 'anapu' and 'akaka'.

All these terms are employed by females as well, except 'anonta', for which categories a woman employs 'akai' and 'aramumi'. There is no difference between a term of address and a term of reference, and a person denotes the personal relationship with each of these relatives by prefixing the possessive pronoun for the first person singular tenti- to all the terms listed above.

326

A.

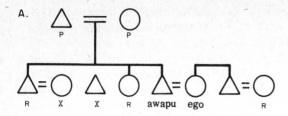

B.

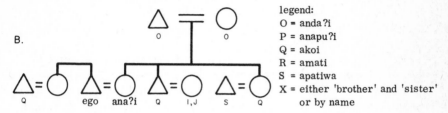

legend:
O = anda?i
P = anapu?i
Q = akoi
R = amati
S = apatiwa
X = either 'brother' and 'sister'
    or by name

Diagram 21. Akunan affinal terminology

An additional set of terms are employed with reference to affines
(Diagram 21).

1. 'Anda?i' — this term is employed by a male when addressing
or referring to his wife's father and wife's mother and is reciprocally
used by them for the daughter's husband.

2. 'Anapu?i' is used by a female to refer to or address her husband's father and husband's mother and is once again reciprocal.

3. 'Akoi' is once again reciprocally used between a man and his
wife's brother and his wife's sister. This is an important term as it
is associated with extreme caution and is translated into Neo-Melanesian as 'tambu'.

4. 'Amati' is also a term which is associated with reserve, but is
not as strict as the previous example. It is used by a woman when
speaking to her husband's brother and husband's sister and is used
reciprocally by them.

5. 'Apatiwa' is used by a man to address or refer to his wife's
sister's husband. It is a term, the significance of which was never
clearly fathomed, but on the basis of behavior patterns observed the
relationship is marked by great freedom and is more or less the
antithesis of the previous two terms, 'akoi' and 'amati'.

The link between the kin and the affines is of course the married
couple, who address each other as husband ('awa.pu') and wife
('ana?i'). Their relationship is a cordial one in which they spend
much of their time together while working on a garden, having their
meals, or simply sitting in the sun or around the hearth at night. At
these times discussion takes place on general village topics or family

matters which are of mutual interest. While unmarried persons hardly ever walk together, it is not uncommon to encounter a man and his wife walking through the bush or looking for a pig in the kunai. This is partially the result of the fact that sexual intercourse between a man and his wife always takes place in the bush or in the kunai, but never in the village. A woman will never have intercourse in her own garden as it will affect the growth and produce of the plants. A man does not mind this in his garden, but he would hardly take another woman there for adultery as his wife might find them.

By and large kinship and affinal terms are employed in everyday village life, but very often personal names will be used, and when questioned regarding this practice an informant will state that so-and-so started the use of the personal name. It always happened that an elder or senior person deviated from the accepted rule, and that it then was taken up by the younger or junior person. Husband and wife rarely use the terms in address, but always for reference. Exceptions to this, namely cases where the terminological system is rigidly adhered to, are those dealing with elder persons — 'anapu', 'akaka', 'apo', 'ano', and 'amamu' (A through E in Table 18), and 'anda?i' and 'anapu?i' (Diagram 21) — and the patterned relations between 'a?o' and 'anonta'. While these persons' names may be used in ambiguous situations, this does not hold for 'akoi' and 'amati', resulting in a major problem during the collecting of genealogical material. Older informants who had placed their trust in the ethnographer would occasionally whisper the name into the recorder's ear, while younger partly-schooled informants found that it did not violate their ethical system to sketch the name in large letters on the ground.

The relationship between a child ('akai' or 'aramumi') and its parents ('apo' and 'ano') is close, for it is within this social universe that the child first recognizes faces and experiences emotions. It is also the parents that introduce the child to the Akunan world of ideas and the behavior patterns which flow from these. Both parents spend considerable time with the infant or care for it during the absence of the other, and while we realize that the man nowadays spends more time with his family than he might have formerly, it is suggested that this is true for traditional times as well. Akunans are very fond of children, and it was instructive to observe old luluai Nori pick up and fondle a child crawling in the village square.

Children are spoken to, scolded or even threatened, but very rarely punished by parents or other adults. The parental attitude seems to be one of "do as I do" rather than one of "do as I say", and children are largely left to learn from life and association with siblings and age mates. A man would sooner beat his wife for a mistake, and vice versa, than beat a child, and physical punishment of children is exceedingly rare. Children are close to the parents until they are fairly independent, age five or six, after which they may spend the day in the village under the watchful eye of a kinswoman or — and this is especially true of boys — start roaming around with their age mates.

From that time on boys are encouraged to become more and more independent, and after the eighth or ninth year it is not insisted that they eat or sleep with the mother. In the years between puberty and betrothal boys live in a carefree world, devoid of strict or binding rules and free to make decisions of their own. This, it is suggested, is causally linked to the freedom of choice regarding a marriage partner and the locality of post-marital residence. In contrast to their brothers, girls remain close to the house and spend much time in domestic duties. A close tie develops between a mother and her daughters, while the latter accompany their mother to the gardens or assist her in the preparation of food and household goods. As there is a taboo on any foodstuffs handled by a woman during menstruation, daughters may also substitute for their mothers at these times, or on occasions when the mother is ill. Their education and training is in the practice of those roles they will soon have to perform.

The relation between siblings, and by extension, all those persons referred to by kinship terms 'apa?i', 'awa?i', 'anano?i', and 'umi' is a close one, especially during childhood. When the child starts to act independently of the mother and spends the day playing in or near the village, it is invariably in the company of an elder sibling. It is within this social context that the child learns the forms of permitted behavior, and while an older sibling may punish a child for unacceptable behavior, such disciplinary action is not welcomed from non-relatives.

A boy learns from siblings and parents the behavior toward a sister and also those acts or topics of discussion which are not permitted. While the relationship between a brother and his sister is reserved, avoiding bodily contact or sexual topics, it is usually more durable than that between two siblings of the same sex. Freedom of association with age-mates and the selection of friends is highly valued, and the result is that brothers usually associate with different individuals who occupy the same age-grouping. This is true for sisters as well.

Marriage betweel cross- or parallel-cousins is unheard of among the Akunans, and their kinship terms intimate this by classifying all persons on ego's age level as brothers and sisters.[20] The fact of

20. Neighboring linguistic groups do permit this tie in marriage and during a cultural survey tour among the Onteno, Oyana, and Ataya Gadsup Outliers, this form of marriage did occur. Among the latter, in fact, the mother's brother's daughter is terminologically distinguished.

seniority is important within this category of persons, and among both sexes a person is careful to distinguish between an elder brother and sister, 'apa?i' and 'anano?i', and a younger brother and sister, 'awa?i', and 'umi'. Relations with junior brothers and sisters are by nature freer and subject to less restrictions than is the reverse relationship.

When a person is very ill and has lost his strength, it is his wife who sits with him all day, resting his head on her lap. Should she at any time leave him for natural relief, his brother may take her place, but never his own sister. Occasionally the wife's sister or even the wife of his 'awa?i.' (younger brother) will take her place, or even a son, a daughter or a son's wife. This strict taboo on physical contact between a brother and sister emphasizes the absolute horror with which Akunans think of incest, and no informant could recall an example of incestuous unions having taken place.

This very close yet reserved relationship between a man and his sister can potentially lead to divided loyalty when the wife comes between them. Since these are agriculturalists, the working of land by a woman for her husband will cause resentment of other women who enjoy these products of her labor, even if such women are the 'anano?i' and 'umi' of her husband, thus posing no sexual threat to her. This ambivalence on a wife's part is once again explained in a cultural tale:

This story is about a woman called Sumeriya. She had a son and the two would come whenever her brother gave some feast, or had some food to eat. When her brother was having a ceremony for the piercing of his son's nose, she arrived. When she came her brother's wife saw her and said: 'Tentimati, you always come running when we are eating some food!' and she was angry at her husband's sister. When her brother's wife was cross at her, she went and sat down on the other side of the door where an old woman was busy making a fire. Now the men and women started to prepare the food and the wood to cook it on. When this had been completed the husband sent his wife to go and tell his sister to come and help. His wife left but did not go all the way; she only went half way and then returned to her husband, informing him, 'I went and told her but she said, "I don't want to go" and she stayed back.'

That night when it was dark they went to sleep and when the day started they went to kill the pigs. As they were starting out, the husband sent his wife, saying 'Go and tell her to come and help with the food.' She left but travelled only half way and then returned to her husband informing him that she had told his sister to come and help, but that she again had replied I do not want to go. After the pigs had been killed and the meat cut, the man again sent his wife, saying: 'Go and

330

tell her to come and help prepare the meat and soup.' Again she went
only half way and returned, saying, 'I told her but she did not want to
come and remained.'

When they started to prepare the earth oven for the meat, the man's
sister was sitting on the other side near the door by the old woman
who was preparing the fire; she was waiting. She told the old one that
her 'timati' had done thus-and-so, and that she was going to leave. 'If
my brother comes tell him that his wife has done thus-and-so and that
I have left.' The old woman remained behind and pretty soon the
woman's brother arrived. The old one told him what his sister had
said, and he realized it and ran after her. When he came to a hill he
saw her and he called out for her to wait, but she did not wish to and
told him to go back. She kept running up the hill until she came to
another hill and there was a hole in the ground and this place is called
Sumeriyampa?i. He called out for his sister to wait, but she threw
her mats and sticks into the hole and before the brother could stop
her, the sister and her son jumped into this deep hole which I have
mentioned. As she was crying, the brother cut off his earlobe and a
finger joint so that blood covered his chest and he wept. At this time
a fog came up over that place which is called the place of Sumeriya.

After he had returned to his house, his wife and the old woman
asked him what was the matter and he told them that he had cut his
finger while working. He said: 'I do not want to exchange pigs and
have a feast,' and he told his wife to place all that meat in the house.
When she had placed it all in the house, he told her to remain there
and to eat it. When the night had come up on that place, he took bush
liana and tied the door of the house; he also tied the secret door of
the house, and he told his wife to eat the meat which she had taken
inside.

When he had done this he asked his wife which part she was eating
and she replied that she was eating the soft meat; he said that that
was good and left. Later he returned, again asking what part she was
eating and she replied that she now was eating the soup of the pigs'
insides. After a while he returned once more and she informed him
that she was eating the meat and skin. 'That is good', he said, and
left. When he returned again she said that now she had started to eat
the seeds of the kopunayun. The man went away. When he returned
she said that all the food was eaten and that she had now started to
eat the string bags and clothes. When he arrived yet another time he
asked her and she said, 'I have eaten all things that were in the house
and now I am starting on my nails and hair.' He said 'yes' and left.
Again he returned and she was crying for want of food and he listened
and departed. Later on he returned again to this house of his and when
he asked her what she was eating, she did not reply. Then he took

some fire and set it to the door and to the secret door. She did not reply and was burned with the house.

Now, if you want your sister to come some place, go and get her yourself and give her food and she can share it, but do not send your wife to call your sister. That is something that a man should not do. Now this story is completed.

This link between kinsmen also extends into the political realm, for hostilities will not develop between kinsmen and especially not between persons who call one another 'brother'. Should a person, in an act of fury or in an accident, kill a kinsman, the matter is considered in a serious light. His brothers will congregate and say, "One of our brothers is dead, let us not all die." They will then go to the culprit and ask him to arm himself. This consists of putting his shield ('yani upekunde?no?u', lit. wood carried on the back) over his shoulder and taking up his bow and arrows. Then he will appear outside his house while his male kinsmen release their arrows, allowing them to pierce and break the shield. Informants explained that with one man already dead, it is not advantageous to kill a second, and this expression of grief and anger clears the air of hostilities, for the matter is not touched on again. If the kinsman is not closely related or closely allied to the deceased and his interest group, matters may naturally take a different course.

An important relationship for every Gadsup is that between 'ainapu' or 'aikaka' and 'anai'. It is a bond which is cultivated from early childhood and serves the adults as an old-age insurance. For this reason it is important that a person be loyal to the members of his interest group, observe his kinship obligations, produce offspring, and assure a special relationship with one of the grandchildren. If these requirements are met, assistance is assured and senescence can be accepted with resignation and without fear.

A grandparent may discipline an 'anai', but is more likely to spoil him. The grandchild may use anything belonging to the grandparent, but may not misuse this right and must report to him regarding the use of the possession. A grandparent is likely to mark a specific tree pig, or even part of a garden for the grandchild — a wish which will be honored when inheritance comes due. While the adult is still strong he/she will rarely return from the bush or another village without something for the child, such as a ripe fruit, or a choice piece of meat. The grandchild remembers this and frequently visits with the aged 'ainapu' and 'aikaka', and on various occasions informants stated, "My father and mother are still strong and will be here for a long time. It is my duty first to look after my grandparents. When they are dead I will look after my parents." The 'anai', then, when as an adult he visits some other village will frequently place a piece of meat in a bamboo container and take it back to an ailing grandmother or grandfather. Occasionally the statement was heard that "so-and-so is poor, he has no grandparent", thus denoting a lack of something valued for every individual, the absence of which causes other persons to feel sorry for him.

Akunans rarely remember relatives beyond the second ascending generation, and will brush off inquiries to this effect with a shrug of the shoulder stating "iyeninde?u peni pukenoyo" (lit.: I do not know, all they are deceased).

The most important relationship is that which exists between the mother's brother and his sister's child. While it is usually the sister's son that is primarily involved in the reciprocal nepotic-avuncular bond, the sister's daughter is not ignored. When Numino, the ten-year-old daughter of Duka and Munoki, died during the field period, it was the mother's brothers who relieved each other during the all-night wailing and mourning ceremony in keeping up the loud and moving expression of grief.

When a child is about four or five years old, the mother's brother performs two ceremonies, though the second was often omitted, especially in girls. The first of these, 'ati?ankarano' ('ati?i', nose), is the piercing of the septum, and the second 'ankamankemi' ('akami', ear), the piercing of the ear lobe. Again at a girl's puberty, 'akintamwaremi' (lit.: the girl she is sleeping, i.e. euphemism for her sojourn in the menstrual hut), the mother's brother plays an important part in killing a tree kangaroo and ritually cooking the meat. From this stage onwards he is of special importance to a boy. The night before a boy's puberty initiation is spent in the young men's house where the initiates are huddled around a roaring fire while the elder males form a circle around them and periodically the mother's brother will relieve his nephew, offering him a period to cool off before he again joins the circle. The following day the ordeal is continued, climaxing in the exposure of the mystic sacred flutes and the bull roarer, the 'upaiyami' (nose bleeding), and finally the cutting and treatment with native salt and 'wimi' or wild ginger (Zingiberaceae alpina species) of the tongue and the glans penis. Once again it is the mother's brother who appears out of the bush, attracted he says, by the pitiful shouts of his nephew. Protectively placing his arms around the boy, he scolds the adult males for ill-treating his 'anonta' in this way and then leads the boy off to the safety of the village and his place in the 'pumara yi ma?i' (young men's house).

This affable and sympathetic relationship continues for the rest of their lives, occasionally leading to avunculocal residence, but always assuring each of them that assistance will be given when needed. It

333

also assures both persons in this relationship, and this is important, that there will be someone to cry at his death.

It remains for us shortly to discuss affines. The parents of the spouse do not require any special avoidance, but caution is called for during betrothal and before marriage. Should a person accidentally encounter the parents of his betrothed, he will utter an excuse like "I'm sorry, how are you?" and not awaiting nor expecting a reply, continue on his way. After marriage persons have freer interaction with their affines, but the man rarely forms close ties with his wife's mother.

Within the group of kinsmen who fall in the category of persons between whom marriage is most likely to take place, namely age peers of the opposite sex, we find the strict institutionalized taboo with prescribed behavior patterns. It will be recalled that 'akoi' is applied reciprocally between ego and the wife's brother and the wife's sister, thus confirming the suggestion raised earlier in this study that the sororate is absent, for the wife's sister is treated very formally, a pattern which borders on avoidance. This applies to both the younger and the older sisters of the wife. A somewhat relaxed relationship is expected between persons who call each other 'amati', and while certain topics are not discussed between them, marriage is permitted — thus allowing for the levirate. This term is applied to both younger and elder siblings of the husband. While discussing the nuclear family, occasions were mentioned when a child might embarrass a parent's sibling by defecating or urinating on him and that in such cases amends had to be made. It seems quite logical that this behavior, namely the disregard in these instances for the father's brother by not paying him to remove the "shame", can be explained in terms of the leviratic practices and the absence of sororate. Though there is a similarity for the child in the kinship terms used to denote mother and mother's sister, and father and father's brother, and the possibility does exist in theory for father to marry mother's sister once mother has deceased, this does not happen frequently; nor is ego's matrilateral kin (thus the affines of father) expected to substitute a wife if mother deceases; nor is father permitted to practice sororal polygyny. The mother's sister, then, is a mother in only a very vague and classificatory sense, and while she may assist the family during mother's absence, thus performing the role of substitute mother, there is little or no chance for her ever to be mother.

Ego's father in fact addresses his own wife as 'tentina?i', but his wife's sister as 'tentikoi' and acts appropriately to the avoidance term. Ego's mother on the other hand calls her husband 'tentiwa.pu', and her husband's brother (junior or senior) 'tentimati', a term, as we have seen, which allows for free and unrestrained, if somewhat for-

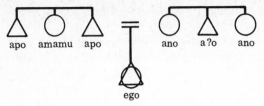

ego

Diagram 22. 'Shaming' and its structural implications

malized relations. The formality, then, is a temporary one, one that
may change should father die and father's brother marry mother under
the rules governing the levirate.

From this discussion it is clear that while mother's sister is called
mother (and usually 'ano?i', as was pointed out), she has little likelihood of becoming father's wife and must thus be treated as mother's
sister and not as mother by making amends for any degrading act of a
child. The father's brother, in addition to being a classificatory father
by virtue of his sibling relationship with father, is a very potential
mother's husband by virtue of the presence of the levirate. The relationship in this context with father's sister and mother's brother is
of a different order and also differs terminologically.

Two males who have married sisters refer to each other as 'apatiwa',
and their relationship is a close one as is the case between Napiwa and
tultul Yerai, or between Torawa and Tapari.

Affinal relations are not transitory, and once these relations have
been established, the terminology and patterned behavior remain.
When an informant who was divorced was asked whether he still called
his previous spouse 'tentina?i', he smiled and said, "No, one doesn't
talk to her."

A child that is adopted is treated as a real child and occupies his
position in the terminological system and relations. He also maintains
these kinship terms and behavior toward his biological kinsmen.

If this terminological system which we have briefly discussed is
analyzed, it serves — with other structural features which have
already become apparent to the reader — to permit a label to be attached to the system. This label is of little importance if the contents
are not clearly described, for labels only apply to general conclusions
rather than to the details of daily living.

1. The princi. of sex is recognized as one of the important features for distinguishing between persons. Thus Akunans distinguish
between males and females on the second ascending generation, among
the parents' siblings, between male and female siblings and counsins,
and between sons and daughters. The exceptions in this respect are to
be found in the second descending generation and in the terms a man

uses to address his sister's children, where the sex differences are ignored.

2. The generational criterion is clearly recognized in the fact that persons of the same sex and the same generational level are grouped together, and the relationship between them and Ego is therefore recognized.

3. The distinction between senior and junior relatives is made in Ego's generation only, by distinguishing between senior siblings and cousins and junior siblings and cousins.

4. Distinction is clearly drawn between cognates and affines in Ego's generation, but not in others. Thus his mother's brother's wife is called mother and the father's sister's husband is called father and this is extended to the grandparental generation. In the same way, Ego speaks of his wife's brother's son as son, and his wife's brother's daughter as daughter.

5. Kinship terms are bilaterally employed as widely as Ego recognizes cognatic ties, but once an elder person has deviated from this principle by using the person's name rather than the term, the junior speaker will follow the pattern which has been established.

6. A total of nineteen terms (compare with the bilateral system of the Tenino, with forty terms, Murdock 1958:307-10) are used for both cognatic relatives and affines.

7. There is no distinction between address and reference.

8. The terms are the same irrespective of the sex of the speaker, and the primary emphasis is upon the connecting link and exact relationship between them. Note that a female Ego addresses her sibling's children as "my son" and "my daughter".

9. We note further that the criterion of bifurcation is observed, thus resulting in a single term being used for mother and mother's sister, but a different one for the father's sister. This same principle is present in niece terms.

10. The principle of collaterality is ignored, the parents and their siblings being merged in the same way as cousins and siblings, or one's own children and nephews and nieces, are merged. Collaterality is recognized only in the special terms for mother's brother and father's sister.

11. A man calls all his female cousins by the same terms he uses for sister, and she in turn uses the terms for "brother" for all male cousins irrespective of the sex of the connecting relative.

12. By noting the presence of bifurcation (see number 9 above) and the cousin terms (see number 11 above) it is quite clear that we are dealing with a Hawaiian kinship terminology (see Murdock 1949:223), which is usually associated with a bilateral system.

PRINCIPAL STRUCTURAL FEATURES

In this study I have outlined the various categories of Akunan social organization, examining how they operate. An attempt was made to show that organizational features are seen in and inferred from personal behavior and that the organization, based on regularities and co-variance, is a dynamic system which changes as the behavior and preferences of persons change. This in general is also true of the nomenclature of kinship, in that the discriminatory features in the terminology are reflections of particular behavior patterns, or rules regarding these. The fact that a person is not allowed to marry his mother's brother's daughter is an established rule in the behavioral realm, but it is given recognition in the terminological system by calling such a person sister. All of this in turn is recognized as an organizational feature, namely the fact that we have here Hawaiian cousin terminology while the marriage rules reflect bilateral tendencies. The same can be said for those features which are summarized as "bifurcate merging", but which are in turn based on the behavior of and the recognition by the people whom we are studying. Furthermore, it is important to keep in mind that these features which we group under the rubric "bilateral tendencies" merely reflect the behavior preferred by those people at that particular time and under the particular conditions under which they are living. When conditions, ecological or cultural, change, and the people readapt themselves, it might result in a change in the structure of that community. Social organization, then, is not a thing sui generis but a description of regularized, patterned behavior and the rules which underlie these. The view is taken here that social structure is an abstraction, a construct, a statistical notion, analyzed and abstracted from the regularity and frequency with which certain features appear in the organizational aspects of a society. It is made by social scientists to describe and categorize organized behavior among a group of persons.

In describing the principal features of Akunan social structure, we are therefore analyzing the organizational system and setting up a series of categories in order that the system may be classified according to one of the accepted structural types. Such a type is in fact a label which marks off various goals or expected behavior patterns operating in the social organization. An act of marriage, then, is individual behavior, but the successive series of marriages in a society constitute patterned behavior which follows the goals or values accepted in that society. In Akuna we happen to be dealing with maximal bilateral extensions of exogamy rules (an organizational feature) which as such is valued and accepted as the goal. We accept this as one aspect of the structure which, with others, allows us in the last analysis to classify and label these features. Now it is not denied that, loosely speaking, "structure influences individual behavior", but only very indirectly so, only because the structure is a "label" attached to certain values in the social realm. It is an accepted thesis in social sciences to hold that values influence the behavior and choices of individuals. That behavior may be affected by external conditions or change of its own accord as the ideas of the people change, cannot be denied, and for that reason social organization is in a continuous condition of change. This dynamism is also present and reflected in the value systems of the group (just as the Akunans in time may come to accept, value, and expect the man to reside with his wife as the village leaders proposed), the value system in turn affects the behavior of other members of the group, also causing us to change our label, reclassifying this structural type.

This study has been marked by the frequent mention and emphasis of such features as individual ownership, individualism, and relative freedom in choice situations. These may very easily create the impression that the end result is chaotic and unorganized, but such reasoning is based on a lineal causal explanation according to which one thing leads to another while maintaining the same direction. The paradigm which underlies such reasoning is:

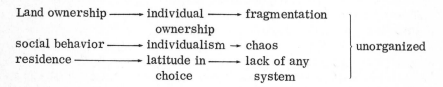

Land ownership ⟶ individual ⟶ fragmentation
               ownership
social behavior ⟶ individualism → chaos   } unorganized
residence ⟶ latitude in ⟶ lack of any
              choice   system

Instead of this directional-cumulative effect, it is suggested here that we are dealing with a system which contains what Bateson (1958:292) calls a "self-corrective characteristic". This means that a cyclical effect is noticed where one aspect leads to another but in turn affects the first. We have then an oscillation between the two tendencies while the cyclical corrective tendency is of overwhelming influence. In the same context Ronald Berndt comments on the Kamano and their neighbor's south of the Arona Valley. He mentions some of the features we have been outlining and then asks: "Is this to be viewed as 'looseness', in the sense of a lack of co-ordination? Or should we identify it as flexibility, which implies a capacity to adjust and be receptive? From the evidence put forward it would seem that the second alternative is the more appropriate, when we consider the apparent ease with which these people have responded to alien demands and adopted and adapted new forms of control. Not all the changes now under way have been forced on them. Many of them have been accepted enthusiastically,

338

even to the extent of anticipating the wishes of the Administration and the Missions" (1962:422).

To return to our paradigm, we still deal with the same characteristics of individualism, private ownership, and latitude in choice, but each is characterized by the self-corrective tendency. People like to have the right and freedom to behave and choose as they prefer, but simultaneously they are in need of help from the local residential group, thus permitting individualism to operate up to a point — that point where it endangers the unity and future of the group. Private land ownership might very well lead to fragmentation of the land holdings to the point at which each child finally would own a minute tract hardly large enough to make a living. Here, too, we have the self-correctiveness that not all persons retain land rights, that land which is not cultivated returns to bush and is free to the claims of any person, that the acculumation of land is not valued, and that inter-village mobility usually leads to a shift in loyalty, thus a shift in land interest. While land rights in many cases are retained, this is not always the case, as people lose interest after a number of years. In this way it can be seen how flexibility exists while not leading to a lack of organization, that the social organization is open but still organized.

The following, then, is a summary of the social organization, typifying certain valued behavioral patterns and finally labelling each in order that we may classify the social structure of Akuna as a certain type (see Murdock 1949:223-31):

1. Descent is patrilineal with a strong bilateral tendency.
2. Cousin terminology is of the Hawaiian type.
3. Post-marital residence is ambilocal with an average percentage of couples residing neolocally, and occasionally avunculocally.
4. Clans are present but of minor importance to Akunans.
5. The 'ankumi' to which each person belongs is an egocentered ramage.
6. The incest taboo has a maximal bilateral extension and the eleven recorded marriages between persons who can trace relatedness are recognized as irregular unions.
7. Marriage on the whole is monogamous, but polygyny is present with an incidence of only 12.1 per cent. The majority of polygynous males have two or three wives and the largest number recorded was five.
8. The independent nuclear family is the primary social grouping in Gadsup society, but a person's interest group usually includes members of the ambilocal extended family.
9. Aunt terms distinguish between the father's sister on the one hand and group the mother and mother's sister, on the other, and are

therefore of the bifurcate merging type. In addition, the mother's brother is terminologically distinguished from father and father's brother.

10. The same is true for niece terms which merge the daughter and brother's daughter, but distinguish the sister's daughter from them, while merging son and brother's son and distinguishing sister's son from them (even though he is merged with sister's daughter).

Based on this analysis of the principal features of the social structure, Akunans are classified as belonging to the Guinea-type of social organization as outlined by Murdock. He explains this type of social organization as follows: "It includes, by definition, all societies with ... cousin terminology of the ... Hawaiian type. It is devised to accomodate those tribes which formerly belonged to one of the stable bilateral types, Eskimo or Hawaiian, and which have evolved patrilineal descent on the basis of patrilocal residence without having yet undergone the adaptive modifications in cross-cousin terms necessary to achieve a more typical patrilineal structure" (1949: 235-36).

The major point with which we would take issue is Murdock's suggestion that we must necessarily be dealing with a structure which shows these bilateral tendencies due to an evolution from an earlier lateral to a more lineal type of social structure. Anthropological data seem to point increasingly at social structure as a response to ecological conditions, and so, when unpredictable or harsh conditions are dealt with the structure allows for a wider social universe on which to draw and on which to depend. This approach of Murdock's may prove to be irrelevant to the New Guinea situation, and perhaps for other peoples who live in these same conditions. This is inherent in Pouwer's discussion of a horizontal versus a vertical stress (1966: 278) in New Guinea; in Goodenough's outline of the "nodal kindred" as a structural type in Lakalai, New Britain (1962:10); in Pehrson's clear picture of perpetual structural rearrangements, alliances and the severances of alliances (1957:108) among the Könkämä Lapps. It seems possible that we are not dealing with an evolution from a previous structural type to another due to contact or integration but with structural features which are part and parcel of the ecological setting of these people.[21]

21. See also the discussion on page 350.

Chapter 11

THE VILLAGE COMMUNITY AND SOCIAL ORDER

In the foregoing pages we have been constructing a quiltwork in which
material of various hues and shapes, sizes and textures has been used.
An attempt was made to fit these all together in such a way that the
product is neat and clear, the result well-rounded, and the effect
esthetically pleasing. Yet there are a number of strands which have
been used to sew the material together which are still loose. We need
to tie them together so that the seams will be united and the product
complete. The community of Akuna with its nearly three hundred
members, who represent different kingroups and different backgrounds,
has been discussed as a community. We have looked at a variety of
contexts in which these people function: the fifty-five households, the
daily round of activity, the kingroup (both clan and 'ankumi'), the
systems of belief, the charter which guides action, and process of
growth and maturation throughout the cycle of a lifetime. The study
as a whole gives a picture of life in Akuna. There seem to be three
areas in which this picture is hazy, three strands which are loose and
need to be tied in. The reason why this village in the New Guinea Highlands is approached in terms of the community study approach is
linked directly to the second question, namely the social structural
features which allow us here to speak of the social organization as
showing features which we have called open, loose, optative and similar terms. Both of these areas are related to the third, namely that
of social order which exists in this community and which is both cause
and effect of Akunan community life.

THE VILLAGE AS COMMUNITY

In the latter part of the previous century the social-philosopher, Ferdinand Tönnies constructed two models by which human aggregates
can be approached and in terms of which these groups can be discussed and compared. These models described general characteristics
and were not based on any particular examples or actual human group.

Tönnies was influenced by Hobbes and Herbert Spencer, and while quoting Adam Smith and Henry Sumner Maine, it is with the latter that his concepts basically agree. Thus Wirth has stated that "the concept community corresponds to Sir Henry Maine's status, while society roughly parallels his contract" (1926:416).

Tönnies explains that every human being stands in a particular relationship to others, and due to these relationships, particular types of associations emerge which mark the group of people, setting one off from another. "The relationship itself, and also the resulting association, is conceived of either as real and organic life — this is the essential characteristic of the Gemeinschaft (community), — or as imaginary and mechanical structure — this is the concept of Gesellschaft (society)" (1940:37). It will be noted that the author speaks of constructs which label particular characteristics or features. The model or construct, however, is based on particular characteristics which may be found in a particular assemblage of people or among whom certain of these features may be present. In his own words, Tönnies describes these as follows: "The Gemeinschaft by blood, denoting unity of being, is developed and differentiated into Gemeinschaft of locality, which is based on a common habitat. A further differentiation leads to the Gemeinschaft of mind which implies only cooperation and co-ordinated action for a common goal. Gemeinschaft of locality may be conceived as a community of physical life, just as Gemeinschaft of mind expresses the community of mental life. In conjunction with the others, this last type of Gemeinschaft represents the truly human and supreme form of community. The first or kinship Gemeinschaft signifies a common relation to, and share in, human beings themselves, while in the second one such a common relation is established through collective ownership of land, and in the third the common bond is represented by sacred places and worshipped deities. All three types of Gemeinschaft are closely interrelated in space as well as in time. They are, therefore, also related in all such single phenomena and in their development as well as in general human culture and its history. Wherever human beings are related through their wills in an organic manner and affirm each other, we find Gemeinschaft of one or another of the three types. Either the earlier type involves the later one, or the later type has developed to relative independence from some earlier one. It is, therefore, possible to deal with (1) kinship, (2) neighborhood, and (3) friendship as definite and meaningful derivations of these original categories" (ibid.:48-).

In contrast to those associations where the will governs and where people are naturally united, Tönnies sees the Geselschaft as an artificial construction, where individualism triumphs, and where there exists in fact a constant state of latent aggression against all other members of this group. "However, in the Gemeinschaft they remain essentially united in spite of all separating factors, whereas in the Gesellschaft they are essentially separated in spite of all uniting factors. In the Gesellschaft, as contrasted with the Gemeinschaft, we find no actions that can be derived from an a priori and necessarily existing unity; no actions, therefore, which manifest the will and the spirit of the unity even if performed by the individual; no actions which, in so far as they are performed by the individual, take place on behalf of those united with him. In the Gesellschaft such actions do not exist. On the contrary, here everybody is by himself and isolated, and there exists a condition of tensions against all others. Their spheres of activity and power are sharply separated, so that everybody refuses to everyone else contacts with and admittance to his sphere, i.e. instrusions are regarded as hostile acts" (ibid.: 74).

The first then describes "reales und organisches Leben" — actual and organic life — while the latter type of association represents "ideele und mechanische Bildung" — an ideal and mechanical structure. Every person due to birth and residence is a member of a community but for utilitarian purposes and by design he incorporates himself with a society. For Tönnies there was a historical trend in human associations where men moved gradually from the pre-modern Gemeinschaft to the modern Gesellschaft.

This position of course was also basic to the approach of Maine who sees the change occuring from kinship to local contiguity. He states: "The history of political ideas begins, in fact, with the assumption that kinship in blood is the sole possible ground of community in political functions nor is there any of those subversions of feeling, which we term emphatically revolutions, so startling and so complete as the change which is accomplished when some other principle — such as that, for instance of local contiguity — establishes itself for the first time as the basis of common political action"(1888:124). Accompanying this change is the change from status to contract (ibid.: 163–165).[1] The fact that Maine and Morgan later denied that primitive peoples united on any basis save that of kinship is well known and is also crucial to this study. In fact, the third lecture on early institutions was titled "Kinship as the Basis of Society" (1893:64) and described the changes which had taken place in India where the village community "is a body of men held together by the land which they occupy: the idea of common blood and descent has all but died out" (ibid.:82). When dealing with the Eastern Village-Community he even

1. Essentially the same position is found in the writings, though evolutionary, of Herbert Spencer (1896:649) and Lewis Morgan (1878:6–7).

states that the "Indian and the ancient European systems of enjoyment and tillage by men grouped in village-communities are in all essential particulars identical" (1889:103).

While many of these ideas were influential in the thinking of Tön-nies, we are concerned here primarily with the latter author's developmental sequence as it was first published in 1887. This "Gemein-schaft und Gesellschaft" trend, which he worked out as a model or tool for the study of human groupings, was used by later writers as being a description of actual human groups. Rather than using the tool as a guide, the various features which constitute each of these conceptual systems were listed and used as a kind of check list to locate societies in this continuum. As some social scientists have tended to use the folk-urban as opposite points on a scale in which the one excludes the other (du Toit 1974) so it has been thought that the gemein-schaft is a pure community where peace, harmony and good will is always present, while the Gesellschaft is the modern society where individualism, feelings of hostility, and suspicion are always present. Most studies along these lines must admit of course that any community and any society (to use the terminological distinction under discussion) contain elements of the other. No traditional or primitive little village community is completely alter-oriented and marked by only friendship and good will and by an absence of thoughts of suspicion or acts of hostility. No modern society contains only egotistical and individualistic characteristics which are continuously tugging at disuniting and disintegrating the members of this society. Both human groupings contain elements of the Gemeinschaft and both show characteristics which mark the Gesellschaft type of human group. They are only present to a greater degree in the various types, as Tönnies certainly would agree.

Somewhat along the same lines, and this similarity was in fact pointed out by the author himself (Weber 1922:22), are the concepts used by Max Weber. While Tönnies showed the primary features which marked each of these types, Weber was interested in the processes which produce each type. He speaks then of "Vergemeinschaftung", or communalization, as against "Vergesellschaftung", or socialization. The first, then, is the communal group or the community, which is "based on a subjectively felt (affectual or traditional) togetherness by those involved" (ibid.:12). This may in fact be based on erotic relations, spiritual brotherhood, pietistic relations, feelings of comradeship, or the bonds of a national group. On the whole these communal groups are opposed to conflict, but conflict marks their relations with other such groups.

Both these authors formulate the concept of the community as a small group of people who are associated due to choice, due to their

344

wills, or due to subjective feelings. The emphasis is upon choice and upon the decision on the part of each member regarding this identification. Without attempting to use these features as guides for our classification, we may then see Akuna as basically a Gemeinschaft which has grown and keeps developing according to the characteristics of "Vergemeinschaftung". The preceding pages are filled with examples of persons who decided to move elsewhere, to develop new ties of comradeship, or to show new loyalties. This is even true at the postpubertal stage when the first choice opens up and the alternatives remain open theoretically for as long as the person is alive. The community of Akuna is basically village-oriented but intravillage hostilities did and even still take place. Within the village there exists a number of subfields of interaction, but in the last analysis it is the household which forms the basic unit of the community life. We pointed at a variety of contexts in which feeding the child (or pig) establishes real bonds between persons. Also people who have shared their food develop certain ties as was the case during the war when a communal village was established by Gadsup and Tairora. In this context Weber also sees the household community as the basic group and the persons with whom one first develops as a person and where all the basic needs are satisfied. But this need not imply that it is in fact the first or universal social group for in some of the hunting and gathering groups it need not be present in this form — Weber's statement does apply, though, to the Akunan situation, the value of eating together or suckling another's child, adoption where we are in fact dealing with sociological parenthood, and similar factors which were mentioned above. The author expresses it in the following way: "The father relationship cannot exist without a producing household unit of father and mother; even where there is such a unit the father relationship may not always be of great import. Of all the communal relationships arising on the basis of sexual intercourse, only the mother-child relationship is fundamental, because it is a household unit whose biologically based cause stability is sufficient to cover the period until the child is able to search for means of subsistence on his own. Thereupon comes the community of experience of siblings brought up together. In this connection, it may be noted that the Greeks spoke of homogalaktes (literally: persons suckled with the same milk; hence, foster brothers or sisters) to denote the closest kin. Here, too, the decisive thing is not the fact of the common mother, but the existence of the extended kinship household as a producing unit. Criss-crossing of communal, sexual, and physiological relationships occurs particularly in the family as a specific social institution" (1961:297).

While Weber felt that hostilities and conflict were not characteristics of the communal group, our community, they are of course partly related to the very existence of the community. The community itself is marked to a major degree by social solidarity and social order — these in fact form part of the basis on which we decided that this is a community — but these are not possible unless there is also some conflict. Conflict of course can be present in the form of latent feelings of aggression which are always present and supposedly underlie Tönnies' Gesellschaft, but it can also be present in other forms and take other avenues of being expressed. It does not seem possible that any group of people, each person with his own personality and peculiarities, can permanently reside together without some form of conflict between them. This is also related in a causal way with mobility.[2] Conflict and social order then are mutually supporting.

The fact that conflict is present need not suggest open acts of aggression among members of a community. While these occured on a number of occasions within the community of Akuna, and also in other similar communities, conflict may take on other forms. On a number of occasions in the text we remarked on the prevalence of sorcery. This in many cases is used to initiate, or balance, hostilities with neighboring village communities, which usually took place in the form of warfare, and an excellent discussion along these lines has been presented by Berndt (1962) for the Eastern Highlands, and also by Murphy (1957) for the Mundurucu. In many cases within the community, however, an open act of aggression might produce schism and so we find what Coser has called the "safety valve" mechanism in a society (1956). This in most cases takes the form of sorcery accusations and a whole process set in operation by such an accusation. With statistical evidence to support their studies a number of anthropologists have concentrated on these forms of conflict (Kluckhohn 1944; Marwick 1965; Wilson 1951) while Krige offered a good discussion of "the social function of witchcraft" (1947). In the same light of functional explanations Kluckhohn has suggested that sorcery and witchcraft may become prevalent when "a tribe's customary outlet for aggression in war is blocked" (1949:267). Unfortunately we do not have enough data to compare the acts of hostility and sorcery in earlier times with the present situation except in so far as comparable data has been extracted from Flierl's observations. In the concluding section of this chapter we will return to this topic of the mutually

2. In a recent study of life in a Tennessee Ridge community Matthews (1966) has suggested that conflict and hostility are major forces which mark this little folk community. In spite of this and the commercial which suggests that "we'll rather fight than switch", du Toit (1974) in a similar kind of community in South Africa, found that persons tended to drift away leaving a basically homogeneous personality type with much harmony among community members.

enforcing effects of conflict and solidarity, or in the words of Georg Simmel, the "Liebe und Hass", attraction and repulsion, thesis and antithesis.

When this discussion is linked to material presented in our third chapter above, it is clear that we are perfectly justified in speaking of Akuna as a community. The fact that most studies of this nature concentrate on other factors merely point at different interests of the authors. It seems that as acculturation continues, as the process of change progresses and these people are drawn further and further into the world culture of the twentieth century we are being forced to look more and more at the binding and disrupting forces of the community. The traditional structural anthropological approach may have a limited contribution to make as we look at interaction and fields of interaction which exist in spite of kin or origin differences.

SOCIAL STRUCTURE OF THE COMMUNITY

The descriptive and statistical data presented in this study permit us to see Akuna as a village community, a Gemeinschaft, composed primarily of like-minded persons. People who have selected this community as their place of residence have done so to a large extent because of kinship ties and because they were born there, but where choice was involved — and this certainly applies also to those just mentioned as to whether they are going to remain or move elsewhere — ties of friendship and amicable associations are important.

As has been pointed out, Akunans distinguish between three clan-groups, or stated differently, genealogical data which includes every person living in the community linked members into three concentrations. I was never able to ascertain a name for either one of these and designated them as A-C in the genealogical analyses. This problem was also encountered by Frantz (1962 personal communication) who spoke the language and had lived in a neighboring community for three years at the time of our study. He states that after repeated questioning he was unable to come up with any named groups. In the ethnographic study and concentration on Akuna as a community, these major kingroups did not appear at any stage as the crucial grouping. Murdock states that "for a group to constitute a genuine clan it must conform to three major specifications. If any one of the three is lacking, the group is not a clan, however greatly it may resemble one in composition and external appearance. In the first place, it must be based explicitly on a unilinear rule of descent which unites its central core of members ... In the second place, to constitute a clan a group must have residential unity. This cannot exist if the residence rule is inconsistent with that of descent, e.g., patrilocal or neolocal when the latter is matrilocal (sic). Nor can it exist if any appreciable degree of individual deviation is permitted from the normal rule of residence. In the third place, the group must exhibit actual social integration. It cannot be a mere unorganized aggregation of independent families like those residing in a block in an American residential suburb. There must be a positive group sentiment, and in particular the in-marrying spouses must be recognized as an integral part of the membership" (1949:68). We have intimated that Akunans state an ideal of virilocal residence but in fact allow a high degree of neolocal residence. They recognize a patrilineal tendency but trace descent through both parents to constitute the 'ankumi'. They show social integration of the kingroup to a certain degree but first and foremost loyalty and integration is at the community level. These patrilineal kinship groups are more extensive than lineages, but lack the significance usually ascribed to them, and instead of cross-cousin marriage we find a bilateral extension of the incest taboo. Akunan kingroupings do not agree with the Chimbu clan which Brown originally described as: "exogamous group is ugually also a territorial, ceremonial, and military unit within which warfare is restricted" (1960:33). It resembles more closely the kin categories among the Chimbu which Brown described after reworking her data (1962).

The grouping which is of basic importance for every Akunan is the 'ankumi'. This, it seems, is a grouping with strongly bilateral tendencies, so much so in fact that I had earlier referred to it as a kindred (du Toit 1963 and 1964(d)). In reworking and rethinking, the conclusion was reached that it would be impossible to explain nominal clan groups and kindred among the same people. But the importance of this group does not diminish by critical analysis. It is still the most important group for every individual, the group to whom bilateral extensions of the incest taboo apply and the kinship category which is the main contributor of members to an ego-oriented interest group. It is this 'ankumi' to whose members the kinship terms are applied, including equally the relatives through the father and through the mother. It is also through the terminological system and by extension, then, through the 'ankumi' that we compared the kinship structure in Akuna with Murdock's structural principles, deciding on a bilateral tendency in the social structure. Is it possible to have lineality and bilateral principles present in the same social structure? Obviougly the nuclear family — or household family which is of singular importance in Akuna — is by its very nature bilateral. In this group the father's kingroup and the mother's kingroup are united in the children, and while an overall lineal tendency might be present in terms of inheritance or clan membership, each generation is united

348

in an ego — or possibly sibling — oriented 'ankuni'. It may be in fact
due to this change by which a man's affines become matrilateral kinsmen for his children which led Nadel (1947:12), Holleman (1952:29),
Geddes (1954:14), and Schapera (1955:12) to employ a definition for
kindred which included affines. Since bilaterality and linealism seem
to be mutually exclusive I searched through anthropological literature
for examples of the simultaneous occurence of these and the clearest
example is Raymond Firth's ramage. Not only is the tree metaphor
used in Tikopia (1965:328), as was the sugarcane metaphor in Akuna
(and so among the Gahuku-Gama (Read 1954:11fn)), but the
Tikopian description of structural principles agrees in a large degree
with the Akunan case. It is best to quote Firth in full and allow the
reader to compare it with the picture which emerged in the previous
chapters. He states: "Looking at the matter from another angle, it
may be said that in each generation complete bilaterality is attained
for each individual, since the houses of his father and his mother are
united around him. But from generation to generation these bilateral
alignments are constantly shifting, a process of focal substitution
takes place, with a fresh individual family as nucleus each time. These
changing bilateral groupings cross and interweave with the unilateral
groups, which persist throughout the generations, changing their personnel almost imperceptibly by a process of unitary substitution. The
point of this distinction is that certain institutions are most appropriately performed by one set of groups, or the other, as they are of
primary interest to the individual per se, or concern him mainly as
a mechanism of cultural continuity; as for instance funeral rites in the
first case, or inheritance in the second" (ibid.:329).

I am not suggesting that the 'ankumi' is a ramage, a kindred, or
anything which has been named before. In the case of the "ramage"
there seems to be a choice where the individual may elect to follow
the father's or mother's group. Akunans have the choice of associating more with one group than another, or of residing with one rather
than the other, but it is not a true choice as Firth has sketched it.
The latter would be too close to ambilineality (Murdock 1960:3-4),
nor does it fit the criteria as discussed by Davenport (1959) for
unilinear and non-unilinear descent groups. Akuna in fact looks more
like Brown's restudied Chimbu (1962), or like the settlement group
among the Bena Bena (Langness 1964) than the Kamano to the south
(Berndt 1962). There is a danger in using any kind of descriptive term or label for this kingroup for labels immediately suggest
extensions and features which might not be intended. By using 'ramage',
or 'paito', or 'hapu', or 'anga?sa' to illuminate the 'ankumi', other
extensions of these terms might be projected onto or read into the
Akunan structure than were intended. Also we would be falling into

the trap of labels. In this study I have attempted to refrain from labels and simply to explain life in Akuna. I have attempted not to arrive at preconceived conclusions of nomenclature and classification but to describe the social system as a set of empirical facts. We are wholly in agreement with Leach when he states: "what we need to understand about a society is not whether it is patrilineal or matrilineal or both or neither, but what the notion of patrilineality stands for and why it is there" (1961:11).

It seems that earlier anthropologists working in New Guinea had in fact tended to label. Barnes discusses this in terms of applying African models to the New Guinea Highlands (1962) while du Toit (1963) pointed at the possible influence of the background and training of researchers as well as ecological factors influencing the community under analysis. Additionally, Langness has made a very valuable survey of problems involved in the conceptualization of social structure in the New Guinea highlands. In this last mentioned survey he also analyzes his ethnographic data from the Bena Bena, questioning in conclusion "why the unilineal principle exists and/or survives at all in groups like Bena Bena?" (1964:181).

A number of anthropologists have ascribed social structure partially to the influence of ecological factors. Thus harsh and unpredictable conditions have been used to explain polyandry (Saksena 1962), but in most cases they are used as part explanation for bilateralism or bilateral tendencies. This has been done for the norther Lapps (Pehrson 1954 and 1957) but more specifically it is inherent in the studies of Pouwer (1961:3) and van der Leeden (1956) while this adaptive adjustment is also mentioned by Boelaars (1959) and Goodenough (1962:10-11). It seems to be too early for us to attempt an overall survey of social structural features in New Guinea[3] or even in the highlands, but a salient feature which nearly every field worker remarks on is an anomaly which has been called "structural looseness","non-unilinear" features, "quasi unilineal groups", based on openness, freedom of choice, and "flexibility". These aspects are closely related and do not necessarily represent a tendency among recent field workers. These same aspects were remarked on by Held (1951), Thurnwald (1916), van Eechoud (1962) — in research conducted during 1939-40 — and van Baal (1954) and even van der Leeden and Pouwer in studies conducted a number of years before Barnes' call for a more critical approach.

Additionally one finds repeatedly in studies of this area the emphasis which is placed on the village community, irrespective of the structural features present. Barnes states that "the New Guinea hamlet is

---

3. A valuable start in this direction appeared recently by Pouwer (1966).

found to be full of matrilateral kin, affines, refugees, and casual visitors ..." (op. cit.: 5). This picture agrees to a large extent with Akuna as we have discussed it. The picture agrees furthermore with descriptions of the restudied Chimbu village (Brown 1962), or a Bena Bena settlement (Langness op. cit.: 166-168), or even the Ngarawapum (Read 1946: 97-98 and 1950: 200-210) and Busama (Hogbin 1963: 34) examples. In nearly every case we are in fact dealing with a group of people — frequently arranged around a core of kinsmen — who reside together for a variety of reasons but not due to kinship alone.[4] Associated with this is the fact of strong kingroup feelings and loyalty toward residential group members while this solidarity is balanced by hostility toward other villages. Village centered institutions then are participated in by sub-groups or individuals who have a great deal in common. While they might not be kinsmen, they share a loyalty and espirit de corps which brought them together through a process of "Vergemeinschaftung" and unites them as a community. Obviously certain individual forces work at disrupting the unity, and interfering with the process of communal living, at causing intra-group hostilities. These are balanced by a social order which emerges and which assures the continuation of the system.

"To sum up the survival of a society depends on the operation if its social system; a social system is comprised of sub-systems which, in turn, are comprised of institutions; the functions of these institutions are served only if their constituent roles are performed. In turn, this requires the recruitment of individuals for the various statuses which comprise the social structure. If these propositions are valid, we are brought back to the central issue of this chapter — the problem of cultural conformity. How does society induce its members to perform roles — those that are instrumental to the attainment of a status, as well as those that are entailed by the occupancy of a status?" (Spiro 1961: 98).

It seems that in a state of permanent warfare with neighboring villages the social structure we have described, and the flexibility of the structure does in fact assure the system's continued operation. Every person is in fact required and enculturated to perform a particular role, but these roles allow a certain leeway. They are statuses but not straitjackets. The person performs a set of roles; he does not follow detailed instructions. of how to act. But different acts also elicit different responses with the result that certain acts and certain roles are more surely rewarded than others.

4. The significance of non-territorial patrilineal descent groups versus residential, localized political units and their importance in alliances is excellently demonstrated for the Dani of the former Western New Guinea by Heider (1970: 62).

In speaking of a community there comes to mind a group of people who exhibit strong sentiments of solidarity toward fellow group members while frequently considering all nongroup members as outsiders, strangers, or even enemies. But as relations of hostility mark the contact of community members with outsiders, the reverse is also true because these relations intensified their feeling of solidarity. The greater and more lasting the pressures from outside, the greater the need for mutual support. As the small community is forced back repeatedly to utilize its own resources, every member is conscious of strains or disrupting forces within the community whether he is directly involved or not.

While it is occasionally possible to point at single lines, which converge and cause cultural changes, this is not true of a condition like social order. It is not even possible to define social order, for equilibrium in one community might not be so considered by the members of another group. Rather than define this concept I will give a number of examples of what would constitute this condition of order. It might be argued that cultural homogeneity, or a strong leader, or some guiding ideal would cause order to prevail in which the members of a community would look beyond the village borders. From this argument, then, the continuous state of warfare in the highlands would cause social order in each community but conflict between village communities. This must certainly have been a factor which influenced not only community relations and attitudes, but also community structure and composition. Discussing Bena Bena warfare, Langness states: "This appears to have been one of the most continuous and violent patterns of warfare on record. It is therefore surprising that, even though emphasis is placed by some writers on the importance of violence and killing for its own sake, what seems to be an obvious further conclusion is never drawn — namely, that the psychological concomitants of such a pattern of warfare must influence the nature of social groups in some way. Presumably the Highlands New Guinea people, as individuals, have no desire to die; they want to survive and to propagate themselves. One way to survive is to align and maintain a strong and large group. Thus, if expedient, one need not be too particular about someone else's genealogy, so long as he can fight. Perhaps indeed, strict unilineal descent was too costly. Strongly unilineal groups would be very difficult to maintain in this social environment. Groups which constantly find it necessary to scatter and re-group, which are decimated by casualties, which must take refuge with friends (who are willing to accept them for the same reason they want to be accepted) cannot, I submit, maintain lineage purity and cannot insist on

352

descent as the sole (or perhaps even most important) criterion for membership" (1964:174).

The reader will recognize in this statement a number of suggestions which would apply directly to Akuna. This same topic is discussed in some detail by Berndt (1964) under a sub-heading dealing appropriately with Allies, Enemies and Refugees. He concludes that "flexibility, allowance for personal initiative within loosely defined limits, was apparent in this sphere of Highlands living no less than in others" (ibid.:203). While warfare is certainly one of the characteristics of Highland New Guinea it would be only a partial explanation for the solidarity at group level; it would be one of the external factors impinging on a community and forcing its members to assist each other.

This social order cannot be explained by warfare, or internal forces alone, for "it is impossible to look for a single force on which social order can be based" (Hogbin 1961:290). There is something which operates between the members of such a small community which assures a minimum of conflict. Something which permits a person to blow a "safety valve" and then resume normal living. It is possible that this "something" is in fact a part of this openness, this flexibility we have spoken of and which is referred to repeatedly in this study. But it would seem that it also is something which flows from the socialization process, flows from that period during which the emerging community member learns what roles are permissible and what latitude each of them entails. This is the period during which values are instilled and preferences taught to cope with those cases in which choices have to be made. This, then, is the period during which the necessary conditions for social order are instilled.

The suggestion which was made above would suggest alternate ways of acting, depending on whether the field of interaction includes community or non-community members. While we were in the community of Akuna, a number of people in Wopepa were badly beaten, but not killed, when two clangroups in the same community exchanged blows. These actions, it seems, were necessary to re-establish social order. No community of this size can exist when open aggression threatens or when pent up feelings of suspicion and latent hostilities exist. There seems to be an additional safety valve which can be blown — the way a pressure cooker lets out some of its steam and then gets back to its job — and which prevents the whole system from being disrupted yet allows the persons involved to act out some hostility. This must already have become obvious in cases which were mentioned in this text above. When a man attempted to commit adultery with Urinanda's wife, the latter physically attacked him and started beating him with a fence post, but it took only two grown boys to hold Urinanda back and remove him from the scene. He had been extremely cross, his face grey and drawn, but these two young fellows simply gripped him by the arms and he walked off and was seated at his hearth. When the two wives of Domandi had built up so much jealousy, frustration, and even feelings of hostility, they attacked each other with fence posts and a burning log snatched from the hearth. It took only one old man, Munowi, to hold them apart and calm them down. In both cases the people involved had blown off steam, had acted out some of their aggression, and had rid themselves of emotions which would have made it impossible to live together in the same community. It seems as if we are dealing with an emotion which differs from what we colloquially call "blind with rage". It is common in the towns and cities of our Gesellschaft living that two persons lose their tempers with one killing the other; frequently the killer cannot be subdued even when overpowered and will attack or strike at any person that comes between him and his goal. This might very well have been the case had Urinanda been confronted by "outsiders" or strangers rather than members of the community.

Social order within the community does not suggest a permanent state of harmony in relationships. In the fourth chapter of his Soziologie (1908), Simmel discusses the fact that a group which is entirely centripetal and harmonious would be an empirical impossibility and be void of that life process which is marked by harmony and disharmony, liking and disliking, association and disassociation. Every community must display "Liebe" and "Hass". Every community comes into existence and continues to operate due to the presence of both these forces. While the "Hass", the repulsion, the antithesis is always present it is subdued and guided between members of the same community and this Simmel does not recognize. He does not apply his theories to the small communities, the Akunans, the Gemeinschafts of the world. His conflict theories would apply more specifically to the Gesellschaft, within which smaller associations and communities are present, or to the inter-community relations we discussed.

But this mechanism which allows steam to escape and thus makes for a fair degree of harmony within the social field we have designated the community also operates in other contexts. When adultery has been committed by a man from a hostile community — or one with whom an alliance does not exist — it would almost certainly lead to open hostilities, and set in motion a series of acts of vengeance. When the culprit belongs to the same community or a friendly one, an arrow is shot into the thigh of the guilty one. The implication here is two-fold: "Those who live together from day to day cannot have hostile emotions toward each other so they have to be vented; but it would certainly be dangerous to the existence of the group if fellow members kill each other" (see also Berndt 1962:404).

But cases were referred to above when a member of the community killed a fellow member — or a kinsman killed one related to him. In these cases the culprit was made to appear with his shield over his shoulder to protect him, and those who were taking revenge would shoot into the shield. Here we have disharmony which is restored to harmony; we have dislike which is replaced by liking, "Liebe" takes over from "Hass". We also mentioned cases of death by suspected sorcery where this act of letting off steam took place. The previous chapter discussed the kinship terminology. It was pointed out that a person always addresses the wife of his brother — both older and younger than himself — by the taboo term 'akoi'. His relations towards her reflected in his form of address and reference is therefore one of extreme caution or even avoidance, though this rarely occurs in Akuna. Yet Akunans practice the levirate which means that when this brother we have just referred to dies, the surviving brother may marry the widow. Here again observed behavior and material in the field suggested that this is a cultural safety device to prevent a man from committing adultery with a brother's wife — real and classificatory brothers — and by so doing causing hostilities within the group. While not structuralized to the same degree, these principles of caution really apply to all members of the interest group and ideally the village community. Thus Berndt states: "One of the strongest influences which keeps decisions and associated behavior within the range of conformity is the expected reaction of others. 'The most significant others', in this context, are those on whom a person depends for present or future support in matters which he considers to be important. Consistent day-to-day interaction takes place with a relatively small circle of persons and as a matter of direct personal contact." (ibid.: 399).

Mention has been made on a number of occasions of the harmony which is reached by common agreement on questions. In our fourth chapter above, court procedure was discussed and Leininger's observation referred to, namely the aim always to reach consensus, for in this is bound up survival value for the members of the community. The temperance in social relations within the community is not matched by inter-community relations for here, too, was survival value. If the community would continue to exist it needed to form a strongly integrated group. Solidarity was one of the prerequisites in this sphere and it is here that the leader, the strong man, operated. His flamboyant speech and gestures were matched by his personal interest in members of the community; his fighting abilities and harshness in warfare, by his compassion for the needs of people in his community.

The importance of the community and the need for solidarity is also found after a death. The high point in sorrow and wailing is reached when the interest group gathers and when they express their grief in a group. This is the time when a person is re-united, re-integrated. There is no burial ceremony and little in the way of mourning since the feast which follows the wake is aimed at integrating the bereaved with the rest of the community.

But it is very important to keep in mind that inter-village relations are not all hostile. Simmel proposed that in many cases warfare is the only social interaction between primitive communities (1955:33-34), but in the New Guinea Highlands this is not true. Through inter-village kinship bonds, affinal relations, alliances, and similar uniting factors the community is in fact extended beyond the village boundary. But this does not apply to every person nor do the same links, and bonds unite the same persons. Here, then, the interest group is of vital importance. One could propose the super-imposition of the interest groups of all members of a number of village-communities and those villages which have the greatest number of links would be most likely to have alliances or pacts of neutrality. The principle of reciprocity, which is of singular significance at the village-community level, also operates here to confirm and keep viable bonds of sentiment and unity which prevent warfare.

This inter-community field extends in most cases to those in the same district, provided they belong to the same linguistic, and, therefore, cultural group. All villages in the Arona Valley differ from each other as regards language and culture. Akuna and Tompena are within shouting distance — a way in which the news spreads throughout the valley when the police or tax collector is on his way — but culturally and linguistically each is an island and differs from the other and those near to it. Yet they are less different than they would be if compared to Agarabe or Tairora. So the linguistic borders are of importance with a distinct border in fact appearing; the social field, the supra-community, reaches a decline in social ties. While the central village in such a linguistic group would theoretically have ties extending in all directions to village-communities spread around it, the border village would have its ties nearly exclusively with linguistic and cultural group-member communities. As a genetic study will show gene clines which reach their apex where there is greatest contact and drop off where the genetic boundaries are reached; so, too, in terms of interpersonal and inter-community relations. With a sociogram in mind, we might want to speak here of social clines which unite all group members but accentuate certain community centers.

SUMMARY

Some years ago Simpson (1937: 74-94) discussed the requirements
for a community to exist and to function normally. He outlined five
major criteria namely the biological, geographical, legal, political,
and one which entails communication or a common meaning. The last
three of these criteria can be subsumed under a common language and
culture. The geographical criterion is an important one for it shows
the center of the community and also the territory which is included
due to the other criteria. But Simpson's emphasis on the biological
community goes further than simply a statement of the priority of the
familial community. We have agreed above on the basic importance of
the family in Akuna and also that for the community the significance
of the kinship unit is of lesser importance than that of culture and loyalty. The community, the Gemeinschaft, is first and foremost composed of persons related through their wills. It is basically an "optative" system (Firth 1957).

The fact of being associated due to choice and decision on their
own part naturally leads to a greater degree of harmony and order
than would be the case of persons had to associate due to some other
bond such as kinship. While a large percentage of persons in any
community are related, are bound by kinship bonds, some of their
kinsmen moved away for reasons of choice, while other members of
the community with whom they live in harmony are not related by kinship ties. This unity of like-minded persons, including kinsmen and
other associates, also marks the community of Akuna.

Addendum A

AKUNAN KINSHIP TERMINOLOGY AND ITS EXTENSIONS

A. anapu — All male cognates of the second ascending generation and the spouses of female cognates in the same generation. This category would include: father's father, mother's father.

B. akaka — All females cognates of the second ascending generation and the spouses of male cognates in the same generation. Included in this category are: father's mother, mother's mother.

C. apo — Father, but included are all male cognates on the first ascending generation except the mother's brother. This category also includes the spouses of female cognates of the same generation level. Included are, father's brother, father's sister's husband, mother's sister's husband.

D. ano — Mother, but included in this category are all female cognates of the first ascending generation and the spouses of male cognates. Excluded is the father's sister. The term then applies among others to the mother's sister, mother's brother's wife, father's brother's wife.

E. amamu — Father's sister and father's female siblings, real and classificatory.

•F. a?o — Mother's brother and mother's male siblings real and classificatory.

G. apa?i

and

H. awa?i — All brothers, real and classificatory distinguished by age seniority relative to ego. This category of persons would include: father's brother's son father's sister's son   mother's brother's son mother's sister's son   husband's sister's husband

358

I.	anano?i	All sisters, real and classificatory distinguished by age seniority relative to ego. Included in this category of persons are: father's brother's daughter father's sister's daughter mother's brother's daughter mother's sister's daughter wife's brother's wife
and		
J.	umi	
K.	anonta	Sister's son and sister's daughter (male speaking). This term is reciprocal with F above.
L.	akai	Males in the first descending generation. Included then are: son, brother's son, husband's brother's son, wife's brother's son, sister's son (woman speaking), husband's sister's son, wife's sister's son.
M.	aramumi	Females in the first descending generation. Included there are: daughter, brother's daughter, husband's brother's daughter, wife's brother's daughter, sister's daughter (woman speaking) husband's sister's daughter, wife's sister's daughter.
N.	anai	All cognates of the second descending generation. Included are: son's son, son's daughter, daughter's son, daughter's daughter.
O.	anda?i	Wife's father, wife's mother, daughter's husband.
P.	anapu?i	Husband's father, husband's mother, son's wife.
Q.	akoi	Wife's brother, wife's sister, sister's husband.
R.	amati	Husband's brother, husband's sister, brother's wife.
S.	apatiwa	Wife's sister's husband.
T.	awapu	Husband.
and		
	ana?i	Wife.

Addendum B

DAILY WORK SCHEDULES

In this section I have tabulated the daily round of activity of four
Akunans. Each person was observed for three days as he went about
his regular duties and associations. It is not suggested that these are
"typical" daily sequences for all Akunans, or even for these persons,
but they were requested to carry on as they normally would while the
observer tagged behind with notebook, pencil, and watch. These observations were made after we had been in Akuna for at least six months
and the people were already used to being observed. My presence in
these cases was less disruptive than it might have been earlier.

CASE 1: NAPIWA

This is a man between 45 and 50 years of age. Though married for
the second time he is a monogamist and personally cares for his trees
while his wife works a garden for tubers, greens, and other foods.
He is the horticulturalist par excellence in Akuna, for adjacent to his
house is a large garden in which most of the trees and plants which
Akunans use, have been transplanted from the bush. Scattered throughout the garden area and bush belonging to Akunans are trees and
palms which belong to him and these are regularly visited and inspected.

February 1, 1962

6:00	Rises
6:00-6:30	Chats with wife at hearth about her work
6:30-9:20	Court session at luluai's house about seduction case
9:20-10:45	Has meal of food prepared in earth oven
10:45-11:40	Walks to new garden with wife
11:40-12:05	Chats and rests
12:05-3:15	Burns brush and clears new garden plot

360

3:15–3:40	Rests and eats cucumber for its juice
3:40–5:25	Clears and burns brush and trees
5:25–6:35	Walks home via bush to pick some ripe betel nuts observed previously
6:35–7:00	Chats with neighbor about pig hunt
7:00–8:35	Relaxes by hearth, sings, drums and waits for food
8:35–9:25	Eats meal
9:25–10:15	Chats with men about pig hunt the next day
10:15–10:35	Returns home, settles in, and smokes in silence
10:35	Retires

February 7, 1962

6:10	Rises
6:15–6:35	Talks with neighbor
6:35–6:55	Sits at home-fire talking to wife
6:55–7:30	Morning meal of toasted sweet potato
7:30–8:05	Chats with sister's husband in upper part of village
8:05–8:15	Gathers tools for work and tobacco
8:15–9:20	Walks to garden to replace fence
9:20–9:30	Rests taking betel and tobacco
9:30–12:05	Builds fence to replace old one
12:05–12:40	Rests, taking betel and tobacco
12:40–3:20	Builds new fence to replace old one
3:20–5:15	Returns to house via pandanus palms in bush to collect ripe nuts
5:15–5:30	Chats with luluai about missionary's activities
5:30–5:55	Smokes and talks with wife at fire
5:55–7:10	Evening meal of roast corn and boiled sweet potato
7:10–9:05	Chats with wife's sister's brother in lower part of village
9:05–9:35	Stands in village square announcing work of Medical Officer and responsibility of villagers to keep village neat
9:35–9:50	Chats at hearth
9:50	Retires

February 20, 1962

5:55	Rises
5:55–6:40	Stands in sun, stretching and talking with neighbor
6:40–7:05	Sits at hearth talking to wife about the tree kangaroo hunt

361

7:05–8:10	Morning meal
8:10–8:20	Gathers tools and tobacco for days work in coffee garden
8:20–3:05	Sets out for garden, but runs into court case regarding pig killing and joins in
3:05–3:25	Walks to coffee garden
3:25–4:55	Weeds and cuts young trees
4:55–5:10	Returns to house
5:10–6:05	Discusses matters with wife while awaiting meal
6:05–7:15	Eats meal
7:15–7:30	Collects bow and arrows, betel, and tobacco for the kangaroo hunt
7:30–8:45	Walks with other men into thick bush by light of full moon
8:45–12:05	Traps, hunts and collects kapul (tree kangaroo)
12:05–1:30	Returns to village
1:45	Retires

CASE 2: ARA?O

This is a widow of 45 to 50 years of age. Her husband was a polygynist but when he passed away a number of years ago her co-wives re-married while she remained single. In the upper part of Akuna live a married daughter and her children, while Ara?o shares her house with her son, a lad of about seventeen and his younger sister of twelve.

February 28, 1962

6:05	Rises
6:05–7:10	Cleans trash from house and sweeps around house in village square
7:10–7:35	Feeds her pigs outside village
7:35–9:05	Prepares meal and eats
9:05–10:00	Sits talking at fire
10:00–10:50	Walks to gardens
10:50–11:05	Rests
11:05–2:50	Does garden work, cleaning sweet potato patches
2:50–3:25	Rests and eats snack of corn and roasted sweet potato
3:25–5:10	Garden work and collection of food
5:10–6:35	Walks back to village via bush to collect fire wood

6:35–6:50	Rests
6:50–8:05	Prepares meal and eats with children
8:05–9:35	Sits, smoking, chatting
9:35	Retires

March 13, 1962

5:55	Rises
5:55–6:45	Sweeps around house and removes trash and excreta from house
6:45–7:05	Feeds her pigs outside village
7:05–9:45	Prepares earth oven and eats food, feeding others
9:45–10:05	Chats
10:05–10:50	Walks to garden
10:50–2:15	Plants peanuts which had been soaking in water at home
2:15–2:35	Rests, feeding grandchild
2:35–4:50	Garden work and collecting of tubers for food
4:50–5:25	Walks into bush to feed "wild pigs"
5:25–6:30	Returns to village with food and firewood
6:30–6:40	Rests
6:40–7:50	Prepares meal and eats with children
7:50–10:05	Chats with female kinsmen

March 15, 1962

6:10	Rises
6:10–6:35	Removes trash and excreta from house
6:35–7:00	Feeds pigs outside village
7:00–8:35	Prepares food for meal and eats
8:35–8:50	Chats
8:50–12:55	Scrapes 'yienni' while working with six other women and places in sun to dry for grass skirt
12:55–1:20	Chats and eats cold sweet potato
1:20–2:05	Walks to garden
2:05–2:20	Rests
2:20–4:05	Gathers sweet potato and other food
4:05–4:40	Rushes home through thunder storm
4:40–5:05	Rests and chats
5:05–6:15	Prepares meal
6:15–7:20	Evening meal
7:20–9:05	Chats
9:05–10:15	Attends mission convert's class
10:20	Retires

CASE 3: TORAWA

This is a young adult man who had been away from Akuna for a number of years on indentured labor and while being trained as a policeman in the western highlands. He was married before leaving but his wife left him and at the time of the study he was monogamously married to a woman who had born him a son of four and a baby daughter. Torawa is also one of the progressive people in Akuna having organized a work team with whom he cuts trees in the forest for sale to the Summer Institute of Linguistics people in Ukarumpa.

January 24, 1962

6:05	Rises
6:05–7:15	Chats in sun with younger men
7:15–7:55	Sits at fire chatting to wife and mother
7:55–9:00	Eats morning meal
9:00–9:35	Discusses days work with his work team
9:35–10:45	Walks into bush for lumbering
10:45–11:10	Rests and chats while smoking
11:10–3:15	Cuts trees and moves them
3:15–3:25	Chats
3:25–5:50	Continues work on cutting trees
5:50–7:05	Walks back to village
7:05–8:15	Eats meal of boiled sweet potato
8:15–9:10	Chats with kinsman in latter's house
9:10–9:55	Sits at fire, smoking
9:55	Retires

February 22, 1962

5:50	Rises
5:50–6:45	Chats in sun with men — receives message that pigs had ruined his garden
6:45–7:35	Goes to gardens near Wayopa
7:35–9:55	Shoots pig and repairs fence
9:55–11:00	Returns to village
11:00–11:40	Washes in stream
11:40–12:05	Rests in house
12:05–12:50	Walks into bush
12:50–1:35	Chats with man making new garden
1:35–4:40	Walks through bush checking for good trees to cut, and collects betel nut

364

4:40–5:55	Returns to village with heavy log for fire
5:55–7:05	Chats to wife about garden
7:05–8:15	Evening meal
8:15–9:10	Makes address in village square about pigs that ruined garden and one killed
9:10–10:25	Chats at fire of wife's sister's husband, Tapari
10:25–10:40	Sits at own fire
10:40	Retires

March 6, 1962

6:00	Rises
6:00–7:05	Sits at own fire
7:05–7:55	Chats with 'line' about lumbering
7:55–8:10	Sits at fire talking to wife
8:10–9:05	Eats morning meal
9:05–9:15	Starts off for bush to cut trees but meets Amamunta people coming for court case re prostitution
9:15–2:10	Attends court case involving kinsman — takes care of his son while wife works in garden
2:10–4:15	Repairs door of house — takes care of son
4:15–4:35	Chats with wife and mother about case regarding kinsman and prostitution
4:35–5:40	Walks to Wayopa for food exchange and feast
5:40–11:05	Chats and partakes of feast
11:05	Retires

CASE 4: WAKANO

This is a widow of about fifty or fifty five years of age. She lives with her son Torawa (mentioned above) and his wife while her daughter of about fifteen also lives in the house with them and assists her in some activities. The fact that she shares the household chores and the production of food for the household means that she does not have quite as much to do as women who must alone produce and prepare foods for a family.

January 30, 1962

6:05	Rises
6:05–6:15	Removes trash and excreta from house
6:15–6:50	Feeds pigs outside village

6:50–8:05	Prepares and cooks meal
8:05–9:00	Morning meal
9:00–9:50	Chats
9:50–12:05	Sits in sun repairing pandanus leaf sleeping mat–rain gear
12:05–1:15	Feeds grandchildren and turns them over to mother
1:15–2:20	Chats
2:20–3:05	Walks to garden
3:05–4:55	Gathers food and removes weeds
4:55–5:40	Returns to village
5:40–6:05	Rests
6:05–7:10	Prepares food with son's wife
7:10–8:15	Evening meal
8:15–9:45	Chats and smokes
9:45	Retires

February 14, 1962

6:15	Rises
6:15–6:30	Cleans out the house
6:30–7:05	Feeds her pigs outside village
7:05–7:40	Chats
7:40–9:10	Prepares and cooks meal in earth oven with number of neighbors
9:10–9:50	Chats
9:50–10:35	Walks to gardens
10:35–10:50	Rests
10:50–1:35	Cleans garden and plants new shoots
1:35–2:10	Rests
2:10–4:50	Works in garden gathering food
4:50–5:45	Prepares some food
5:45–6:10	Gathers greens and breadfruit leaves
6:10–7:00	Walks back to village
7:00–7:55	Rests and chats
7:55–8:45	Cooks some corn for herself
8:45–10:25	Chats with visitors who came over
10:25–10:50	Relaxes while son beats handdrum
10:50	Retires

March 8, 1962

6:00	Rises
6:00–7:15	Cleans house and feeds pigs while throwing out trash
7:15–7:50	Chats

7: 50- 8: 45	Prepares meal with son's wife
8: 45- 12:0 5	Remains in house chatting while it is raining
12:0 5- 12: 50	Walks to gardens
12: 50- 1: 20	Rests
1: 20- 3: 45	Works in garden
3: 45- 4: 20	Rests
4: 20- 5: 35	Gathers food to return with
5: 35- 6: 25	Returns to village
6: 25- 6: 50	Rests and chats
6: 50- 8:00	Prepares meal
8:00- 10:0 5	Chats while enjoying meal
10:0 5	Retires

Addendum C

TIES BETWEEN AKUNA HOUSEHOLDS AND BETWEEN THESE AND GADSUP VILLAGES

Table 19. Sociometric ties between Akuna households and between these and Gadsup villages

Household No.:	1	2	3	4	5	6	7	8	9	10	11	12	13	14	15	16	17	18	19	20	21	22	23	24
No. of occupants:	1	6	7	5	1	2	8	1	5	4	3	2	5	8	9	6	2	7	6	8	6	4	4	3

Cognatic ties: Households to which each is connected through househead and at least 1 member in other household.

HH	1	2	3	4	5	6	7	8	9	10	11	12	13	14	15	16	17	18	19	20	21	22	23	24
	3	25	1	3	3			5	8	5	35		10	5	24	4	1	19	18	8		4	8	35
	4	4		8				10	10	8	41		14	8	26	18	3			10		16	10	38
	5	5		10				14	13	14				10	38	19	5			14		18	14	
	26	10		15				15		15				15		20	8			23		20	20	
	28	17		16				20		17				17		30	10			30				
	30	26		17				23		23				20			15							
	35	28		20				31		31				23			20							
	39	35		23				38		38				31			23							
	52	39		31										38			28							
		52		38													30							
																	31							
																	38							
																	39							
																	52							

Affinal ties: Households to which each is connected through househead and at least 1 member in another household.

HH	1	2	3	4	5	6	7	8	9	10	11	12	13	14	15	16	17	18	19	20	21	22	23	24
	8	3	2	1	1	5	5	6	1	7	5		5	1	7	35	11	16	27			7	14	25
	9	28	9	5	3	8	8	7	3	11	8		8	7	8		35	20	44			16	42	35
	10	52	25	26	35	10	10	29	13	13	10		14	9	10		41	22				18	44	42
	13	35	28			14	14	44	14	29	14		15	10	14							22		
	30					15	15	35	15				17	13	17							26		
	35					20	20						17				20	29				36		
	39					23	23						18				23					37		
	46					31	31						20				31							
	52					38	38						23				38							
													31											
													38											

Adoption ties:	1	2	3	4	5	6	7	8	9	10	11	12	13	14	15	16	17	18	19	20	21	22	23	24
		44						?			41			38			22				D	44	16	
																	25							

Villages to which each household is cognatically connected through househead and/or spouse.

HH	1	2	3	4	5	6	7	8	9	10	11	12	13	14	15	16	17	18	19	20	21	22	23	24
	A	A	A	D	A	G	C	A	A	A	B	B	A	A	C	A	A	A	A	D	A	C	C	C
	B	C	B		B			B	B	B			B			C	B	C	C	C	C	D		D
	C	D	C		D											D	E			D	D			
					F											F								

Villages to which each is connected affinally, through househead and/or spouse.

HH	1	2	3	4	5	6	7	8	9	10	11	12	13	14	15	16	17	18	19	20	21	22	23	24
	A	C	A	A	A	A	A	A	A	A	B	B	C	A	B	A	A	E	E	A		C	A	G
	B		D	B	C	D	B	C	B	C	E				C	C	B	G	G	C		D	C	
			E	C	E		E	D		D					D	D	D					E		
							F			F						F	F							

```
25 26 27 28 29 30 31 32 33 34 35 36 37 38 39 40 41 42 43 44 45 46 47 48 49 50 51 52 53 54 55
11 12  3  8  3  5  6  2  1  1  8  3  6  4  5  7  5  2  3  4  5  2  5  2  3  1  3  1  3  2  6

52 36 19  1  7  1       31       1  1 26 26 15  1    22 14 22 14  4 19          45       1 45
      37    3 13  3      38       3  3 37 36 24  3    44                         53       3 50
            4 41  4               19  5          26 28                 42                  28
           30     5               28 30          30                   44                  30
           35    35               30             35  ?                 53                  35
           39                     35             52                                        39
           52                     39
                                  52


28  8 21  2  5 28  7 14  1 27  9  8  8 14  2    11 19 32  6 43 30       1 42 41 41  2 19 35
10 28 25  8 39 38        3           9 10 10 31  9    21 25    19 50 31    3 51 50 25 25    37
14 30    14 46 39              35       14 14 32       51 32    25 53 32   30            42
23 34    15 52                          23 23          46       42         35            44
39 52    17                                           53       46         39            46
   52    20                                                    46
         23                          unknown                   53
         26
         31
         36
         37
         38

                        14          11                    2                              B
                                                         19
                                                       ( 21)

C  B     A  C     D  C  A  A  B  B  B  C  C     E  D  A  B  C  D  D  E  C  C  D  A  C  A  A
D        C        D     C        C              C  E  D  E     G              C  D  C  E
         E                                      D                                  E
                                                E

A  A  A  A  A  D  A  A  C  A  A  A  A  C  G  C  B  B  A  C  E  C  C  C     C  A  A  E  C
B  D  E  B  E  B  D     C  C  B  B  C  D        F  D  E              D     C  D
C           F        D  D  C  C  D                 E                       D
                                                  G
```

BIBLIOGRAPHY

Adam, Leonard, "The Discovery of the Vierkantbeil or Quadrangular Adze Head in the Eastern Central Highlands of New Guinea", Mankind, 4, 1953.

Arensberg, C. and Kimball, S.T., Culture and Community. New York: Harcourt, Brace & World, Inc., 1965.

Aufenanger, Heinrich and Georg Holtker, Die Gende in Zentral Neuguinea. Wien-Modling: Missionsdruckerei, 1940.

Bank, T.P., "Report of the Subcommittee on Ethnobotany", The Proceedings of the Ninth Pacific Science Congress 1957, Vol. 4, 1962.

Barnes, J.A., "The Collection of Genealogies", Rhodes-Livingstone Journal, 5, 1947.

Barnes, J.A., "African Models in the New Guinea Highlands", Man 62, 1962.

Barnett, H.G., Innovation: the Basis of Cultural Change. New York: McGraw-Hill, 1953.

Barth, Frederick, Nomads of South Persia. Boston: Little, Brown and Co., 1961.

Bateson, Gregory, Naven. 2nd ed. Stanford, Calif.: Stanford University Press, 1958.

Belshaw, Cyril S., "The Identification of Values in Anthropology", The American Journal of Sociology 64(6), 1959.

Benedict, Ruth, "Continuities and Discontinuities in Cultural Conditioning", Psychiatry, 1, 1938.

Berndt, Catherine H., "Socio-Cultural Change in the Eastern Central Highlands of New Guinea", Southwestern Journal of Anthropology 9(1), 1953.

Berndt, Ronald M., "Kamano, Jate, Usurufa, and Fore Kinship", Oceania 25, 1955.

Berndt, Ronald M., Excess and Restraint. Chicago: University of Chicago Press, 1962.

Berndt, Ronald M., "Warfare in the New Guinea Highlands", American Anthropologist 66(4), 1964.

Boelaars, J., Papoea's aan de Mappi. Utrecht: de Fontein, 1959.

Booth, Doris R., Mountains, Gold and Cannibals. Sydney: 1929.

Brandel, Mia, "Urban Lobola Attitudes", African Studies 17(1), 1958.

Brass, L.J., "Results of the Archbold Expeditions No. 86. Summary of the Sixth Archbold Expedition to New Guinea (1959)". Bulletin of the American Museum of Natural History 127, Article 4, 1964.

Brongersma, L.D. and G.F. Venema, To the Mountains of the Stars. London: Hodder & Stoughton, 1962.

Brown, Paula, "Chimbu Tribes: Political Organization in the Eastern Highlands of New Guinea", Southwestern Journal of Anthropology 16(1), 1960.

Brown, Paula, "Non-Agnates Among the Patrilineal Chimbu", Journal of the Polynesian Society 71(1), 1962.

Bulmer, Susan, "Radiocarbon Dates from New Guinea", Journal of the Polynesian Society 73(3), 1964a.

Bulmer, Susan, "Prehistoric Stone Implements from the New Guinea Highlands", Oceania 34(4); 1964b.

Bulmer, Susan & Ralph Bulmer, "The Prehistory of the Australian New Guinea Highlands", American Anthropologist 66(4), Part 2, 1964.

Carstens, Peter, The Social Structure of a Cape Coloured Reserve. Cape Town: Oxford University Press, 1966.

Chinnery, E.W.P., "Stone-Work and Goldfields in British New Guinea", Journal of the Royal Anthropological Institute 49, 1919.

Chinnery, E.W.P., "Mountain Tribes of the Mandated Territory of New Guinea from Mt. Chapman to Mt. Hagen", Man 34(140), 1934.

Conklin, Harold C., "The Study of Shifting Cultivation", Current Anthropology 2(1), 1961.

Coser, Lewis A., The Functions of Social Conflict. Glencoe: The Free Press, 1956.

Dalton, George, "Economic Theory and Primitive Society", American Anthropologist 63, 1961.

Davenport, William, "Nonunilinear Descent and Descent Groups", American Anthropologist 61, 1959.

Davis, W.A. and R.G. Havighurst, "Social Class and Color Differences in Child Rearing", American Sociological Review 11, 1946.

De Josseling de Jong, J.P.B., Levi-Strauss' Theory of Kinship and Marriage. "Mededelingen van het Rijksmuseum voor Volkenkunde 10". Leiden: 1952.

Detzner, Hermann, Kreuz — und Querzuge in Kaiser-Wilhelmsland (Deutsch Neuguinea) während des Weltkrieges. "Mitteilungen aus den deutschen Schutz-gebieten 32". 1919.

Detzner, Hermann, Vier Jahre unter Kannibalen: von 1914 bis zum Waffenstillstand unter deutscher Flagge im unerforschten Innern von Neuguinea. Berlin: August Scherl, 1920.

Dexter, David, The New Guinea Offensive Australia in the War of 1939-1945. Vol. VI. Canberra: Australian War Memorial, 1961.

du Toit, Brian M., "Structural Looseness in New Guinea", Journal of the Polynesian Society 71(4), 1962.

du Toit, Brian M., Organization and Structure in Gadsup Society. Unpublished doctoral dissertation. University of Oregon: Eugene, Oregon, 1963.

du Toit, Brian M., "Gadsup Culture Hero Tales". Journal of American Folklore 77(306), 1964a.

du Toit, Brian M., "Die Somer Taalkundige Instituut". Die Kerbode 116, 1964b.

du Toit, Brian M., "Review of 'Etnografie van de Kaowerawedj' by J.P.K. van Eechoud", Journal of the Polynesian Society 73(4), 1964c.

du Toit, Brian M., "Filiation and Affiliation Among the Gadsup", Oceania 35(2), 1964d.

du Toit, Brian M., Beperkte Lidmaatskap. Cape Town: John Malherbe Publishers, 1965a.

du Toit, Brian M., "Gewoonte en Reg in Antropologie", Acta Juridica, 1965b.

du Toit, Brian M., "Die Lutherse Sending in Nieu-Guinee", Die Krygsbanier 5(1), 1966.

du Toit, Brian M., "Misconstruction and Problems in Communication — A New Guinea Case". American Anthropologist, 71(1), 1969.

du Toit, Brian M., People of the Valley: Life in an Isolated Afrikaner Community in South Africa. Cape Town: A.A. Balkema, 1974.

du Toit, Brian M., Configurations of Cultural Continuity. (Unpublished Manuscript).

Eliade, Mircea, Rites and Symbols of Initiation (translated from French by Willard R. Trask). New York: Harper Torchbooks, 1958.

Ember, Melvin, "The Nonunilinear Descent Groups of Samoa", American Anthropologist 61(4), 1959.

Embree, John, "Thailand — A Loosely Structured Social System". American Anthropologist 52(2), 1950.

Field, M.J., Religion and Medicine of the Ga People. London: Oxford University Press, 1937.

Firth, Raymond, We, the Tikopia, Chicago: American Book Company, 1936.

Firth, Raymond, Elements of Social Organization. New York: Philosophical Library, 1951.

Firth, Raymond, "A Note on Descent Groups in Polynesia", Man 57(2), 1957.

Firth, Raymond, We, the Tikipia. Boston: Beacon Press, 1965.

Flierl, Johannes, Forty-five Years in New Guinea, (translated). Columbus, Ohio: 1931.

Flierl, Leon H., Unter Wilden: Missionarische Anfangsarbeit im Innern von Neuguinea. Neuendettelsau, 1932.

Fortes, Meyer, "Time and Social Structure", in Meyer Fortes (ed.), Social Structure. Oxford: Clarendon Press, 1949.

Fortes, Meyer, "Introduction", in Jack Goody (ed.), The Developmental Cycle in Domestic Groups. "Papers in Social Anthropology". Cambridge, Mass: Cambridge University Press, 1958.

Fortes, Meyer, "Some Reflections on Ancestor Worship in Africa", in African Systems of Thought. Studies presented and discussed at the Third International African Seminar in Salisbury, December 1960: Oxford University Press, 1965.

Frantz, Chester I., "Grammatical Categories as Indicated by Gadsup Noun Affixes", Oceania Linguistic Monographs 6, 1962.

Frazer, Sir James G., The Golden Bough — A Study in Magic and Religion, abridged edition. London: MacMillan & Co., 1954.

Freeman, J.D., "On the Concept of the Kindred", Journal of the Royal Anthropological Institute 91(2), 1961.

Frerichs, A.C., Anutu Conquers in New Guinea. Ohio: The Wartburg Press, 1957.

Geddes, W.R., The Land Dayaks of Sarawak. "Colonial Studies No. 14". London: Colonial Office, 1954.

Gitlow, Abraham L., Economics of the Mount Hagen Tribes, New Guinea. New York: J.J.Augustin Publishers, 1947.

Goodenough, Ward H., "Kindred and Hamlet in Lakalai, New Britain", Ethnology 1, 1962.

Goodenough, Ward H., "Residence Rules", Southwestern Journal of Anthropology 12(1), 1956.

"Government in New Guinea". The Bulletin, September 28, 1960.

Gulliver, P.J., The Family Herds. London: Routledge, & Kegan Paul, 1955.

Havighurst, R.J. and A. Davis, "A Comparison of the Chicago and Harvard Studies of Social Class Differences in Child Rearing", American Sociological Review 20, 1955.

Heider, Karl G., The Dugum Dani. Chicago: Aldine Publishing Co., 1970.

Heim, Roger, "Note succincte sur les champignons alimentaires des Gadsup", Cahiers du Pacifique 6, June, 1964.

Held, G.J., Papoea's van Waropen. Leiden: E.J. Brill, 1947.

Held, G.J., De Papoea, Culturimprovisator. 's-Gravenhage: Vanhoeve, 1951.

Henry, Jules, "The Personal Community and Its Invariant Properties", American Anthropologist 60, 1958.

Hoebel, E. Adamson, The Law of Primitive Man. Cambridge, Mass: Harvard University Press, 1954.

Hogbin, H. Ian, "Sex and Marriage in Busama, North-Eastern New Guinea", Oceania 17, 1964.

Hogbin, H. Ian, Transformation Scene: The Changing Culture of a New Guinea Village. London: Routledge and Kegan Paul, 1951.

Hogbin, H. Ian, Law and Order in Polynesia. Hamden: Shoe String Press, Inc., 1961.

Hogbin, H. Ian, Kinship and Marriage in a New Guinea Village, "Monographs on Social Anthropology No. 26". London: London School of Economics, 1963.

Hogbin, H. Ian and Camilla H. Wedgwood, "Local Grouping in Melanesia", Oceania 23(4), 1953.

Holleman, J.F., Shona Customary Law. London: Oxford University Press, 1952.

Homans, George C., "Social Behavior as Exchange", American Journal of Sociology 62, 1958.

Hsu, Francis L.K., Under the Ancestor's Shadow. New York: Columbia University Press, 1948.

Klatskin, E.H., "Shifts in Childcare Practices in Three Social Classes Under an Infant Care Program of Flexible Methodology", American Journal of Orhtopsychiatry 22, 1952.

Kluckhohn, Clyde, Navaho Witchcraft. "Papers of the Peabody Museum of American Archaeology and Ethnology 22". Cambridge, Mass: Harvard University, 1944.

Kluckhohn, Clyde, Mirror for Man. New York: McGraw-Hill Book Co., 1949.

Knak, D. Siegfried, German Protestant Missionary Work; its characteristic features in practice and theory. International Missionary Council, 1938.

Krige, J.D., "The Social Function of Witchcraft", Theoria 1, 1947.

Kroeber, A.L., "The Societies of Primitive Man", Biological Symposia 8, 1942.

LaBarre, Weston, "Some Observations on Character Structure in the Orient: The Japanese", Psychiatry 8, 1945.

Langness, L.L., "Some Problems in the Conceptualization of Highlands Social Structure", American Anthropologist 66(4), Part 2, 1964.

LaPiere, Richard T., A Theory of Social Control. New York: McGraw-Hill Book Co., 1954.

Lawrence, Peter, Land Tenure Among the Garia. Canberra: Australian National University, 1955.

Leach, E.R., Social Science Research in Sarawak. "Colonial Research Studies I", London: Colonial Office, 1950.

Leach, E.R., Political Systems of Highland Burma. Cambridge: 1954.

Leach, E.R., Pul Eliya: A Village in Ceylon. Cambridge: University Press, 1961.

Leahy, Michael, "The Central Highlands of New Guinea", Geographical Journal 87(3), 1936.

Leahy, Michael and Maurice Crain, The Land that Time Forgot. New York: Funk and Wagnalls Co., 1937.

Leininger, Madeleine, "A Gadsup Village Experiences Its First Election", New Guinea's First National Elections — A Symposium. Journal of the Polynesian Society 73(2), 1964.

Le Roux, C.C.F.M., De Bergpapoea's van Nieuw-Guinea en hun Woongebied II, Leiden: Brill, 1950.

Levi-Strauss, Claude, "Social Structure", in A.L. Kroeber (ed.), Anthropology Today. Chicago: University of Chicago Press, 1952.

Levi-Strauss, Claude, "The Family", in Harry L. Shapiro (ed.), Man, Culture and Society. New York: Oxford University Press, 1956.

Levi-Strauss, Claude, Structural Anthropology. New York: Basic Books Inc., 1963.

Linton, Ralph, The Study of Man. New York: Appleton-Century-Croft, 1936.

Linton, Ralph, The Cultural Background of Personality. New York: Appleton, 1945.

Lowie, Robert H., "Social Organization", in Encyclopedia of the Social Sciences. New York: MacMillan Co., 1935.

Lowie, Robert H., The History of Ethnological Theory. New York: Rinehart and Co., Inc., 1937.

Lowie, Robert H., Social Organization. New York: Rinehart and Co., (first published 1948), 1953.

McCarthy, J.K., Patrol Into Yesterday. Canberra: F.W. Cheshire, 1963.

McKaughan, Howard, "A Study of Divergence in Four New Guinea Languages". American Anthropologist 66(4), Part 2, 1964.

McMillan, N.J. and E.J. Malone, The Geology of the Eastern Central Highlands of New Guinea. "Report No. 48, Bureau of Mineral Resources". Australian Department of National Development, 1960.

Maine, Henry Sumner, Ancient Law. New York: Henry Holt & Co., (fifth ed.), 1888.

Maine, Henry Sumner, Village-Communities in the East and West. New York: Henry Holt & Co., 1889.

Maine, Henry Sumner, Lectures on the Early History of Institutions. London: John Murray, (6th ed.), 1893.

Mair, L.P., Australia in New Guinea. London: Christophers, 1948.

Malinowski, Bronislaw, Argonauts of the Western Pacific. London: Routledge and Kegan Paul, 1922.

Malinowski, Bronislaw, The Sexual Life of Savages. New York: Halcyon House, 1929.

Marwick, M.G., "Some Problems in the Sociology of Sorcery and Witchcraft", in African Systems of Thought. "Studies Presented and Discussed at the Third International African Seminar in Salisbury December, 1960". Oxford University Press, 1965.

Matthews, Elmora Messer, Neighbor and Kin. Nashville: Vanderbilt University Press, 1966.

Mead, Margaret, Coming of Age in Samoa. New York: William Morrow, 1928.

Mead, Margaret, Sex and Temperament in Three Primitive Societies. New York: Mentor Books, 1959.

Michener, Charles D., "Observations on the Nests and Behavior of Trigona in Australia and New Guinea (Hymenoptera, Apidae)". American Museum Novitates 2026, May 1961.

Morgan, Lewis H., Ancient Society. New York: Henry Holt and Co., 1878.

Murdock, George P., Social Structure. New York: MacMillan Co., 1949.

Murdock, George P. (ed.), Outline of Cultural Materials, New Haven: Human Relations Area Files, 1950.

Murdock, George P., "Social Organization of the Tenino". Miscellanea Paul Rivet. Universidad Nacional Autonoma de Mexico, 1958.

Murdock, George P. (ed.), Social Structure in Southeast Asia. "Viking Fund Publications in Anthropology 29". Chicago: 1960.

Murphy, Robert F., "Intergroup Hostility and Social Cohesion". American Anthropologist 59(6), 1957.

Nadel, S.F., The Nuba. London: Oxford University Press, 1947.

Nadel, S.F., "Social Control and Self-Regulation". Social Forces 31, 1953.

Nadel, S.F., The Foundations of Social Anthropology. London: Cohen & West Ltd., 3rd Impression, 1958.

Newman, Philip, Supernaturalism and Ritual Among the Gururumba. Unpublished doctoral dissertation, University of Washington, Seattle, Washington: 1962.

Notes and Queries on Anthropology. London: Routledge, Kegan, Paul, (sixth edition), 1954.

Official Handbook of the Territory of New Guinea administered by the Commonwealth of Australia under Mandate from the Council of the League of Nations. Canberra, 1937.

Oosterwal, G., "The position of the bachelor in the Upper Tor Territory". American Anthropologist 61(5), 1959.

Oosterwal, G., People of the Tor. Assen: Royal van Gorcum Ltd., 1961.

Opler, M.E., "An Outline of Chiricahua Apache Social Organization", in Fred Eggan (ed.), Social Anthropology of North American Tribes. Chicago: University of Chicago Press, 1937.

Park, Robert E. and Ernest W. Burgess, Introduction to the Science of Sociology. Chicago: University of Chicago Press, 1921.

Pehrson, Robert N., "Bilateral Kin Grouping as a Structural Type", Journal of East Asiatic Studies 3(2), 1954.

Pehrson, Robert N., The Bilateral Network of Social Relations in Könkämä Lapp District. Bloomington: 1957.

Pelzer, K.J., Pioneer Settlements in the Asiatic Tropics. "Special Publication No. 29 of American Geographical Society". New York: Institute of Pacific Relations, 1945.

Pospisil, Leopold, Kapauku Papuans and Their Law. "Yale University Publications in Anthropology 54", Cambridge, Mass:1958.

Pospisil, Leopold, The Kapauku Papuans. New York: Holt, Rinehart and Winston, 1965.

Pouwer, J., Enkele Aspecten van de Mimika-Cultuur. 's-Gravenhage: Government Printers, 1955.

Pouwer, J., "Loosely Structured Societies in Netherlands New Guinea", Bijdragen Land-, Taal-, Volkenk 66, 1960a.

Pouwer, J., "Social Structure in the Western Interior of Sarmi", Bijdragen Land-, Taal-, Volkenk 66, 1960b.

Pouwer, J., "New Guinea as a Field for Ethnological Study", Bijdragen Taal-, Land-, en Volkenkunde 117, 's-Gravenhage: 1961.

Pouwer, J., Toward a Configurational Approach to Society and Culture in New Guinea", Journal of the Polynesian Society 75(3), 1966.

Prins, A.H.J., East African Age-Class System. Groningen: Wolters, 1953.

Radcliffe-Brown, A.R., "Preface" to Meyer Fortes and E.E. Evans-Pritchard (eds.), African Political Systems. London: Oxford University Press, 1958.

Rappaport, Roy A., Pigs for the Ancestors. New Haven: Yale University Press, 1967.

Read, K.E., "Social Organization of the Markham Valley", Oceania 17, 1946.

Read, K.E., "Political Systems of the Ngarawapum", Oceania 20, 1950.

Read, K.E., "Cultures of the Central Highlands, New Guinea", Southwestern Journal of Anthropology 10(1), 1954.

Reay, Marie, The Kuma. Melbourne University Press, 1959.

Reay, Marie, "Mushroom Madness in the New Guinea Highlands", Oceania 31, 1960.

Rhys, Lloyd, Highlights and Flights in New Guinea. London: Hodder & Stoughton, Ltd., 1942.

Rowley, Charles, The New Guinea Villager: The Impact of Colonial Rule on Primitive Society and Economy. New York: Frederick A. Praeger, 1966.

Ruhen, Olaf, Mountains in the Clouds. Adelaide: Rigdy Ltd., 1963.

Saksena, R.N., Social Economy of a Polyandrous People (2nd ed.). London: Asai Publ. House, 1962.

Salisbury, R.F., From Stone to Steel Economic Consequences of a Technological Change in New Guinea. Melbourne University Press for Australian National University, 1962.

Schapera, I., A Handbook of Tswana Law and Custom. London: Oxford University Press, 1955.

Schindler, A.J., "Land Use by Natives of Aiyura Village, Central Highlands, New Guinea", South Pacific 6(2), 1952.

Schoorl, J.W., Kultuur en kultuurveranderingen in het Moejoe-gebied. Den Haag: Voorhoeve, 1957.

Sears, Robert R., Maccoby, E.E. and Levin, H., Patterns of Child Rearing. Evanston: Row, Peterson & Co., 1957.

Sherwin, V.H., "Ancient Carved Stone Objects, Watut River, Territory of New Guinea", Man 38(69), 1938.

Simmel, Georg, Soziologie. Leipzig: Duncker & Humbolt, 1908.

Simmel, Georg, Conflict (translated by Kurt Wolff). Glencoe: The Free Press, 1955.

Simpson, George, Conflict and Community. New York: Liberty Press, 1937.

Soderstrom, J., "Die Rituellen Fingerverstummelungen in der Sudsee und in Australien", Zeitschrift für Ethnologie 70, 1938.

Souter, Gavin, New Guinea the Last Unknown. Sydney: Angus and Robertson, 1964.

Spencer, Herbert, The Principles of Sociology, II. New York: D. Appleton & Co., 1896.

Spiro, Melford E., "Social Systems, Personality, and Functional Analysis", in Bert Kaplan (ed.), Studying Personality Cross-culturally. New York: Row Peterson & Co., 1961.

Taylour, H. and Morley, I., The Development of Gold Mining in Morobe, New Guinea, "Proceedings of the Australiasian Institute of Mining and Metallurgy 89". Melbourne: 1931.

Thurnwald, Richard, Barnaro Society. "Memoirs of the American Anthropological Association III; 1916.

Tönnies, Ferdinand, Fundamental Concepts of Sociology Gemeinschaft und
Gesellschaft, (translated by C.P. Loomis). New York: American Book Co.,
1940.

van Baal, J., Over wegen en drijfveren der religie. Amsterdam: Noord-Hol-
landsche Uitgevers Maatschappij, 1947.

van Baal, J., "Volken" in W.C. Klein (ed.), Nieuw Guinea. Vol. II. 's-Gravenhage: Staatsdrukkerij, 1954.

van der Leeden, A.C., Hoofdtrekken der sociale structuur in het westelijke bin-
nenland van Sarmi, Leiden: Ijdo, 1956.

van der Leeden, A.C., "Social Structure in New Guinea", Bijdragen Land-, Taal-
Volkenk. 66, 1960.

van Eechoud, J.P.K., Etnografie van de Kaowerawedj. "Verhandelingen van de
Koninklijk Instituut voor Taal- Land- en Volkenkunde 37", 's-Gravenhage:
1962.

van Gennep, Arnold, Les Rites de Passage. Paris, 1910.

Verschueren, J., "Het mensen offer op de zuidkust van Nederlands Nieuw
Guinea", Indonesië 1(5), 1948.

Vicedom, Georg F. and Herbert Tischner, Die Mbowamb, II, Hamburg: 1943-48.

Watson, James B., "A New Guinea 'Opening Man'", in J.P. Casagrande, In the
Company of Man. New York: Harper and Brothers, 1960.

Watson, James B., "A General Analysis of the Elections at Kainantu", "New
Guinea's First National Elections: A Symposium". Journal of the Polynesian
Society 73(2), 1964a.

Watson, James B., "Introduction: Anthropology in the New Guinea Highlands".
American Anthropologist 66(4), Part 2, 1964b.

Watson, Virginia. "Pottery in the Eastern Highlands of New Guinea", Southwestern Journal of Anthropology 11(2), 1955.

Weber, Max, Wirtschaft und Gesellschaft, 2 vols. Tubingen: Mohr (Siebeck),
1922.

Weber, Max, "The Household Community", in Talcot Parsons et al (ed.),
Theories of Society. Vol I. Glencoe: The Free Press, 1961.

White, J. Peter, "Archaeological Excavations in New Guinea: An Interim Report",
Journal of the Polynesian Society 74(1), 1965.

White, J. Peter, "New Guinea: The First Phase in Oceanic Settlement", Pacific
Anthropological Records 12, 1971.

White, M.S., "Effects of Social Class Position on Child-rearing Practices and
Child Behavior", (Abstract), American Psychologist 10, 1955.

Whiting, J.W.M., "The Cross-Cultural Method", in G. Lindzey (ed.), Handbook of Social Psychology. Vol. I, Cambrdige, Mass: 1954.

Whiting, J.W.M. and Child, Irvin L., Child Training and Personality. New
Haven: Yale University Press, 1962.

Williams, F.E., Orokaiva Society. London: Oxford University Press, 1930.

Williams, F.E., "Natives of Lake Kutubu, Papua", The Oceania Monographs 6,
1940.

Wilson, Monica, "Witch Beliefs and Social Structure", The American Journal
of Sociology 56, 1951.

Wirth, Louis, "The Sociology of Ferdinand Tönnies", The American Journal of
Socioloty 32, 1926.

Wurm, S.A., "The Changing Linguistic Picture of New Guinea", Oceania 31(2), 1960.

Wurm, S.A., "The Question of Language and Dialect in New Guinea", paper presented at 35th meeting of the Australian and New Zealand Association for the Advancement of Science. Brisbane: 1961.

Wurm, S.A., "Australian New Guinea Highlands Languages and the Distribution of Their Typological Features", American Anthropologist 66(4), Part 2, 1964.

Wurm, S.A. and D.C. Laycock, "The Question of Language and Dialect in New Guinea", Oceania 32(2), 1961.

Yalman, Nur, "The Structure of the Sinhalese Kindred", American Anthropologist 64, 1962.

Young, Frank W., Initiation Ceremonies: A Cross-Cultural Study of Status Dramatization. New York: Bobbs-Merrill Company, Inc., 1965.

Zeitschrift der Gesellschaft für Erdkunde zu Berlin, 1932.

INDEX